PRACTICAL
ELECTRICAL
WIRING

Practical
Electrical
Wiring

RESIDENTIAL, FARM, AND INDUSTRIAL

Based on the 1968 National Electrical Code

H. P. Richter

*Member, International Association of Electrical Inspectors
and National Fire Protection Association*

EIGHTH EDITION

McGRAW-HILL BOOK COMPANY

**New York   St. Louis   San Francisco   Dusseldorf   London
Mexico   Panama   Sydney   Toronto**

Preface

In preparing this book it has been the author's aim to make it simple enough for the beginner, yet complete enough so that it will be of value also to those already engaged in electrical work. It is intended to be, not a manual that merely recites the methods used in wiring buildings for the use of electricity, but rather a book that explains the subject in such fashion that the reader will learn both the *way* things are done and *why* they are done in that particular way. Only in this manner can the student master the subject so that he can solve his own problems as they arise in actual practice, for no book can possibly cover all the different problems that are likely to arise.

Since this book is not intended to include the subject of electrical engineering, only so many basic engineering data as are essential have been included, and these so far as possible have been boiled down to ABC proportions.

All methods shown are in strict accordance with the National Electrical Code, but no attempt has been made to include a detailed explanation of *all* subjects covered by the Code. The Code is written to include any and all cases that might arise in wiring every type of structure from the smallest cottage to the largest skyscraper; it covers ordinary wiring as well as those things that come up only very rarely.   The scope of this book has been limited to the wiring of structures of limited size and at ordinary voltages, under 600 volts.   Skyscrapers and steel mills and projects of similar size involve problems that the student will not meet until long after he has mastered the contents of this book.

The book consists of three parts:

Part 1 presents the fundamentals of electrical work; terminology; basic principles; the theory behind general practices.   Part 2 deals with the actual wiring of residential buildings and farms.   Part 3 covers the actual wiring of nonresidential buildings, such as stores, factories, schools, and similar structures.

The science or principles of electricity do not change; the art or method of application does change.   That portion of this book having to do with *principles* has in this eighth edition been revised and amplified to present such principles more clearly.   That portion concerned with *methods* has been revised or rewritten as required by Code changes, to describe new materials and methods, and to outline the methods more clearly.

The author will welcome practical suggestions for the improvement of the content or the presentation of this book.

H. P. RICHTER

Contents

PART 2   Actual Wiring: Residential and Farm

PART 3   Actual Wiring: Nonresidential Projects

Theory and Basic Principles

Part 1 of this book is the introduction to practical electrical work, the ABC of the science and the art of electrical wiring.   In order to master the art, naturally you must clearly understand the science or the principles involved.   The terms used in the measurement of electricity, the names of the devices used in wiring, must be at your finger tips.

For that reason a considerable portion of the material presented in Part 1 of this book emphasizes the "why" more than the "how."   If you understand the "why," the "how" becomes obvious.   Master both, and Parts 2 and 3 of this book will be relatively simple.

1

Underwriters and Codes

This book will deal with the electrical materials that are installed in a building so that electric power can be safely used for all intended purposes.   It will also deal with *how* these materials are installed and *why* they are installed in a certain way.

In any discussion of electrical work you will often see the phrase "listed by Underwriters" and also references to the "Code."   You must understand clearly who the Underwriters are and what the Code is.

Electricity is a powerful force.   Under control, it safely performs an endless variety of work for us; uncontrolled, it can lead to great harm and damage.   It is under control when the right kinds of materials are installed in the right way.   It becomes uncontrolled when the wrong kinds of materials are used or when the right kinds of materials are wrongly installed.

Uncontrolled electric power can cause fires, can kill people, and can lead to a multitude of other costly results.   When these things happen, individuals are the greatest sufferers.   Insurance companies also suffer losses, which can result only in higher insurance rates which individuals are called upon to pay.   It is then only natural that insurance companies have led the way in work which leads to the setting up of minimum standards of quality in electrical materials, as well as uniform methods of installation—standards and methods that experiment and experience have shown lead to a maximum of usefulness with the least amount of danger.

Underwriters' Laboratories.   This is an independent not-for-profit testing organization with testing stations at Chicago, Ill., Melville, N.Y., and Santa Clara, Calif.   Manufacturers who wish to do so submit samples of their product to these laboratories before going into production. In the laboratories the product is given exhaustive tests in accordance with established standards, and if it in every way comes up to the minimum requirements, it is listed in the Underwriters' official published list and is then known as "Listed by Underwriters' Laboratories, Inc."

It should be interesting to know that the tests by the Underwriters, and frequently also the tests that the manufacturers are required to make, are many times more severe than any conditions that are likely to be imposed upon the merchandise in actual use. For example, consider the tests required of the manufacturers by the Underwriters for ordinary wire used in residential wiring.   The manufacturers have several choices of procedure.   One of them is to submerge every foot of the wire in a tank of water for 6 hr, then test at 1,500 to 4,000 volts (depending on the size and type of wire) against breakdown while still submerged.   Only after the wire meets this test is it labeled with the Underwriters' label, which indicates that the wire is suitable for use at not over 600 volts.

An example of the extreme severity of tests at the Underwriters' Laboratories is that applied to all plug fuses when submitted by manufacturers for listing.   Such fuses are never rated at more than 30 amp.   In the laboratory test, the fuse is shortcircuited across a circuit capable of delivering 10,000 amp.   The fuse will naturally blow, but it must cause no damage external to itself.   See Fig. 1-1, which shows two fuses after test.   The one shown at the bottom of the picture has blown, but it is still otherwise intact, and the fuseholder is not damaged; it was listed. The one shown at the top of the picture also has blown, but in the process it demolished both itself and the fuseholder; it was not listed. Used in a home, such a fuse could easily start a fire or cause an injury.

In order that the product may remain listed, the safety level must

Unlabeled fuse

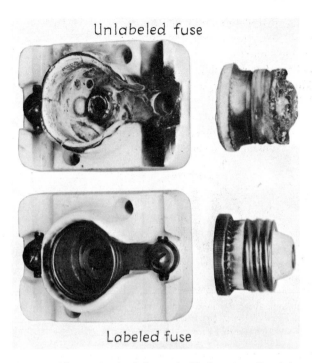

Labeled fuse

Fig. 1-1 After testing both fuses, the Underwriters listed the one at the bottom. The fuse at the top did not have a chance of becoming a listed item. (*Underwriters' Laboratories, Inc.*)

remain at least as high as the Underwriters' minimum specifications. To make sure that it does, the Underwriters send traveling inspectors to the factory from time to time to see that the required factory tests are made and that every effort is made to preserve adequate safety. Inspectors are more or less permanently stationed at some of the larger factories. The Underwriters also regularly buy samples in the open market from merchants stocking the item, and repeat the original test. If the safety level is not maintained, listing is withdrawn. There are two basic types of follow-up service by the Underwriters: reexamination and label.

Reexamination Service. This type of service is used when it is relatively simple to maintain the safety of the original sample of the product. Many devices fall into this classification: sockets, receptacles, outlet boxes, porcelain insulators, most appliances, etc. On such merchandise it is not necessary to test and inspect a large percentage of merchandise manufactured.

It is not essential that items falling into the reexamination classification be individually labeled to the effect that they are listed, although practically all manufacturers are eager to state this fact so that the purchasing public will not overlook it. For this purpose the uniform markers shown in Fig. 1-2 have been developed and are in general use, both on the merchandise and on containers.

Fig. 1-2 The markers above on merchandise indicate that the material is listed by Underwriters. (*Underwriters' Laboratories, Inc.*)

Label Service. This type of service is applicable to products where each piece of merchandise is individually labeled with a label that reads "Underwriters' Laboratories, Inc. Listed" or an abbreviation like "Und. Lab. Inc. List." Under the label service, representatives of the Underwriters' Laboratories make frequent inspections at the factory to inspect and test samples of the labeled products. Into this classification fall wire and cable of all kinds, conduit, switches of all kinds, wiring devices in general, lighting fixtures, and many similar items. Examples of the labels used on the merchandise are seen in Fig. 1-3.

Flexible cords sold over the counter by the foot, formerly were labeled with a "bracelet" label shown in Fig. 1-4, applied every 5 ft on the cord. It was easy for the purchaser to determine quickly that the cord in question was a listed item. However, the label was misleading to the public, for many people assumed that because the cord *on an appliance* bore the bracelet label, the appliance itself was a listed item: a totally wrong conclusion.

Beginning in 1969, the "bracelet" label was discontinued. When buying cords by the foot over the counter, look for a label similar to those shown in Fig. 1-3, on the spool, or on precut lengths of 6 to 30 ft.

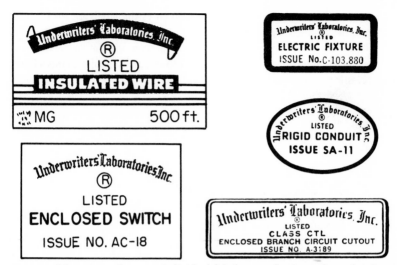

Fig. 1-3 Many items have an individual label on each piece of merchandise. Above are shown a few samples. (*Underwriters' Laboratories, Inc.*)

Replacement cord sets, with a plug or similar device on one or both ends, bear either a "doughnut" or a "flag" label shown in Fig. 1-5, indicating that the cord assembly is a listed item. As to appliances, look for the symbol of Fig. 1-2 on the carton, on the appliance, or a statement on the name-plate stating that it is listed.

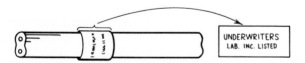

Fig. 1-4 Listed flexible cords formerly bore the label shown above, every 5 ft.

In certain classes of devices, such as, for example, toggle switches, the words "Und. Lab. Inc. List." are molded or stamped or otherwise made an integral part of the merchandise, and constitute the label. Look for these words molded or stamped into bakelite, steel, or porcelain parts of such devices.

How Underwriters Are Supported. Underwriters' Laboratories is a nonprofit organization, and its cost of operation is absorbed by the manufacturers who submit merchandise. There is a fee for testing merchandise, and when a manufacturer buys the labels for his product from the Underwriters, he pays more than the bare cost of the label, the difference being a service charge which pays for the inspectors' expenses in checking at the factories, and which also supports the laboratories in general.

Fig. 1-5   Cord assemblies, if listed, have one of the above two types of labels.

What "Listed by Underwriters" Means.   Listing assures the public that the manufacturer of the listed item has submitted samples to the Underwriters, who have tested them and found that they meet the required minimum *safety* standards.   The presumption is that the manufacturer in his future production will maintain that quality.   The public is assured that the listed item probably meets minimum safety standards; the manufacturer knows his competitors are not skimping; insurance companies suffer smaller losses than would be the case if substandard materials were used.   This is a happy arrangement whereby everybody benefits.

Listing does not mean that all listed items are of equal quality, or indeed that a listed item is necessarily of good quality for its purpose; it merely means that the listed item meets the minimum *safety* requirements of the Underwriters.   For example, one listed switch may outlast another by five times, but both meet the minimum safety requirements. Exercise the same judgment in buying listed merchandise that you would in buying other kinds of merchandise with which Underwriters are not concerned.

An automobile tire designed for a passenger car will no doubt be suitable when used for that purpose but would be totally unsafe if used on a 3-ton truck.   So also with electrical materials, "Listed by Underwriters" implies that the part or material in question is safe when used for the purpose for which it was designed.   For example, armored cable is suitable for use only in dry locations, for which reason the electrical inspector will not give approval for its use in wet locations such as barns. It isn't suitable for that purpose and can't be used even if it is listed by Underwriters. Flexible cord is listed for use in connecting floor lamps, clocks, and similar devices, but may not be used for the wiring of circuits in a house, even if the cord is listed by Underwriters.

Identification. In any event, regardless of whether a product is listed under reexamination or label service, it must always be identified by the manufacturer's name or trademark, by a number that has been assigned by the Underwriters, or by some arbitrary symbol, and in addition by a type or catalogue description or by a "listed" label, so that it can be recognized as listed merchandise.

How to Recognize Listed Merchandise. How do you recognize listed merchandise when you see it? If the merchandise is labeled "Underwriters' Laboratories, Inc.," or with an abbreviation indicating the same, it is listed. If the merchandise or its carton bears the UL insignia of Fig. 1-2, it is listed. If it does not bear the manufacturer's marking or other designation which will identify it, it is not a listed item. Reference to the Underwriters' list of listed devices will establish in any event whether the merchandise is listed. The Underwriters' list contains tabulations showing manufacturers' names as well as catalogue numbers, thus making it easy to check any device. The safest policy is to buy electrical merchandise only from concerns that have a reputation to maintain and that cannot afford to jeopardize it by selling unlisted products.

It does not follow that just because an item is not listed it cannot be good merchandise, for some manufacturers have an exaggerated idea of the cost of having an item inspected and listed, and others just do not care; for these reasons they do not submit their products for listing. The presumption, however, is against merchandise that is not listed, especially if a different brand of the same item is listed.

In fairness it must be stated that the Underwriters do not concern themselves at all with certain devices on which it is a simple matter to maintain good quality or which present no hazard whatever. In this class are, for example, low-voltage devices, such as doorbell wires and doorbells (but doorbell transformers *are* listed). Strangely, electric motors, except those of the explosion-proof type, are not listed.

National Electrical Code. If approved electrical devices of high quality are used, but installed in a haphazard fashion and with no regard to the relation of one device to the other or the total load they may be called upon to carry, the complete installation may still be dangerous. It is necessary therefore that standardized methods be set up which have been found in practice to be safe.

These standardized methods that experiment and experience have shown to be correct are set down in a form which has come to be known as the National Electrical Code. Whenever in this book the word "Code" is used it refers to this National Electrical Code, abbreviated NEC. In this book, excerpts from the NEC are published with the permission of the National Fire Protection Association. Any further reproduction of this material is not authorized except with the permission of that Association.

The purposes of the Code are outlined in the following quotation from the Introduction of the Code:

> *a.* The purpose of this Code is the practical safeguarding of persons and of buildings and their contents from hazards arising from the use of electricity for light, heat, power, radio, signaling, and for other purposes.
> *b.* This Code contains basic minimum provisions considered necessary for safety. Compliance therewith and proper maintenance will result in an installation essentially free from hazard, but not necessarily efficient, convenient, or adequate for good service or future expansion of electrical use.
> *c.* This Code is not intended as a design specification or an instruction manual for untrained persons.

From the above you can readily understand that the chief objective of the Code is to promote *safety* in electrical installations. Moreover, its specifications are the *minimum* requirements, again for safety. An electrical installation made in strict accordance with the Code will be safe, but it might still be impractical from many standpoints. As the quotation above from the Code points out, the installation might not be efficient, convenient, or adequate for good service. Aside from complying with the Code requirements, the designer of an installation and those who do the actual work must still use their own good judgment and skill to produce an installation that will be completely satisfactory to the occupant of the building.

Do note that the Code is "not an instruction manual for untrained persons." Study of the Code is very necessary to learn the science and art of wiring, but the Code alone is not sufficient.

The Code is reissued every three years. Once a Code is issued, interim amendments appear that modify the original issue. In due course of time, a new Code appears, incorporating all the amendments and other new material. The Code consists of several dozen "Articles" each covering one major division of the over-all subject, each divided into as many "Sections" as required. Each new Code of course carries changes in the rules, as compared with the previous issue. Thus new Sections or even new Articles are introduced from time to time.

In studying the Code, it is essential that you pay particular attention to Art. 90, Introduction, and Art. 100, Definitions. Unless these basic areas are well understood, you will have difficulty understanding the remainder of the Code. Also important: The Code uses both the terms "shall" and "should." "Shall" means that whatever is prescribed *must* be done. "Should" means just what it says: While not required, it should be done. Often the "should" of one Code becomes the "shall" of a later Code.

The Code is not written by one man or by a group of just a few people. Each phase of the work is handled by those people in the industry who are most competent to handle it. These people are members of permanent committees established for the purpose. There is a Correlating Committee of about a dozen members, plus 17 subcommittees called "Code Making Panels," each with six to twelve members. To these committees falls the tedious responsibility of listening to endless suggestions for revisions of the Code, debating the merits of the suggestions, and finally revising the Code as seems desirable.

"Approved." You will hear, or see in print, the words "Approved by Underwriters." That is a wrong usage of words, for the Underwriters do not *approve* an item. They *list* the item, which indicates that the item, so far as can be determined, meets their minimum *safety* standards.

The Code, however, does use the word *approved*, and in Art. 100 defines it as "Acceptable to the authority enforcing this Code." This in turn generally speaking means "acceptable to the electrical inspector." But inspectors rarely have adequate equipment or the necessary specialized knowledge to pass on all materials. The inspector then proceeds per Code Sec. 90-8, which suggests that he depend on the findings of "organizations properly equipped and qualified for experimental testing." The obvious organization is Underwriters' Laboratories. If the Underwriters *list* the item, the inspector *approves* it; if not listed, generally he will not approve it.

The inspector is also the final authority with respect to installation method. There are many situations that cannot be 100% covered by the Code, and qualified inspectors can be depended on to render fair judgments. Indeed, it is the combined experience of inspectors, and their consensus of opinion with regard to new materials and new methods, that often lead directly to changes in the Code.

History of the Code. The first National Electrical Code was adopted in 1897. Before that time there were many other collections of rules or codes, not national in scope. For example, there was the 1890 "Rules and Requirements of the Michigan Inspection Bureau for Electric Lighting." Be thankful that today you have no such rule as one contained in that 1890 code: "All circuits shall be tested at least twice a day with a suitable magneto or other approved device, in order to discover any escapes to ground that may exist."

Enforcement of the Code. By this time you should logically be asking, "What jurisdiction do insurance companies or their laboratories or a group of manufacturers have over me—why can't I use such merchandise as I please and in such ways as I please?" The answer is that these bodies have no jurisdiction over you whatsoever; as far as they are concerned, you may do as you please. However, you are still obligated to obey the laws of your state or city or other municipality; with few exceptions these lawmaking bodies pass laws or ordinances which require that the provisions of the National Electrical Code must be observed in the territory involved. Another important consideration is that fire-insurance companies may refuse to issue policies to cover buildings that are not properly wired.

Aside from all legal considerations, common sense suggests that you and I take advantage of the experience of those who know more about the subject than we do. The experience of such experts collectively has resulted in the National Electrical Code, which then becomes a guide for all.

Unfortunately many parts of the Code are interpreted differently by different people; the words of the Code mean different things to different people. The interpretations in this book are the opinions of the author, based on his experience and the opinions of others. There is no "official interpretation"[1] of the Code, and the Code in Sec. 90-7 makes it entirely clear that the local electrical inspector has the final word in any situation.

Local Codes.   The National Electrical Code defines in a broad way what may and may not be done in the line of wiring, the different methods that are permitted, and so on.   Frequently local ordinances limit the National Code, so that only a portion of what it permits is then permitted locally.   For example, the National Code specifies that under certain conditions a conduit system, an armored-cable system, a non-metallic system, or one of several other systems may be used in wiring a house.   The local code or ordinance may specify that all these systems may be used locally except armored cable.   It is therefore important that you be familiar not only with the National Electrical Code[2] but also with the local codes or ordinances that apply in your locality.

Permits.   In many places it is necessary to get a permit from city, county, or state authorities before a wiring job can be started.   The fees charged for permits generally are used to pay the expenses of electrical inspectors, whose work leads to safe, properly installed jobs.   Power suppliers usually will not furnish power until an inspection certificate has been turned in.

Licenses.   Many cities, counties, or states have laws which require that no person may engage in the *business* of electrical wiring without a license.   That leads to the question: When you do electrical work *on your own premises*, are you engaged in the *business* of wiring?   Of course, if the interpretation of the local laws says that in doing so, you are indeed engaged in the business, then your procedure is "illegal." But it raises the larger question of whether such laws in themselves are valid.   For a final answer, expensive legal procedures are necessary, but so far as the author knows, all cases carried to higher courts have led to answers that in doing electrical work on your own premises, you are not engaged in the *business* of wiring, and that therefore you are violating no law.   Many localities recognize this.

---

[1] There are indeed "Official Interpretations" of specific parts of the Code, but not of the Code as a whole.   Very few such interpretations are handed down.   The Code itself in its "Rules of Procedure" outlines how such interpretations may be obtained; it is not a simple procedure.

[2] A copy of the National Electrical Code can be obtained by sending $2 to the National Fire Protection Association, 60 Batterymarch St., Boston, Mass. 02110.   Every student is urged to obtain and study a copy.

The author is not qualified to give legal advice, and will not do so. You will have to be guided by local circumstances. But before proceeding to do your own wiring, remember that a permit, if required, must still be obtained. Before applying for the permit, be sure to understand completely all problems in connection with the job, so that your wiring will in every way meet Code and local requirements. If it does not, the inspector will be required to turn down the job, until all errors are corrected.

2

Electricity: Basic Principles and Measurements

In the study of electricity you will meet the many different terms that have to do with measurement: "volts," "amperes," "watts," and others. It will be much easier to learn how one is related to the other than to get an idea of the absolute value of each. That is because electric power is measured in units which cannot be compared directly with feet, pounds, quarts, or any other measure familiar to you. If you consult your dictionary, you will find definitions like these:

Volt: the pressure required to force one ampere through a resistance of one ohm.

Ampere: the electric current which will flow through one ohm under a pressure of one volt.

Ohm: the resistance through which one volt will force one ampere.

These definitions show a clear interrelationship between the three items, but unfortunately define each item in terms of the other two.

How "big" is each unit?   It is as confusing as the beginner's first encounter with the metric system: 10 millimeters make a centimeter, 100 centimeters make a meter, 1,000 meters make a kilometer; further, 1,000 cubic centimeters make a liter; a liter of water weighs a kilogram. But these terms are meaningless unless you can translate them into familiar units such as inches, yards, miles, pounds.

After using these metric terms for a while, you will indeed begin to see some relationship between them and the more familiar terms you are accustomed to using.   So also in electrical work; after a while you will begin to see some meaningful relationship among the various terms such as volts, amperes, ohms, and others.

Unfortunately in electrical work, there is no way of translating an ampere, a volt, or an ohm into something that is familiar to you. Therefore we shall have to compare these terms with other measures which behave something like electrical measures. The best of comparisons or analogies are not very good, but they are better than none.

Gallons of Water.   We can measure water in pounds or cubic feet or acre-feet or in many other ways.   The measure known to most people is no doubt the gallon.   A gallon of water is a specific *quantity* of water.

Coulombs of Electricity.   When we come to measure electric power, the term that corresponds directly to the gallon in the case of water is the "coulomb."   Ask all the people you know in the electrical business, "How much is a coulomb of electric power?" and 99% or more will answer, "Coulomb? I vaguely remember the term from way back when, but I don't know what a coulomb is."   That being the case, why should we talk about it here?   The only answer is that, while very few people indeed remember the definition of the term, it is nevertheless very helpful in getting to understand other electrical terms.   We shall use it as a temporary tool, just as a child learns how to ride a tricycle before he learns how to ride a bicycle.   Just accept it as a fact that a coulomb is a very definite *quantity* of electric power; do not bother trying to understand how big that quantity is.

Water in Motion.   Gallons of water standing in a tank are just quantities of water.   But if there is a small hose connected to the tank, water will flow out of it.   If there is a big hose water will flow out of it faster than out of the small hose.   If we want to talk about how much faster it flows out of the big hose than out of the small one, we must use some measure to denote the *rate* of flow, so usually we talk about "gallons per minute."   This phrase tells us about the quantity (gallons) and the time (per minute) and collectively indicates what we call the rate of flow.

Electric Power in Motion. Instead of a tank of water, let us now consider a battery, a generator, or other source of electric power. Instead of a hose connected to the tank, let us think of wire through which electric power will flow. Coulombs of electric power will flow through that wire, just as gallons of water flowed through the hose. And just as in the case of water we talk about "gallons per minute," so in the case of electric power we can talk about "coulombs per minute." Only in the case of electric power, it has become the custom to talk about "coulombs per second." Instead of saying that water flowed at the rate of "10 gallons per minute," we shall be talking about electric power flowing at the rate of "10 coulombs per second." But instead of using an awkward, long phrase "10 coulombs per second," we say "10 amperes," because an ampere[1] long ago was defined as a flow of one coulomb per second.

You must never say that current is flowing at 10 amperes *per minute*, for that would be the same as saying that it is flowing at the rate of 10 coulombs *per second per minute*, which does not make sense. Just remember that a coulomb is a *quantity*, an ampere a *rate*. Once you clearly understand that amperes measure the rate of flow, you can forget the coulomb completely and think in terms of amperes.[2]

Water under Pressure. A gallon of water standing in a tank is an inert, static quantity. Water dribbling out of your not-quite-shut-off garden hose at a gallon per minute is just a nuisance. Water coming

[1] The ampere is named after André Marie Ampère (1775–1836), one of the great scientists of the early nineteenth century, who discovered many of the fundamental laws concerning the flow of electric current.

[2] In these days of electronics and atomic and nuclear bombs, a word may be in order to those who are interested in the purely scientific aspect of the subject. A flow of one ampere (one coulomb per second) is equivalent to a flow of 6,280,000,000,000,000,000 electrons per second past a given point. However, it is entirely safe to forget all about coulombs and electrons per second, except for those who intend to delve very thoroughly into basic electrical engineering, and even for those the exact figure is of more academic than practical interest. For those who are interested in comparing numbers as such, it may be interesting to note that the big number shown above, representing the number of electrons moving past a given point every second when a current of one ampere flows, is approximately two million times greater than the number that represents the number of seconds that have elapsed in the more than 1,900 years since the beginning of the Christian era.

out of your sprinkler on the lawn at a few gallons per minute waters your lawn nicely and can be a delightful shower for children playing under it.   The same gallons-per-minute flow coming out of the same hose in the form of one tiny stream and directed at one of those same children who a moment ago thought it was fun will be considered painful because it seems to hurt.   The same gallons-per-minute flow coming out of a high-pressure hose can cause real discomfort or perhaps even injury.

Suppose there are three water tanks located 10, 20, and 100 ft above ground level.   A pipe runs from each tank to the ground level, and a pressure gauge is connected *at the ground level.*   The gauges on the three tanks will show pressures of 4.3, 8.6, and 43 lb per sq in.   (This disregards such things as, for example, pipe friction; let us not complicate the problem by going into details that are for the finished engineer, details that theoretically should be taken into consideration but which may be overlooked for the sake of simplification.)   If a tank were located 1,000 ft above ground level, the pressure would be 430 lb per sq in.

One gallon per minute running out of the first tank would be a nuisance; a gallon per minute out of the last tank at 430 lb per sq in. could do a lot of damage.   The difference lies in the difference in pressure between the two.   The difference is measured in pounds per square inch, and the various pressures spell the difference between nuisance, convenience, and danger.

Electric Power under Pressure.   Electric power is also under pressure, but instead of being measured in pounds per square inch, it is measured in volts.   One volt[3] is a very low pressure; in commercial work higher pressures or voltages are used.

An ordinary flashlight cell or dry-cell battery (regardless of size) will if fresh develop approximately $1\frac{1}{2}$ volts.   Four of them connected in series, as in the so-called "hot-shot" battery, will develop $4 \times 1\frac{1}{2}$, or 6 volts.   Connect 30 of them in series, as in the B battery used in some portable radios, and you have $30 \times 1\frac{1}{2}$, or 45 volts.   A single cell of a storage battery develops about 2 volts when fully charged; the six cells

---

[3] The volt is named after Count Alessandro Volta (1745–1827), one of the great pioneer scientists who had much to do with the early research in electricity.   For example, he discovered that when two dissimilar metals are immersed in an acid, an electric current will flow through a wire connecting the two metals.   In other words, he discovered the principle underlying batteries of all kinds.

of an automobile battery develop 6 × 2, or 12 volts.   An ordinary house-lighting circuit operates at about 115 or 230 volts.   The high-voltage lines feeding the transformers found in alleys of city streets operate at 2,300 to 7,200 volts.

Refer back to footnote 2 on page 17.   The number of electrons flowing past a given point in one second define the number of amperes flowing.   The *speed* with which they move defines the voltage.   Compare the electrons with rifle bullets: If I toss one of them against you, no harm is done because the speed is so slow (the voltage is low).   If one is fired at you from a rifle, it will kill you because the speed is high (the voltage is high).

Water Once More.   Suppose that each of the water tanks we have talked about has a capacity of 100 gal and that each tank is empty. You must fill each tank using a hand pump (pumping out of a source of water at pump level, just to simplify the problem).   You will have to work quite hard to fill the tank located 10 ft above ground level in, say 5 min.   You will have to work twice as hard to fill the one located 20 ft above ground level *in the same time*.   To fill the one located 100 ft above ground level in the same 5 min, you will have to be "Superman"; to fill the one located 1,000 ft above ground level would be a fantastic job to do by hand.

It takes you 5 min to fill the 100-gal tank located 10 ft above ground level.   Then to fill a tank ten times as large (1,000 gal) located at the same level should take you ten times as long, or 50 min.   Then by like token, to fill the original 100-gal tank but now located ten times as high (100 ft) should take you ten times as long, also 50 min.   In other words, it should take you 50 min to fill a 1,000-gal tank located 10 ft above ground level or a 100-gal tank located 100 ft above ground level.   All this is probably easier to grasp in tabular form:

Capacity of tank, gallons	Height above ground, feet	Approximate pressure, pounds per square inch	Time to fill, minutes
100	10	4.3	5
1,000	10	4.3	50
100	100	43.0	50

From this discussion it is easy to see that neither gallons nor gallons per minute determines how much work is involved in filling the tank; pressure alone does not determine it; the two in combination with each other do determine it.   In other words, gallons per minute times pressure per square inch measures the amount of work being done at a specific moment in pumping.

Watts.   In measuring electric power, neither the rate (amperes) nor the pressure (volts) tells us the energy flowing in a circuit at any moment.   A combination of the two does tell us the answer very simply, for amperes × volts = watts.   Watts measure the total energy flowing in a circuit at any given moment, just as horsepower measures the power developed by an engine at any given moment.   Indeed, horsepower and watts are merely two different ways of measuring or expressing rate of work or power; 746 watts are always equal to 1 hp.

If a lamp consumes 746 watts, you would be entirely correct if you called it a 1-hp lamp, although that method of designating lamps is never used.   Likewise you would be entirely correct if you said that a 1-hp motor is a 746-watt[4] motor, although that method of designating power of motors is not used (except in the case of "flea-power" motors which are sometimes rated, for example, "approximately 1½ watts output," instead of being rated as "$1/500$ hp").

Again note that the electrical term is just "watts," not "watts *per hour.*"   You would not say that the engine in your automobile delivers 250 hp *per hour;* at any given moment it delivers 250 hp, the power that 250 horses would or could deliver at any one moment if working simultaneously.   Both watts and horsepower denote a *rate* at which work is being done, not a *quantity* of work being done in a given time.

From the above you can see that a given wattage may be a combination of any voltage whatsoever and the correspondingly correct amperage.   For example:

$$3 \text{ volts} \times 120 \text{ amp} = 360 \text{ watts}$$
$$6 \text{ volts} \times 60 \text{ amp} = 360 \text{ watts}$$
$$12 \text{ volts} \times 30 \text{ amp} = 360 \text{ watts}$$
$$60 \text{ volts} \times 6 \text{ amp} = 360 \text{ watts}$$
$$120 \text{ volts} \times 3 \text{ amp} = 360 \text{ watts}$$
$$360 \text{ volts} \times 1 \text{ amp} = 360 \text{ watts}$$

---

[4] Note, however, that a motor which delivers 1 hp or 746 watts of power actually consumes more nearly 1,000 watts from the power line.   The difference between the 1,000 watts consumed and the 746 watts delivered as useful power is consumed at heat in the motor, to overcome bearing friction, to overcome air resistance of the moving parts, and similar factors.

Carrying the illustration further, a lamp in an automobile headlight consuming 5 amp from a 12-volt battery consumes a total of 5×12, or 60 watts; a lamp consuming ½ amp from a 120-volt lighting circuit in a home consumes a total of $\frac{1}{2} \times 120$, also 60 watts. The voltages and the amperages differ widely, but the wattages of the two lamps are the same.

This simple formula is not correct under all circumstances; the exceptions will be covered in a later chapter.

Kilo-, Mega-, Milli-, Micro-.   A watt is a very small amount of energy; it is only $\frac{1}{746}$ hp. The Greek word "kilo" means thousand, so when we say "one kilowatt," it is just another way of saying "1,000 watts." The abbreviation is "kw," and 25 kw then means 25,000 watts. In the spoken word, the two letters of the abbreviation are usually pronounced (25 kay-double-u) instead of the words they stand for. By like token 20 kilovolts (20 kay-vee) means 20,000 volts, and so on.

Other prefixes that you will meet from time to time are the following:

Mega-: a million. Examples: megawatts, megacycles. Thus 25 megacycles means 25,000,000 cycles.

Milli-: one-thousandth. Examples: milliamperes, milliwatts. Thus 25 milliamperes means 25/1,000 amp.

Micro-: one-millionth. Examples: microvolts, microamperes. Thus 25 microvolts means 25/1,000,000 volt.

Watthours.   The watt merely indicates the total amount of electrical energy that is flowing at a given moment; it tells us nothing about the total quantity of electrical energy that has flowed during a period of time, just as the fact that a man earns $3 per hour tells us nothing about his earnings per year unless we know how many hours he works. Multiplying the watts flowing at one time by the number of hours during which this number of watts flowed gives us *watthours*, which definitely measure the total amount of electrical energy consumed or flowing during a given time. For example:

$$10 \text{ watts} \times 1,000 \text{ hr} = 10,000 \text{ watthours}$$
$$100 \text{ watts} \times 100 \text{ hr} = 10,000 \text{ watthours}$$
$$1,000 \text{ watts} \times 10 \text{ hr} = 10,000 \text{ watthours}$$
$$5,000 \text{ watts} \times 2 \text{ hr} = 10,000 \text{ watthours}$$
$$20,000 \text{ watts} \times \tfrac{1}{2} \text{ hr} = 10,000 \text{ watthours}$$

Kilowatthours. One kilowatthour (kwhr) is 1,000 watthours, 20 kilowatthours is 20,000 watthours, etc. Power is paid for by the kilowatthour.

Reading Meters. Reading meters of the type shown in Fig. 2-1 needs no explanation. Some modern kilowatthour meters have a cyclometer dial, as shown in Fig. 2-2, and they are the most easily read. Other

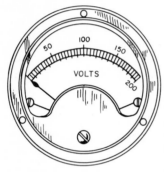

Fig. 2-1 Meters of this type are simple to read.

Fig. 2-2 This type of register is found on some kilowatthour meters. (*General Electric Co.*)

meters, like that shown in Fig. 2-3, have a register, shown in approximately full size in Figs. 2-4 and 2-5, with two different readings. There are four dials, on two of which the figures read from left to right in clockwise fashion, on the other two in counterclockwise fashion. Simply write down the last number the pointer has *passed* on each dial, considering the direction in which the pointer moves.

Refer to Fig. 2-4, and write down the proper numbers, as indicated by the pointers. The first (left-hand) pointer is between 2 and 3, so write down 2; the second is between 7 and 8, so write down 7; the third is between 4 and 5, so write down 4; the last is between 6 and 7, so write down 6. That makes 2,746 and indicates that 2,746 kwhr of energy have been used since the meter was installed.

Now write down the readings in Fig. 2-5, which represents the same meter a month later. The first pointer presents no problem; write down 3. The second pointer, however, points directly at 2. Should you write down 1 or 2? If your watch had no minute hand but only an hour hand and that pointed to 2, how would you know whether the time was 1:58 or 2:02? You would have no way of knowing. However, the watch does have a minute hand, and if it points to 2 min before 12, you know it is 1:58, but if it points to 2 min after 12, you know it is 2:02.

Fig. 2-3 The ordinary meter is a little harder to read. (*General Electric Co.*)

In other words, the minute hand tells you whether the hour hand has not quite reached 2 or has just passed 2.   So also on the kilowatthour

Fig. 2-4 Enlarged register of a kilowatthour meter.   Read the figure the pointer has passed.   The reading above is 2,746 kwhr.

meter, look at the pointer on the dial to the right of the one that points directly to a figure.

If it has not reached 0, the pointer to the left has not reached the figure to which it *apparently points.* In this particular case, the second pointer points directly at 2, but since the third pointer is between 0 and 1, the second one must point a little after the 2, so we write down 2, not 1.   The third and fourth pointers present no problem; write down 0, 7, making 3,207 kwhr.   The difference between 3,207 and 2,746, or 461, naturally represents the number of kilowatthours used since the previous reading.

Fig. 2-5   The meter now reads 3,207 kwhr.

Some meters used for measuring very large quantities of power may have a notation on the register "× 10," "× 100," "multiply by 10," or some similar words.   In that case the indicated kilowatthours must be multiplied by the "multiplier" to arrive at the correct reading.

Power Consumed by Various Devices.   It will be well for you to know the approximate power consumed by everyday devices used in homes. The wattages are as follows:

		Watts
Lamps:		
Incandescent	10 upward	
Fluorescent	15 to	60
Lights, Christmas tree	20 to	150
Clock	2 to	3
Radio	40 to	150
Television	200 to	400
Sun lamp (ultraviolet)	275 to	400
Heat lamp (infrared)		250
Heating pad	50 to	75
Blanket, electric	150 to	200
Razor	8 to	12
Projector, slide or movie	300 to	500
Heater:		
Portable, household type	1,000 to	1,500
Wall type, permanently installed	1,000 to	2,300

Fan, portable . . . .   50 to   200
Air conditioner, room type . . . .   800 to 1,500

Sewing machine. . . . .   60 to   90
Vacuum cleaner. . . . .   250 to   800
Refrigerator, household. . . . .   150 to   300
Freezer, household . . . .   300 to   500

Iron, hand (steam or dry). . . . .   660 to 1,200
Hot plate, per burner.   600 to 1,000
Range (all burners and oven "on"). . . . . 8,000 to 14,000
Range top (separate) . . . . 4,000 to 8,000
Range oven (separate) . . . . 4,000 to 5,000

Toaster . . . .   500 to 1,200
Coffee maker (percolator) . . . .   500 to 1,000
Waffle iron . . . .   600 to 1,000
Roaster . . . . 1,200 to 1,650
Rotisserie (broiler) . . . . 1,200 to 1,650
Fryer, deep fat . . . . 1,200 to 1,650
Frying pan. . . . . 1,000 to 1,200

Blender, food . . . .   500 to 1,000
Knife, electric   100
Food mixer . . . .   120 to   250
Dishwasher . . . . 1,000 to 1,500
Garbage-disposal unit . . . .   500 to   900
Washing machine. . . . .   350 to   550

Washer, automatic . . . .   600 to   800
Dryer, clothes . . . . 4,000 to 8,000
Water heaters . . . . 2,000 to 5,000

Motors:
  ¼ hp . . . .   300 to   400
  ½ hp . . . .   450 to   600
  Over ½ hp, per hp. . . . .   950 to 1,100

Figuring an Electric Bill.   The cost of electricity varies a great deal between localities and also with the amount used per month.   For residential and farm use it seldom falls as low as 1 cent and very rarely reaches 10 cents per kilowatthour.   Usually there is a step rate; the more power you use per month, the lower the cost per kilowatthour. For example, a typical rate is as follows:

First 50 kwhr . . . . 4½ ¢ per kwhr
Next 150 kwhr . . . . 2½ ¢ per kwhr
Over 200 kwhr   . . . . 2 ¢ per kwhr

Assuming a monthly consumption of 461 kwhr, the total bill would
be:

 50 kwhr @ 4½ ¢ . . . . $ 2.25
150 kwhr @ 2½ ¢ 3.75
261 kwhr @ 2 ¢ 5.22
  Total 461 kwhr. $11.22
  Average per kilowatthour . . . . 2⁴⁄₁₀¢ (approx)

To determine the cost of operating any electrical device for 1 hr multiply the watts consumed by the rate in cents per kilowatthour and point
off five decimal places, giving the cost directly in dollars per hour.   For
example, assume a 1,000-watt flatiron at a rate of 5 cents.   Multiplying
$1,000 \times 5 = 5,000$; pointing off five decimals gives 0.05000, or 5 cents per
hour.   For a 40-watt lamp the figures are $40 \times 5 = 200$; pointing off five
places gives 0.00200, or $\frac{2}{10}$ cent per hour, 5 hr for 1 cent.   The oven on
a range may consume 5,000 watts, and at 2 cents per kilowatthour the
figures are $5,000 \times 2 = 10,000$; pointing off five decimal places gives
0.1000, or 10 cents per hour.

To determine the number of hours any device can be operated while
consuming 1 kwhr, simply divide 1,000 by the wattage of the device.
Obviously a 1,000 watt lamp can be used just 1 hr; a 50-watt lamp, 20
hr; a large motor consuming 2,000 watts, ½ hr; and so on.

Conductors and Nonconductors.   When you connect a 1½-volt lamp
through a piece of wire across the two terminals of a dry cell, as shown
in Fig. 2-6, current flows for the lamp lights.   Yet no current flows
through the wax that is poured over the top of the dry cell, nor does it
flow through the paper carton that makes contact with the terminals of
the cell during shipment.   If a material will permit current to flow
through it, it is known as a "conductor"; if it will not permit current to
flow, it is an "insulator."   There is no perfect conductor, nor is there a
perfect insulator.

Resistance.   Number 10 copper wire has a diameter of 0.1019 in.
Experiment shows that, if the two ends of a piece of this wire 1,000 ft
long are connected to a source of electricity developing exactly 1 volt, 1
amp will flow through the wire.

If, however, you substitute aluminum wire of exactly the same diameter and length, only about $\frac{6}{10}$ amp flows. Still using aluminum wire of the same diameter but reducing the length from 1,000 to about 600 ft, 1 amp again flows.

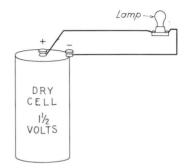

Fig. 2-6 Dry cells, no matter how large or small, develop 1½ volts when new.

If you substitute iron wire of exactly the same diameter and the same 1,000-ft length, only about $\frac{1}{6}$ amp flows. If the length of the wire is reduced from 1,000 to about 167 ft, 1 amp will again flow.

If you substitute a column of mercury of the same diameter and the same 1,000-ft length, only about $\frac{1}{60}$ amp flows. Reduce the column to about 17 ft, and 1 amp will again flow.

In these examples you will note that different metals under otherwise identical conditions permit different amperages to flow; stating it in another way, different metals resist the flow of current to different degrees. Resistance is measured in ohms. A given length of conductor is said to have a resistance of one ohm[5] if it permits exactly one ampere to flow when connected to a source of electricity delivering exactly one volt.

In these examples, the 1,000 ft of No. 10 copper wire and the shorter lengths of other metals each have a resistance of 1 ohm. The precise resistance depends on details such as the temperature, the chemical purity of the metal, and many other factors.

In these examples, note that the wire is connected directly across the source of the electrical power. It is not connected to a lamp, a motor, or other device. The piece of wire constitutes a "resistive load" across the source of power; the power flowing does no work except to heat the wire.

---

[5] The ohm is named after George Simon Ohm, a German scientist of the early nineteenth century, who discovered the basic laws concerning resistance.

Ohm's Law. In the preceding paragraph it was shown that 1 volt forces exactly 1 amp to flow through different lengths of wire each made of a different material but each with exactly 1 ohm of resistance. Further experiment will show you that, if the voltage is doubled, 2 amp will flow; if it is 5 volts instead of 1 volt, 5 amp will flow. *As long as other conditions remain constant, the amperage is in direct proportion to the voltage.*

Still further experiment will show you that, if the voltage remains constant and the material in the wire is not changed but its cross-sectional area (not its diameter) is doubled, twice the amperage will flow. If the wire is increased to five times its original cross-sectional area, five times the amperage will flow. *As long as other conditions remain constant, the amperage is in direct proportion to the cross-sectional area of the wire.* Remembering that doubling the diameter of a circle increases its area four times, tripling the diameter increases the area nine times, it is evident that, as long as other conditions do not change, doubling the diameter of a wire increases the amperage four times, increasing the diameter three times increases the amperage nine times, etc. The reason should be obvious: Increasing the area of a wire by four times reduces its resistance to one-fourth of what it was, and so on.

The data given above pertain only when the wire in question is connected directly across a source of electricity. The conclusion should not be reached that doubling the cross-sectional area of the wire used to connect a motor, for example, will double the amperage flowing through the motor. The wire used for connections is only a small portion of the total wire in the circuit; the wire inside the motor must also be taken into consideration.

The basic principles and definitions outlined in the preceding paragraphs make it easy to recreate the well-known formula, known as Ohm's law, for calculating resistance, voltage, and amperage. If two of these factors are known, it is easy to calculate the missing one. The formula is

$$\frac{\textbf{Volts}}{\textbf{Amperes}} = \textbf{ohms}$$

For brevity, it is customary to use the standard symbols for these three factors:

E for voltage
I for amperes
R for ohms

The same formula then becomes **E/I = R.** If by measurement the voltage is 10 and the amperage is 2, the resistance must be 5 ohms. Likewise, if the voltage is 110 and the amperage is 22, the resistance is 5 ohms.

The formula can be transposed easily so that instead of

$$\frac{E}{I} = R$$

it becomes

$$\frac{E}{R} = I$$

or the third form

$$E = I \times R$$

Divide the voltage by the amperes to find the ohms. Divide the voltage by the ohms to find the amperes. Multiply the amperes by the ohms to find the volts.

Careful inspection of the formula will confirm the facts pointed out in the previous paragraphs:

Voltage being constant, reduce resistance (increase size of wire) to increase the amperage.

Resistance (size and length of wire) being constant, increase voltage to increase amperage.

Amperage being constant, reduce resistance (increase size of wire) to permit lower voltage to be used.

Other Equations Derived from Ohm's Law. If we introduce watts into the formula (using the abbreviation W), the formula can be transposed into many forms. The 12 usual equations follow:

$W =$	EI	I^2R	$\dfrac{E^2}{R}$			
$E =$		IR		$\sqrt{WR}$		$\dfrac{W}{I}$
$I =$			$\dfrac{E}{R}$	$\sqrt{\dfrac{W}{R}}$	$\dfrac{W}{E}$	
$R =$	$\dfrac{E}{I}$				$\dfrac{E^2}{W}$	$\dfrac{W}{I^2}$

$I =$ current, amperes $E =$ voltage
$R =$ resistance, ohms $W =$ watts

Dangerous Voltages. Whether any given voltage is dangerous to human life depends on a great many factors.   Sometimes a voltage as low as 115 is fatal, yet at other times individuals come in contact with much higher voltages and survive.   The sensible course to follow is *safety first*—assume that any voltage of 115 and upward is dangerous. In specific cases, everything depends on such factors as the health of the individual, whether he is in contact with a grounded object, the amperage available along with the voltage, and many other factors. For example, the voltage involved when you touch a spark plug in a car is of the order of several thousand volts yet results only in an unpleasant shock.   It is a very brief shock; if the current were continued, it would perhaps be fatal for at least some individuals.

Voltage Drop.   All conductors have resistance; it requires energy to force a current through them.   Assume a motor connected through a *long* length of wire to a source of electricity.   Connecting a voltmeter directly across the start of the circuit of Fig. 2-7 will indicate the full

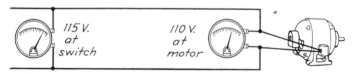

Fig. 2-7 This circuit illustrates voltage drop.

voltage, probably 115 volts. The same meter connected directly across the motor will indicate a lower figure, probably 110 volts.   The difference of 5 volts has been consumed in forcing the current through the wire.

The voltage that is lost in forcing current through the wire is known as "voltage drop." It is wasted power as far as useful purpose is concerned; it merely heats the wires. Excessive voltage drop also is responsible for many other bad effects, which will be discussed further in Chap. 7.

Voltage drop is like lost pressure in a water hose. If a short hose is attached to a water faucet at the side of a house, you can squirt water, say, 50 ft. If a much longer length of hose is attached to the same faucet, you can squirt only a lesser distance, say, 35 ft. A pressure gauge might indicate 35 lb pressure at the faucet but 20 lb pressure at the end of the hose. The difference is expended in forcing water through the hose, and it corresponds to voltage drop in a wire.

The actual amount of voltage drop is easily calculated if the resistance of the wire and the amperage are known, using Ohm's law, $E = I \times R$. For example, if the amperage is $7\frac{1}{2}$ and the resistance is 2 ohms, the drop is $7\frac{1}{2}$ (amp) $\times 2$ (ohms), or 15 volts.

From this it is equally simple to calculate the amount of power wasted. Remembering that watts = volts × amperes, the loss in the example of the previous paragraph is 15 (volts) × $7\frac{1}{2}$ (amp) or $112\frac{1}{2}$ watts.

Assume now that in the same example the amperage is doubled so that 15 amp flows instead of $7\frac{1}{2}$ amp. The voltage drop now is 15 (amp) $\times 2$ (ohms), or 30 volts. *Doubling the amperage doubled the voltage drop.* Calculating the wattage loss involved, 30 (volts) × 15 (amp) gives 450 watts, instead of $112\frac{1}{2}$ watts at $7\frac{1}{2}$ amp. *Doubling the amperage increased the wattage loss four times.* The wattage loss in a circuit is proportional to the *square*[6] of the amperage. Current cannot be made to pass through a wire without some loss; but doubling the amperage of the load without increasing the wire size (which is the same as doubling the wattage so long as the starting voltage does not change) increases the wasted watts by four times; tripling the amperage increases the wasted watts by nine times; etc.

The formula for wattage loss has already been given as

Wattage loss = amperes × voltage drop

However, since voltage drop = amperes × ohms, substituting "am-

---

[6] The square of a number is that number multiplied by itself. Thus 16 is the square of 4($4 \times 4 = 16$).

peres × ohms" for "voltage drop" in the first formula gives this new formula:

Wattage loss = amperes × amperes × ohms = I²R

This merely puts into formula form the statement of the previous paragraph that the wattage loss is proportional to the *square* of the amperage.   All this concerns wattage loss in a given circuit with a given size of wire.

Operating Voltage.   Previous paragraphs explained the fact that the greater the amperage in a wire, the greater the voltage drop and the greater the wattage lost in the form of heat.   From this it should be obvious that, in order to carry high amperages without undue loss, large sizes of wire are required.   The greater the distance, the heavier the wire must be.   Therefore it is distinctly advantageous to keep amperages as low as is practical.

This, at least in theory, is simple, for any given wattage may consist of a low voltage with high amperage or of a high voltage with low amperage.   Therefore for low amperages, relatively high voltages must be used, automatically giving correspondingly low amperages.

In practice, the actual voltage depends on the amount of power to be transmitted and the distance.   In an automobile, while the wattage is very heavy at times, a battery of only 12 volts is used, even if the amperage flowing through the starting motor when it is cranking the engine is often over 250 amp; this is practical only because the distance is so short.

On an old-fashioned farm lighting plant, usually a battery of 32 volts was used.   The distances were seldom over a couple of hundred feet, and the wattages were reasonably small, leading to a relatively small amperage.   In spite of all that, relatively large wires had to be used.

For ordinary residential lighting the voltage is usually 115,[7] while for ranges and water heaters it is 230 volts.   For industrial purposes, where the wattages are great, 460 and 575 volts are usually used.   The distribution lines that run down city alleys and from farm to farm are usually 2,300 to 13,800 volts, but the main distribution lines are at still

[7] The voltage is often referred to as "110 volts" although today actually it is usually either 115 or 120 volts, with the trend toward 120 volts.   Throughout this book, whenever the terms "115 volts" and "230 volts" are used, they will mean the common voltages, whether they happen to be 110 and 220, 115 and 230, or 120 and 240.   In many foreign countries the standard voltage is 220.

higher voltages until, for long-distance cross-country distribution, the voltages are often 345,000 volts or more.

Since it is advantageous to keep amperages as low as possible in order to reduce voltage and wattage losses in the wires and to do away with the necessity of buying large size wire when a smaller size will do, and since this can be done by making the voltage higher, it would appear entirely logical to use a high voltage for all purposes. You might well ask: "Why not use, for ordinary house wiring, 230 volts or 500 volts or higher?"

First of all, the higher voltages require heavier insulation, so that wire becomes more expensive; the higher voltages are more dangerous in case of accidental contact. Another important consideration is the fact that in the manufacture of devices consuming relatively low wattage, under 100 watts, the wire used inside the device is often of almost microscopic dimensions, even when the device is for a voltage as low as 115 volts. For example, the tungsten wire in the filament of a 60-watt 115-volt lamp as manufactured today is only 0.0018 in. in diameter; in a 3-watt lamp it is about 0.00033 in.[8] in diameter. If the device were for 230-volt use or for an even higher voltage, the wire would have to be still smaller, making factory production and uniformity decidedly difficult. The device would also be more fragile, and it would burn out more easily. The present common level of 115 volts is a compromise for lowest over-all cost of installation, operation, and purchase of devices to be operated.

However, since the same home usually has small devices consuming from 5 to 500 watts, also appliances like electric ranges which may consume over 10,000 watts, it would be desirable to have available two different voltages, one relatively low for low-wattage devices and one relatively high for high-wattage devices. Fortunately this is practical.

Three-wire Systems. The 3-wire system in common use in homes today provides both 115 and 230 volts. Only three incoming wires are used and only a single meter. The 3-wire 115/230-volt system constitutes the ordinary system as installed in practically all houses and farms. The higher voltage is usually used for any single device consuming 1,650 watts or more.

Figure 2-8 shows two generators,[9] each delivering 115 volts; the two

[8] To cover a space of 1 in., 3,000 such filaments would have to be laid side by side.

[9] Two generators are not actually used. This will become clear in a later chapter concerning transformers.

combined deliver 230 volts. Any device connected to either wires *A* and *B* (or wires *B* and *C*) will be connected to 115 volts. Any device connected to wires *A* and *C* will be connected to 230 volts. In actual wiring the central or *neutral* wire *B* is white; the outer two or "hot"

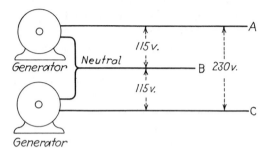

Fig. 2-8 With only three wires, two separate voltages are available.

wires are black or some other color, but never white or green. Connect any device operating on 115 volts to one black and one white wire, and any device operating on 230 volts to the two black wires. (These colors are correct only on grounded-neutral systems; this will be explained in the chapter on grounding. Practically all installations today have a grounded neutral.)

Effects of Electricity. The endless assortment of things that electricity does can, in great part, be broken down into forms or combinations of three basic effects: thermal, magnetic, chemical.

The thermal effect of electricity is simply heat. A current cannot flow without causing some heat. Sometimes heat is not desired, as, for example, in the case of the unavoidable wattage loss referred to in the examples in this chapter. In an ordinary lamp over 90% of the current is wasted as heat, and less than 10% is converted into light, but the light is not possible without the heat. In a toaster or flatiron the heat only is desired.

The magnetic effect can be stated very simply. When a current flows through a wire, the wire is surrounded by a magnetic field—the area immediately around the wire becomes magnetized. Bring a small compass near a wire that is carrying current and the needle will move just as it will when you bring it near an ordinary horseshoe magnet. Wrap a wire a number of times around a piece of soft iron that is not in the least magnetic; during the time that a current flows through the wire, the soft iron becomes a magnet, weak or powerful depending upon such factors as the number of turns of wire and the number of amperes flowing.   The moment the current stops flowing, the iron ceases to be magnetic.   It is this magnetic effect that causes doorbells to ring, motors to run, and telephones and radio speakers to operate.

The chemical effects are of great variety, including the electroplating of metals, the charging of storage batteries, and the electrolytic refinement of metals.   In a dry-cell battery we have the reverse effect: A chemical action produces an electric current.

3

AC and DC; Power Factor; Transformers

As you read about electrical subjects, you will frequently meet the words "direct current" and "alternating current," also "cycles"; you will read about "single-phase," "2-phase," "3-phase," and "polyphase." These terms, while at first formidable and not at all understood by many people, really are fairly simple and easily understood if only you pay close attention to their explanation.

Direct Current.   If an ordinary *direct-current* voltmeter (such as is used for testing dry cells or radio batteries) is connected to a battery, the pointer will swing either to the right or to the left, depending on how the two terminals on the meter are connected to the corresponding terminals of the battery.   Inspection will show that the two terminals of the meter are marked " + " and " − ," "P" and "N," or "pos." and "neg.," all indicating positive and negative; the battery terminals are similarly marked.   Only when the positive terminal of the meter is connected to the positive terminal of the battery will the pointer swing in the right direction.   If on any source of electricity, whether battery or generator or other device, one terminal is positive, the other negative, *and they never change,* the current is known as "direct current," or "DC." Current from any type of battery is *always* direct current.

Alternating Current.   Instead of an ordinary voltmeter which has the zero at one end of the scale, a zero-center voltmeter of the type shown in Fig. 3-1 may be used.   This meter is the same as the first except that

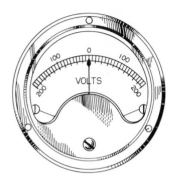

Fig. 3-1 The voltmeter above is the same type as that in Fig. 2-1, except that the zero is at the center of the scale.

the terminals are not marked "pos." and "neg."   Connect the terminals of this meter to the two terminals of a battery and note which way the needle swings.   Then reverse the two leads to the battery, and the needle will swing in the opposite direction.   The meter is equally easy to read whether the pointer swings to the right or to the left, and it provides the additional convenience that it is not necessary, before connection is made, to investigate carefully which is the positive and which is the negative terminal.

We are now ready to perform an experiment using this voltmeter with the zero in the center.   Its two terminals are connected to a source of electricity, the nature of which is unknown to us.   The pointer of the voltmeter performs in a peculiar fashion.   It never comes to rest, but keeps on swinging from one end of the scale to the other, and back again, with great regularity.   Let us watch that pointer carefully, starting from the zero in the center.

It starts swinging toward the *right,* first rapidly, then more slowly, until it reaches a maximum of about 162½ volts in exactly 15 sec. Then it starts dropping back toward 0, first slowly, then rapidly, until in 15 sec more it is back at 0.   It does not stay there but keeps on swinging toward the *left,* and in 15 sec more it reaches the extreme left at

162½ volts, the same relative position as it originally had at the right. Again it swings back toward the right, and in 15 sec more, 1 min from the starting point, it is back where it started from—the zero.

It repeats this same procedure indefinitely, every minute. From observing the pointer it is evident that each wire is first positive, then negative, then positive, then negative, and so on, alternating between positive and negative continuously. The voltage is never constant, is always changing from 0 to a maximum of 162½ volts, first on the positive side, then on the negative. Current in which any given wire regularly changes from positive to negative, not suddenly but gradually as outlined above, is known as "alternating current," or "AC." If the data just observed are plotted, the actual voltage against the time, they will produce a chart such as is shown in Fig. 3-2. This portrays one cycle of alternating current.

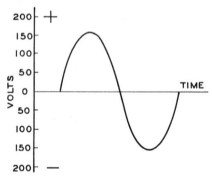

Fig. 3-2 This represents one cycle of alternating current. The voltage fluctuates regularly and continuously from zero to maximum to zero, and each wire alternates regularly between positive and negative.

Alternating current if considered *at any fixed moment,* rather than over an interval of time, is still direct current. *Alternating current may be defined as a current of regularly fluctuating voltage and regularly reversing polarity.*

Frequency. Alternating current which takes a full minute to go through the entire cycle (from no voltage to maximum voltage on the positive side, back to 0, to maximum voltage on the negative side, back to 0) would be known as "1 cycle *per minute.*" There is no such current in actual practice; actually in ordinary 60-cycle alternating current, as used in over 99% of all wired American homes and industrial establishments, all this change takes place at the rate of 60 times *per second*, much too fast to be observed by an ordinary voltmeter. Such current is then said to have a "frequency" of 60 cycles, the "per second" being understood. In the United States, practically all current is 60-cycle, and the few remaining 25-, 40-, and 50-cycle installations are rapidly being changed over to 60-cycle. In foreign countries, most installations are 50-cycle, but various other frequencies are also in commercial use, some as high as 133-cycle. In the United States, 180-cycle current is in use in a few industries, and fluorescent-lighting installations operating at 400 cycles are coming into use; military equipment also often operates at 400 cycles. The advantages of the higher frequencies lie chiefly in the fact that motors, transformers, and similar equipment become smaller and smaller in physical size as the frequency increases.

It will be well to remember that "kilo" means "thousand." When your radio receiver is tuned to a station operating at 1,250 kilocycles (abbr. kc), it means that the signal coming into the receiver is alternating current of 1,250,000 cycles per second. If the receiver is tuned to an FM station operating at 90 megacycles ("mega" is a Greek word that has been adopted to designate "million"), it means that the signal is alternating current of 90,000,000 cycles per second.

Voltage of Alternating Current. In the curve of Fig. 3-2 the voltages range between 0 and 162½ volts. If a lamp rated at 115 volts is connected to a circuit of such varying voltage (1 cycle *per minute*), it will burn far more brightly than normal while the voltage is above 115 volts, less brightly than normal while the voltage is under 115, and part of the time the lamp will not light at all, because the voltage is very low, even zero twice during the cycle. Flickering would be extreme and unendurable. However, in the case of the ordinary 60-cycle alternating current, all this change of voltage takes place twice per cycle, 120 times *every second*. The filament of a lamp does not have time to cool off during the very short periods of time when no voltage is impressed on it, which is the reason for lack of observable flicker. In the case of very small lamps which have very thin filaments that can cool off quickly, operated on 25-cycle current which is still found in a few localities, a noticeable and annoying flicker is present.

The rated voltage of an alternating-current circuit is a value between

0 and the peak voltage and in the case under discussion is 115 volts.[1] An alternating-current voltmeter connected to the circuit will read 115 volts. A 115-volt alternating-current source will light a 115-volt lamp to the same brilliancy as a 115-volt direct-current source.

Alternating Current and Motors.   Alternating current as discussed up to this point is "single-phase" alternating current.   In foreign countries it is frequently designated "monophasic" current.   When applied to a motor, remember that it magnetizes the steel poles of the motor every time it builds up from zero to peak voltage, or in other words 120 times per second, as shown in Fig. 3-3, which shows three consecutive cycles

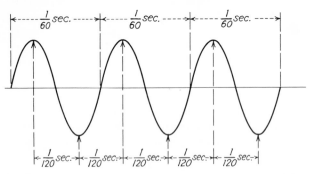

Fig. 3-3   Three cycles of 60-cycle alternating current.   All the changes shown take place in $\frac{3}{60}$ or $\frac{1}{20}$ sec.

of 60-cycle current.   At the top is indicated the time between cycles, or $\frac{1}{60}$ sec.   At the bottom is indicated the time between alternations, or $\frac{1}{120}$ sec.   One might say that the motor is given a push 120 times a second, just as a gasoline engine is given a push every time there is an explosion in the cylinder.   Offhand, 120 times per second may seem fast enough for any purpose, but remember that an ordinary motor runs at 1,800 rpm, which means that the rotor (the rotating part) makes 30 revolutions every second.   In turn this means that the 120 pushes per second become only 4 pushes per revolution; if the motor is a large one, the rotor or armature may be 12 in. in diameter, over 36 in. in circumference, which in turn means that a point on the rotor has to turn about 9 in. between pushes.   Do not imagine that these pushes are abrupt

[1] The rated voltage is 0.707 of the peak voltage; the student will recognize 0.707 as $\frac{1}{2} \sqrt{2}$.   From this it is evident that the peak voltage is rated voltage times $\sqrt{2}$.

sudden impacts.   They are gradual pushes that start slowly and build up to a maximum as the voltage builds up to a maximum value.

In an ordinary 1-cylinder 4-cycle gasoline engine, running at 1,800 rpm, there is an explosion in the cylinder every other revolution, or 900 times every minute, or 15 times every second.   The crankshaft gets a push 15 times every second.   If more pushes are needed every second to secure smoother operation, or more power, it is simply done by using more cylinders: two or four or as many as needed.   How is this to be done in the case of an electric motor?   Fortunately it is rather simple.

Three-phase Alternating Current.   It can be done by putting into the motor three separate windings not connected to each other in any way,

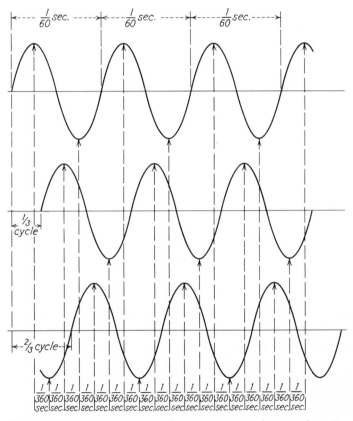

Fig. 3-4   Three separate single-phase currents combine to form 3-phase current.

but each connected to one of three separate sources of single-phase alternating current. The three separate sources of current must be so designed that the peak voltage of one does not coincide with the peak voltage of another. The voltages in the three sources come to their peaks in very regular fashion, one after the other. Then the motor receives three times as many pushes as before. Figure 3-4 shows the voltage curves of the three separate sources. At the top is indicated the time between cycles *in each separate source:* $\frac{1}{60}$ sec. At the bottom is indicated the time between pushes *from the three separate sources combined:* $\frac{1}{360}$ sec. That is 3-phase current, and it is nothing more or less than three separate sources of single-phase alternating current so arranged that the peaks of voltage follow each other in a regular, repeating pattern.

Do note that for each phase, the duration of a cycle is $\frac{1}{60}$ sec, but there are two pushes per cycle, so that the time between pushes is $\frac{1}{120}$ sec. But look at the bottom of the diagram and you will see that the pushes from the three phases combined are only $\frac{1}{360}$ sec apart. The pushes are imparted by each of the three windings in turn, as shown by the dotted lines from the peaks to the bottom of the diagram.

Figure 3-5 shows this diagrammatically: generator *A* and (inside the

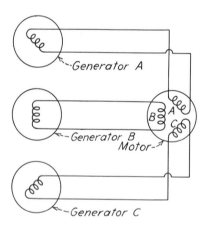

Fig. 3-5 Three single-phase generators form a 3-phase circuit, here connected to a 3-phase motor.

motor) winding *A*, also generator *B* and winding *B*, also generator *C* and winding *C*. In practice it would be impossible to make three separate generators run in such precisely uniform fashion that the peaks of voltage would come at precisely the right time, and it would also be a most uneconomical method.   A single generator is used with three separate windings so that the peak and the zero voltage of each winding come at precisely the right time.   This is shown in Fig. 3-6, which also shows how the six wires shown in Fig. 3-5 become only three wires in actual practice.

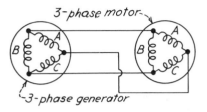

Fig. 3-6 The three separate single-phase generators of Fig. 3-5 have now been combined into one 3-phase generator. Only three wires are used instead of six.

Two-phase Alternating Current.   Instead of three phases, 2-phase power operating on the same principle can also be used.   However, this system is used in so few localities that it does not warrant space in this book.

Polyphase Current.   When a current is either 2 phase or 3 phase, it is known as "polyphase" current (from the Greek word *poly* meaning "many").   It is also known as "multiphasic" current.   With a few rare exceptions, polyphase current is never found in homes but is the general type of current employed for commercial and industrial uses for operating motors and similar devices.   Even in those establishments, single-phase current is of course used for lighting purposes and operating small miscellaneous devices.

Abbreviation.   The word "phase" is usually abbreviated ϕ, the Greek letter *phi.*

Volt-amperes.   The   previous   chapter   contained   the   formula volts $\times$ amperes = watts.   That formula is always correct in connection with direct current, but in connection with alternating current it is not correct because of power factor, which will be discussed in the next paragraph.   Before that can be discussed, it is necessary that the term "volt-amperes" be understood.

In *single-phase work:*

Volt-amperes = volts $\times$ amperes

In 3-*phase work:*

Volt-amperes = 1.73 $\times$ volts $\times$ amperes

In 3-phase work you will frequently meet the multiplier 1.73.   Note that $1.73 = \sqrt{3}$.

Kilovolt-amperes.   One thousand volt-amperes is one kilovolt-ampere, abbreviated kva.   If the power factor of the load happens to be 100%, then and only then is one kilovolt-ampere the same as one kilowatt (kw).

Power Factor.   To explain just what power factor is must be left to any good book on electrical engineering.   An explanation of how to measure it and a general idea of its importance are simple enough to be covered here.

In a direct-current circuit consisting of the load, together with an ammeter, a voltmeter, and a wattmeter, the product of the volts and the amperes is *without exception* equal to the reading of the wattmeter. If, however, the same experiment is made on an alternating-current circuit, sometimes the same is true, sometimes not.

In an alternating-current circuit, whenever measurements show that the product of the volts and the amperes *is exactly equal* to the wattmeter reading, the device that constitutes the load is said to have a power factor of 100%.   Into this classification fall lamps, most appliances that generate only heat, and in general all noninductive devices, that is, those that do not involve windings of wire around a steel core.

If the product of volts times amperes *is greater than* the reading of the wattmeter, then the power factor is less than 100%.

Power factor is abbreviated PF; it is referred to also as "$\cos \theta$." [2]   It is defined as the proportion between the real or measured watts (also known as "effective power") and the volt-amperes (also known as "apparent watts").   The formula is simply

$$\text{Power factor} = \frac{\text{watts}}{\text{volt-amperes}}$$

Measuring Power Factor.   To measure power factor we need only a voltmeter, ammeter, and wattmeter.   Assume a small single-phase motor on a circuit of 115 volts, consuming 5 amp as indicated by the ammeter and 345 watts as indicated by the wattmeter.   The formula then becomes

$$\text{Power factor} = \frac{345}{5 \times 115} \text{ or } \frac{345}{575} \text{ or } 60\%$$

[2] Cosine theta.

In the case of a 3-phase 230-volt motor consuming 12 amp, the volt-amperes are $1.73 \times 230 \times 12$, or 4,775. If the wattage as indicated by the wattmeter is 3,950,

$$\text{Power factor} = \frac{3,950}{4,775} \text{ or } 82.3\%$$

Generally speaking, the power factor of a motor improves (increases in percentage) with the increase in horsepower of the motor and also varies considerably with the type and quality of the motor in question. It may be as low as 50% for small fractional-horsepower motors, and over 90% for a 25-hp motor.

Watts in Alternating-current Work. The correct formula for use in connection with alternating-current work is

Watts = volt-amperes × power factor

In using this formula do not overlook the fact that, if the power is 3-phase, the product of the voltmeter and ammeter readings must be multiplied by 1.73.

Desirability of High Power Factor. Assume that a small factory is using 100 amp of single-phase power at 230 volts, a total of 23,000 volt-amp, or 23 kva. If the power factor is 100%, this is equivalent to 23 kw. At 5 cents per kilowatthour, the power company receives $1.15 per hour for the total power.

Now assume a second factory also using 100 amp at 230 volts, but with a power factor of only 50%. That is still 23 kva but only 11.5 kw, and at 5 cents per kilowatthour the power company now receives only 57½ cents per hour.

Since it is the kilovolt-ampere load that determines wire size, transformer and generator size, and similar factors, and since each factory uses the same 23 kva, the power company must furnish wires just as big for the factory where they are paid 57½ cents per hour as for the one where they are paid $1.15 per hour; they tie up just as much transformer capacity, generator capacity, and all other equipment for the one as they do for the other.

It is natural, therefore, that power companies, when furnishing power to establishments where the power factor is low, not only charge for the kilowatthours consumed but also make an extra charge based on the kilovolt-amperes used during the month or period in question, as compared with the kilowatthours used. Since with a constant load in watts the volt-amperes decrease as the power factor increases, it is definitely in order to watch the power factor very carefully. Few installations attain 100% power factor, and rarely does one fall as low as 50%. The over-all power factor in an industrial establishment is generally determined by the electric motors in use, although other devices also contribute their share.

Power-factor Correction. The theory covering power-factor correction is entirely beyond the scope of this book but can be found in any good book on electrical engineering. The actual correction is accomplished by means of capacitors or synchronous motors; the required calculations should be made by one thoroughly familiar with the subject. Correcting the power factor not only reduces the charges for power consumed but carries with it many other advantages, including higher efficiency of electrical machinery because of reduced voltage drop.

Transformers. When it is necessary to transmit thousands of kilowatts of electrical energy over a considerable distance, wire large enough to transmit it at 115 or even 230 volts would have to be so big that the cost would be entirely prohibitive. If a relatively small wire and a much higher voltage are used, the voltage will be so high as to be dangerous in the final consumption and there will be many other disadvantages.

It would be most convenient, therefore, to have a way of changing current from one voltage to another as required. In the case of direct current there is no simple, efficient device available, but for alternating current there is fortunately a simple and efficient device that does just that—the transformer.

If an electric current flows through a wire that is wrapped around a soft iron rod or core, the core becomes a magnet as long as the current flows. The experiment can be simply made by wrapping a couple of dozen turns of insulated wire around an iron bolt; connect a dry cell to the two ends of the wire, and the bolt becomes a magnet as long as the dry cell is connected (see Fig. 3-7). Magnetic lines of force surround the wire and build up in the iron core. The moment the dry cell is disconnected, the bolt loses practically all its magnetism.

With a good galvanometer (which is a *very* sensitive direct-current voltmeter), this next experiment is simply made. Discard the dry cell, and connect the two ends of a coil with at least several hundred turns of wire to the two terminals of the galvanometer (see Fig. 3-8). Push the iron bolt suddenly into the coil or pull it suddenly out; nothing happens.

Then use a permanent magnet of any type; a 10-cent horseshoe magnet will serve the purpose.   The more powerful the magnet, the easier it will be to make this demonstration.   Push one leg of the magnet suddenly into the coil, and the needle on the galvanometer, if sufficiently

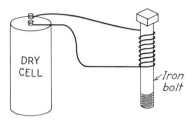

Fig.   3-7 When   current   flows through a wire wrapped around an iron bolt, the bolt becomes a magnet, but only as long as the current flows.

Fig. 3-8 Pushing a magnet into a coil generates a current indicated by the voltmeter. This is the simplest possible electrical generator.

sensitive, will move, then drop back to zero.   As long as the magnet is stationary inside the coil, nothing happens. Pull the magnet out suddenly, and the galvanometer needle will again move, but in a direction opposite to that taken when the magnet was pushed into the coil.   The more turns of wire in the coil, the stronger the magnet, the easier it will be to perform this experiment.

This demonstrates that, whenever there is a *change* in the magnetism in the space occupied by a coil of wire, electricity flows in the wire; if magnetism of a constant nature is there, nothing happens.

Consider now what will happen if 60-cycle alternating current is connected to the two ends of the coil of wire around the bolt.   The voltage in the alternating current applied to the coil changes 120 times every second from zero to maximum to zero.   Therefore 120 times every second the bolt becomes a magnet, and 120 times every second it loses its magnetic power, as the voltage in the circuit builds up from zero to maximum and then drops back to zero.   Consider now a contraption like that in Fig. 3-9, where again an iron bolt is used and on it two coils of wire, *A* and *B*, not connected to each other in any way.   To coil *A* is connected a source delivering 60-cycle alternating current; a lamp is connected in series with it to limit the current.   The other coil is connected to a very sensitive *alternating-current* voltmeter.   Remember that, whenever a magnet was moved inside the coil in the first experiment, electricity flowed in the coil. Remember also that, with 60-cycle current flowing through coil A, the bolt becomes a magnet, then becomes just a plain iron bolt, 120 times every second. That is exactly the same as inserting and removing a magnet into coil B 60 times per

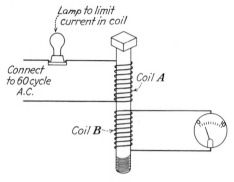

Fig. 3-9   A very crude transformer.

second; accordingly, alternating current will flow in coil B, and the voltmeter, if it is sensitive enough, will show it. The device is a transformer, which transfers current from one coil to another coil not connected to it.

The device described is exceedingly crude. If a longer bolt is used and bent into the form of a complete circle, as shown in Fig. 3-10, it will be much more efficient and more current will flow in coil B.

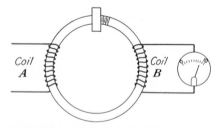

Fig. 3-10   Still a crude transformer, but much more efficient than the one shown in Fig. 3-9.

In commercial use, solid cores like the bolt would be impractical because solid cores heat excessively. Instead, thin sheets of a special grade of steel are used, cut in U sections, which can later be stacked into a core, usually of rectangular shape, generally with the two coils over opposite legs. Many variations are possible, but for showing the principle, the type shown in Fig. 3-11 will serve the purpose. The coil to which the power is applied is called the "primary"; the other coil from which the power is taken is called the "secondary."

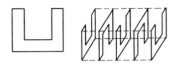

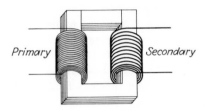

Fig. 3-11 This shows the general construction of transformers.

Primary

Secondary

Transformer Ratios. If the primary of a properly designed transformer is connected to a source of alternating current (of the voltage for which the primary was designed) and of practically unlimited power, but if the two ends of the secondary winding are *not* connected to anything or to each other, practically no current will flow in the primary. If, however, a load (lamps, appliances, motors, etc.) is connected to the secondary, then just as much current will flow in the primary as is required to deliver the required wattage to the secondary, but no more (assuming, of course, that the capacity of the transformer is adequate to the load connected to it).

Experiment shows that, if the primary has as many turns of wire as the secondary, the voltage and the amperage that can be made to flow in the secondary will be exactly the same as the voltage and the amperage in the primary, minus a small percentage because a transformer is not 100% efficient. It would be more correct to say that the voltage of the primary and the secondary will be the same, and the amperage flowing in the primary from the power line adjusts itself to the amperage demanded in the secondary by the nature of the particular load connected to it.

If the secondary has twice as many turns as the primary, the voltage in the secondary will be twice that of the primary but the amperage will be only half as great.   If the secondary has ten times as many turns as the primary, the voltage in the secondary will be ten times that of the primary but the amperage will be only one-tenth as great.   By reversing the proportions and having fewer turns in the secondary than in the primary, it is equally simple to step the voltage down, instead of up; the amperage, of course, will go up as the voltage goes down.   The volt-amperes in the secondary are always equal to the volt-amperes in the primary minus a few per cent, depending on the efficiency of the transformer.

The minimum number of turns must be kept within the limits that good engineering has shown lead to the greatest efficiency, and wire sizes in both primary and secondary must be chosen to carry the amperages involved.   The smallest transformer usually found is the ordinary doorbell type, which steps 115-volt alternating current down to about 8 volts for operating doorbells and similar equipment; the largest are so big that there is difficulty finding railway cars sturdy enough to transport them.

Well-built transformers are very efficient, and, generally speaking, the larger the transformer, the greater the efficiency.   In very large transformers it is possible to recover from the secondary over 99% of the power applied to the primary.

Practical Use of Transformers.   In a large generating station, the power is generated at various voltages; 13,800 volts is typical.   The power is fed into transformers and stepped up to a much higher voltage, from 23,000 to 345,000 volts.   Where the power is to be consumed, it is stepped down to lower voltages at local substations, often 13,800 volts. There it is again stepped down to still lower voltages, usually 2,300 or 4,000 volts, at which voltage it is transmitted in the lines running down city alleys or from farm to farm.   At strategic points it is again fed through transformers and stepped down to 115/230 volts, at which value it is used.   See Fig. 3-12.

The maximum voltage at which power is transmitted depends on many factors, including the amount of power involved, and the distance.   Several 500,000-volt lines, and one 765,000-volt line, are in use, and still higher voltages are being considered.

Series-parallel Connections.   It is the usual practice in power and lighting transformers to have both the primary and the secondary con-

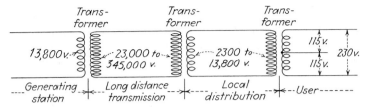

Fig. 3-12 Power is generated at a relatively low voltage, then stepped up to a much higher voltage for transmission over a distance, then stepped down to a working voltage.

sist each of two separate coils. When the two primary coils are connected in series, as shown in A of Fig. 3-13, the primary will be suitable for connection to a 4,600-volt line; reconnected in parallel, as shown in

Fig. 3-13 By connecting the windings of transformers in series or parallel, one transformer serves for two different voltages.

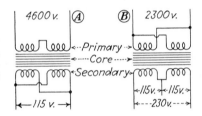

B, the transformer becomes suitable for 2,300 volts. Likewise, the two secondary sections can be connected in parallel to deliver 115 volts, as shown in A, or in series to deliver 230 volts. More usually the secondary coils are connected in series, with a tap at the midpoint, forming the common 3-wire 115/230-volt system, as shown in B. Any given transformer, when the secondaries are connected in parallel to deliver 115 volts, will deliver twice the amperage that it will on 230 volts.

Use on Alternating Current Only. Considering the discussion in earlier paragraphs, it should be superfluous to state this, but let it be repeated: A transformer operates only on *alternating* current.

Three-phase Transformers. A 3-phase transformer bank consists of three separate single-phase transformers, one for each phase. In the diagrams that follow, only the secondaries of the transformers are shown. They may be connected in either the delta fashion shown in Fig. 3-14, or the wye or Y (sometimes called "star") fashion shown in Fig. 3-15. Three wires must be run to any 3-phase load such as a 3-phase motor, and the current is of course automatically 3-phase.

But remember that the current from any *two* wires of a 3-phase system is automatically *single-phase*. From the standpoint of both the user and the power supplier, it is desirable to have a supply that will deliver both single-phase and 3-phase power. It is a simple matter.

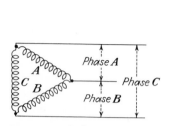

Fig. 3-14 Delta method of connecting 3-phase transformers.

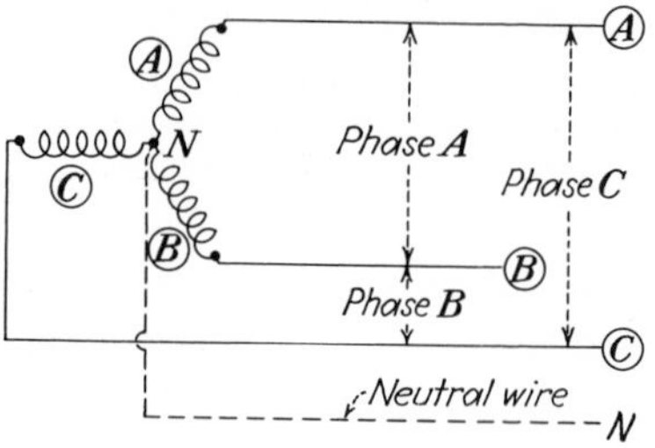

Fig. 3-15 Star or Y method of connecting 3-phase transformers.

In the delta system of Fig. 3-14, the power may be at 230, 460, or 575 volts (or more often at 240, 480, or 600 volts). The single-phase power then available is one of those voltages. Single-phase power at only 230 volts isn't very practical for most purposes; we want 115/230-volt single-phase power. See now Fig. 3-16, which shows the same transformers as in Fig. 3-14, but with one of the secondaries tapped at the midpoint. Assuming the basic voltage is 230, that secondary tapped at the midpoint will deliver 115/230-volt single-phase power, which may be used at the same time as the 3-phase power.

In the Y system shown in Fig. 3-15, a neutral wire is run from the junction of the three secondaries. The 3-phase voltage of any circuit connected to all three wires, A, B, and C, is usually 208 volts. The single-phase voltage between wires A and B (or B and C, or A and C) is

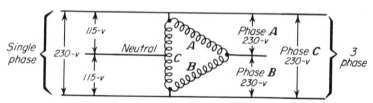

Fig. 3-16 A delta-connected transformer can deliver 3-phase power, and dual-voltage single-phase power, at the same time.

208 volts.   You would then expect the voltage between the neutral wire *N* and either *A, B,* or *C* to be half of 208, or 104 volts, but that is a wrong conclusion.   The voltage between the neutral and any hot wire of a 208-volt 3-phase system connected in Y fashion is 120 volts.

At first glance this may seem all wrong for, if the voltage between wires *A* and *B* in Fig. 3-16 is 208 volts, the voltage between the neutral wire and either *A* or *B* might be expected to be one-half of 208, or 104 volts, instead of 120 volts as previously stated.   Remember, however, that in a 3-phase circuit, the voltage comes to a peak or maximum at a different time in each phase.   At the instant that the voltage in secondary *A* is 120 volts, that in *B* is 88 volts, so that across wires *A* and *B* there is a voltage of 120 + 88, or 208 volts.[3]   The system therefore has the advantage of making it possible to transmit over only four wires (including a grounded neutral) 3-phase power at 208 volts, single-phase power at 208 volts, and single-phase power at 120 volts.   Occasionally in a home, instead of providing the usual 115/230-volt 3-wire system, three wires of the star-connected system (the neutral and any other two wires of Fig. 3-16) are provided, thus furnishing 120 volts for lighting and 208 volts (instead of the usual 230 volts) for water heaters and similar large loads.

Instead of 120/208 volts, newer installations in commercial and industrial establishments provide power at 277/480 volts.   More will be said about this later.

Autotransformers.   An autotransformer can be defined as a transformer in which a portion of the turns are common to both primary and secondary (see Fig. 3-17).   Let there be a tap at the midpoint of the

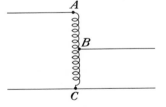

Fig. 3-17 In an autotransformer, some of the turns are common to both primary and secondary.

coil so that, although there are, for example, 1,000 turns of wire between *A* and *C*, there are only 500 between *B* and *C*.   The entire coil *A* and *C* may then be considered the primary, those turns from *B* to *C* the

---

[3] Note that $208 = 120 \sqrt{3}$.

secondary.   Whatever the voltage across *A* to *C*, the voltage across *B* to *C* will be exactly half.   The tap may be at any point in the coil.   The voltage across *B* to *C*, as compared with the total voltage across *A* to *C*, will always be proportional to the number of turns from *B* to *C*, as compared with the total number of turns from *A* to *C*.   This type of construction is somewhat less expensive than the usual two-coil type, but is rarely employed, as its use in general is prohibited by the Code with certain exceptions, notably in connection with motor-starting devices.

High-voltage Direct Current.   Throughout this chapter it has been emphasized that transformers operate only on alternating current.   Yet you may read about high-voltage *direct*-current lines that are in use mostly in foreign countries, at pressures up to 1,000,000 volts or more. How is this possible, when transformers won't work on direct current?   The power is generated as alternating current, stepped up to a high voltage by transformers, and then rectified or changed to direct current.   At the point where it is to be used, it is reconverted to alternating current.   The exact procedure had best be left to books concerning electrical engineering.

There are many advantages to long-distance transmission of direct-current power, but the method is practical only if very large amounts of power are to be transmitted from one point to one particular distant point, without need for tapping off any of the power at intermediate points.   While this method is not yet in common use in the United States, one million-volt line is under construction in the western states. It seems probable that over a period of many years, direct-current transmission will very slowly become more common.

4

Basic Devices and Circuits

In order properly and intelligently to assemble the great number of available electrical devices to form a complete wiring system, you must understand the basic principles regarding electrical devices and electrical circuits.

If the electric current is to produce an effect, it is not enough that the current merely flow up to the device that is to be operated; the current must flow *through* it. In other words, there must be two wires from the starting point (the source of power) to the device. The electric current can be compared with a series of messengers who start from some given point (the generator of an electrical system), make a trip to their destination (the device to be operated), and return to the starting point before their errand is completed. The wires can be considered the streets over which they travel, only they must be considered one-way streets; the messengers must go out on one, return over a different street (wire), because there are millions of them. As a matter of fact, an electric current can be considered as consisting of many millions of billions of such messengers per second for every ampere flowing.[1]

Lamps. The most common electrical device is probably the "light bulb." The correct name is "lamp"; the glass part of the lamp is the bulb. The lamp consists essentially of a filament which is a wire made of tungsten, a metal having a very high resistance and a very high melting point. This makes it possible to heat the wire to a very high temperature (over 4,000°F in ordinary lamps) without it burning out. The filament is suspended on supports inside the lamp, from which the air has been exhausted and into which, in most sizes, usually some inert gas like argon has been introduced to prolong the life. The ends of the filament are brought out to a convenient base, which makes replacement simple. In the base the center contact is insulated from the outer brass part of the base, thus providing two terminals for the two wires leading up to the lamp. The cross section of a lamp shown in Fig. 4-1 should make this clear.

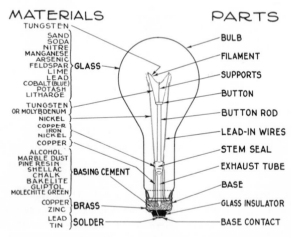

Fig. 4-1 Cross section of an ordinary lamp. (*General Electric Co.*)

Lamps are held in sockets, the simplest being the screw-shell receptacle shown in Fig. 4-2; Fig. 4-3 shows a cross section of it. One terminal *A* is connected to the center contact corresponding to the center

---

[1] See footnote 2, p. 17.

contact on the lamp base; the other terminal *B* is connected to the screw-shell terminal (which is carefully insulated from the center contact and terminal *A*), corresponding to the outer shell of the base on the lamp. When a lamp is screwed into such a socket, the current will flow in at one terminal, through the filament, and out again at the other terminal.

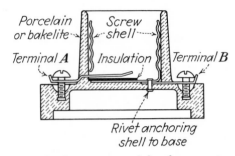

Fig. 4-2 A "cleat receptacle"—the simplest form of socket. (*General Electric Co.*)

Fig. 4-3 Cross section of the cleat receptacle shown in Fig. 4-2.

Circuits.   Used in a general sense, as it will be in this chapter, a circuit is any combination of wires and devices which will permit electric power to do its work.   Perhaps the words "hookup" or "wiring diagram" would be more descriptive.   Only the basic devices necessary to make the combination of devices work will be included in this chapter; the supplementary devices such as conduit, outlet boxes, switch plates will be deferred to a later chapter.

Outlets.   Every point where electric power is taken from the wires *and consumed* is an outlet.   Receptacle (plug-in) outlets in themselves use no current, but since current-consuming devices like radios and lamps are plugged into them, they are considered outlets.   A switch uses no current; therefore it is not an outlet.   Sometimes the term "outlet" is loosely and improperly used to indicate also any point where a device such as a *switch* (which *consumes* no current) is connected to the wires, this being commonly done in contracting work, when estimating the cost of a job on a "per-outlet" basis.

You will often see the term "wiring device."   According to the Code, a wiring *device* is a component that carries current, but does not consume it; examples are switches and receptacles already mentioned, also sockets, push buttons, and so on. Anything that *consumes* power (such as a lamp, a toaster, a motor, and so on) per the Code is "utilization equipment" and constitutes the "load" on the circuit.

Source. In all the diagrams in this book where the word SOURCE appears, it will mean the generator, the battery, or wherever the current comes from—the SOURCE of supply. Actually, it may be the point where the wires enter the building or the point where the particular circuit under discussion begins.

Basic Circuit. Figure 4-4 shows a wire running from SOURCE to the socket with the lamp and another wire from the socket back to SOURCE. The current flows outward through one wire, through the lamp, and

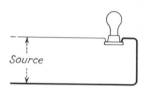

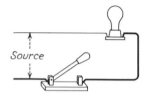

Fig. 4-4 The most simple circuit. There is no way of turning the lamp on or off.

Fig. 4-5 A switch has been added to the circuit at the left to control the lamp.

back through the other wire. This makes a complete circuit, and as long as SOURCE furnishes power, the lamp will light. It is not at all a practical circuit, since it is necessary to disconnect one of the wires from the socket or to cut a wire whenever the light is to be turned off. Such a circuit would not be very sensible, so a switch must be included. This has been done in Fig. 4-5, the switch being the open porcelain-base type. Opening the blade is the same as disconnecting or cutting a wire, or, comparing it with the one-way street, it is the same as opening a drawbridge in the street and thus allowing no way of going ahead on that street until the bridge is again closed.

Toggle Switches. In actual wiring we would not use a clumsy porcelain-base switch of the type shown in Fig. 4-6. Instead, we use a neat toggle switch of the type shown in Fig. 4-7, concealed in the wall, with only the handle showing. It has two terminals just like the knife switch shown in Fig. 4-6. The mechanism is small and compact, but it does exactly what the knife switch does; in one position of the handle the switch is open, in the other position it is closed. Any switch that merely opens one wire is known as a "single-pole" switch. A single-pole toggle switch is identified by its two terminals, and the words ON and OFF on the handle. Obviously this style of switch is much safer than one with an exposed mechanism.

Fig. 4-6 The switch opens one wire.

Fig. 4-7 A toggle switch. The mechanism is completely enclosed. It does exactly what the switch shown in Fig. 4-6 does—it opens one wire. (*General Electric Co.*)

Series Wiring. The circuit of Fig. 4-5 controls only one lamp; often one switch must control two or more lamps. In drawing a diagram for this, most beginners will connect several sockets as shown in Fig. 4-8. The current can be traced from the SOURCE along the one-way street

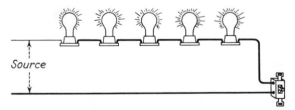

Fig. 4-8 This type of wiring is known as "series" wiring.

(wire) to the first lamp, to the second, to the third, to the fourth, to the fifth, and then along the other one-way street (wire) back through the switch to the SOURCE; consequently the lamps should light. They will light if the correct sizes are used. However, assume that each lamp is a different size; since all the current that flows through one must also flow through the other, the smallest lamps will carry more current than they should and will burn more brightly than normal.   The biggest ones will carry less current than they should and will burn less brightly than normal.   Medium-size lamps may burn at normal brilliancy.   So far the scheme does not seem very practical.   Burning out one lamp or removing it from its socket, as shown in Fig. 4-9, is equivalent to opening

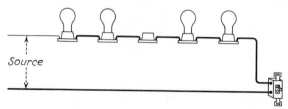

Fig. 4-9   In series wiring, when one lamp goes out, all go out.

a switch in the circuit.   All the lamps go out.   This type of wiring is known as "series" wiring and is impractical for ordinary purposes.[2]

Instead of a picture of a lamp in a socket being used as in past diagrams, from this point onward the arbitrary symbol of Fig. 4-10 will be used to denote a lamp and its socket.   Note also the diagrams of Fig. 4-11, indicating whether wires that cross each other in diagrams are connected to each other or not.

Fig. 4-10 In illustrations from this point onward, the symbol above will be used to indicate a lamp and its socket.

Fig. 4-11 Note carefully the designations above, which show whether crossing wires are connected to each other, or not.

Parallel Wiring.   The scheme used in ordinary wiring is known as "parallel" or "multiple" wiring, shown in Fig. 4-12.   When one lamp

[2] The series circuit was used on old-style Christmas-tree outfits, where eight identical lamps were used and consequently all burned at the same brilliancy.   Each lamp was rated at 15 volts; they could be used on a 115-volt circuit because each lamp received one-eighth of the total of 115 volts, or about 15 volts each.

burns out or is removed, the current can still be traced from the source directly to *each* of the lamps whether there are five as shown, or a dozen or more. From the other terminal of each lamp the current can be

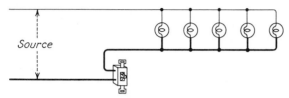

Fig. 4-12   One switch here controls five lamps.

traced back along the wire through the switch to the source. Try it; cover one or more of the lamps with a narrow strip of paper, leaving the wires exposed; the circuit will operate, regardless of the number of lamps in place, and the switch will always turn all the lamps on and off. This is the way the sockets in a five-light fixture are wired, operated by a single switch in the wall.

Using Several Switches.   The circuits covered up to this point might serve well in a one-room summer cottage or an outbuilding on a farm, but all the lights in an entire house would never be controlled by one single switch. It is equally simple to wire a number of sockets with separate switches. Figure 4-13 will be recognized as the same as Fig.

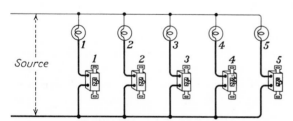

Fig. 4-13   Now each lamp is controlled by a separate switch.

4-12, except that in place of one switch there are now five switches; these have been numbered 1, 2, 3, 4, and 5, and likewise the lamps have been numbered 1, 2, 3, 4, and 5. Cover with a piece of paper both lamps and switches 2, 3, 4, and 5, leaving 1 exposed; immediately it becomes the simple circuit of Fig. 4-5.   Cover lamps and switches 1, 2, 3, and 4, and again it becomes Fig. 4-5.   Cover *any* four switches and lamps, and it becomes Fig. 4-5.   Trace the current from the SOURCE to *any* lamp; it can be traced through the lamp to the switch for that lamp and back to the SOURCE.   This can be done whether one or two or all the switches are on; each one is independent of the others.

Turn now to Fig. 4-14, where a *group* of lamps has been substituted for each single lamp, so that there are now five *groups* of lamps and five

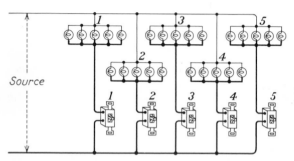

Fig. 4-14   This is the same as Fig. 4-13, except that each switch controls five lamps.

switches, numbered 1, 2, 3, 4, and 5.   Cover with a piece of paper groups 2, 3, 4, and 5 with their switches, and immediately the simple circuit of Fig. 4-12 appears—five lamps controlled by a single switch. Cover any four groups, and in each case the current can be traced from SOURCE to any one of the lamps and through the switches controlling the group (if switches are turned on) back to SOURCE.

Figure 4-14 is the basic wiring diagram for a five-room house with a five-light fixture in each room, controlled by one switch for each room. Actually the wires would run more as shown in Fig. 4-15, which is more pictorial, with wires coming into the basement, then running to two rooms on the first floor and three rooms on the second floor.

Receptacles.   Radios, toasters, floor lamps, and similar devices must be portable; receptacles are used to plug in these devices as required. The basic idea is shown in Fig. 4-16: a pair of metal clips or contacts, one attached to each of the two wires from SOURCE; a plug which has two corresponding clips or contacts which can be brought into connection with the first pair, and a pair of wires leading from them to the lamp or appliance.   Figure 4-17 shows the finished product ordinarily known as a "duplex receptacle" (because it has *two* pairs of openings which will accommodate two plugs at the same time).   The old-fashioned "single receptacle" of Fig. 4-18 is seldom used in new installations today.

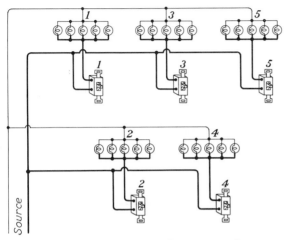

Fig. 4-15   The circuit of Fig. 4-14, but rearranged.

In any wiring diagram, a receptacle can always be substituted for a socket; if, however, the socket is controlled by a wall switch, then whatever is plugged into the receptacle substituted for the socket will also be

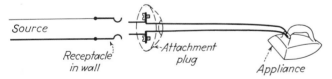

Fig. 4-16   The principle of a plug-in receptacle outlet.

turned on and off by the switch.   In any diagram or circuit, connect the receptacle in such a way that, if it were a lamp, it would always be on. If in doubt, go back to the one-way-street idea, and see if the messengers can go from SOURCE to the receptacle and back again to SOURCE even if all switches are in the open or off position.

Double-pole Switches.   While opening one of the two wires to a lamp turns it on and off, still both wires can be opened if desired, as is done in

Fig. 4-19. The porcelain-base switch there shown is known as the "double-pole single-throw" type, and the corresponding flush toggle switch of the type shown in Fig. 4-7 is usually referred to simply as a

Fig. 4-17 A duplex receptacle permits two different devices to be plugged in at the same time. (*General Electric Co.*)

Fig. 4-18 Single receptacles are little used today. (*General Electric Co.*)

"double-pole" switch.   It is identified by the fact that it has *four* terminals for wires *and* the words ON and OFF on the handle.

Double-pole switches are required by the Code when one of the two conductors is not grounded.   Grounding will be discussed later in this book.   In practice, this means that you must use double-pole switches when lamps operate at 230 rather than 115 volts.

Three-way Switches.   Often it is convenient to be able to turn a light

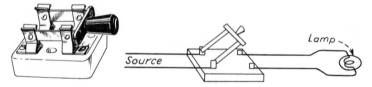

Fig. 4-19 Both wires are disconnected when a lamp is turned off with a double-pole single-throw switch.

on and off from two different places, for example, a hall light from upstairs and downstairs or a garage light from either the house or the garage.   Fortunately this is easily done by the use of switches of the

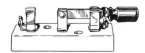

Fig. 4-20 A single-pole double-throw switch. An enclosed toggle switch that performs the same operation is called a "3-way" switch.

type known as "single-pole double-throw," pictured in the porcelain-base type in Fig. 4-20.   Figure 4-21 shows the diagram; call the two switches A and B.   Tracing the circuit will show that, when the handles of A and B are both *up,* the lamp will light; when they are both

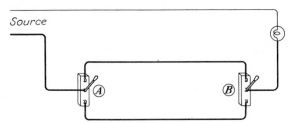

Fig. 4-21 The basic diagram for 3-way switches, which are used to control a light from two different points.

down, the lamp will also light.   If either one is up and the other down, the lamp cannot light.   Careful study will show also that if the light is on (regardless of whether the handles of the two switches are both up or both down), it can be turned off by throwing the handle of either A or B to the opposite position; likewise, if the light is off, it can be turned on by throwing the handle of either A or B to the opposite position.   The light can be controlled by either switch A or switch B, regardless of the position of the other switch of the pair.

In actual wiring, a switch that looks like the switch in Fig. 4-7 is used, except that it has *three* terminals instead of two and the words on and off do *not* appear on the handle.   Switches of this kind are known as 3-way switches, a name which is misleading because it seems to imply that by the use of such switches a light can be controlled from three points instead of only two.   The name is no doubt derived from the three terminals on the switch.   The terminal that corresponds to the center terminal of the porcelain-base switches *A* and *B* of Fig. 4-21 is usually marked by being of a different color, usually a dark or oxidized finish.   Analyzing Fig. 4-21 carefully shows that the wiring of these switches is really very simple.   On one of a pair of such switches, run the wire from source to the marked or "common" terminal; on the other switch, run a wire from the light to the marked terminal.   Then run two wires from the two remaining terminals on one switch to the two remaining terminals on the other.   The wires that start at one switch and end at another are called "runners" or "travelers."

The mechanical construction of 3-way switches varies among manufacturers, so that the marked terminal is sometimes alone on one end of the switch, sometimes alone on one side.   Therefore the pictorial diagram will be either that of Fig. 4-22 or that of Fig. 4-23, depending on

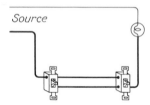

Fig. 4-22   If the "common" terminal on 3-way switches is alone on one *side*, use this diagram.

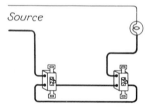

Fig. 4-23   If the "common" terminal on 3-way switches is alone on one *end*, use this diagram.

the brand of switch.   Fortunately no harm is done if the wrong terminals are selected, except that the circuit will not work, and if there is any doubt as to which are the correct terminals, proceed by trial and error until a combination is found that works properly.   For the purposes of this book, whenever a pictorial diagram involves 3-way switches, the terminal that is alone on one *side*, as in Fig. 4-22, is the common or marked terminal.

Four-way Switches.   The preceding paragraphs showed how to control a light from two different points.   What about three different points?   It is a bit more complicated, although still relatively simple. At the point nearest the source, and also at the point nearest the light, use the 3-way switches just described.   At the in-between point use what is known as a 4-way switch, the construction of which is such that it performs the operations shown in Fig. 4-24. In one position of the handle the terminal K is connected to the terminal L; also the terminal M is connected to the terminal N. When the handle is thrown, K is connected to N, and M is connected to L, as the diagram shows.

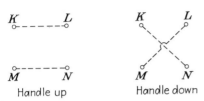

Handle up   Handle down

Fig. 4-24 This shows what happens inside a 4-way switch when the handle is thrown from one position to the other.

With this operation clearly in mind, now note Fig. 4-25, which shows a light with three switches: a 3-way at A, another at B, and a 4-way at C in the center. As long as the 4-way switch C is in the position shown, the current flows through the switch from K to L and from M to N. The wires from A to B might just as well be continuous wires without the switch C. In this picture the handles of switches A and B are both in the up position, and of course, the light is then on. If, then, the wires from A to B are considered as continuous wires (forgetting for the moment that switch C is there), Fig. 4-25 becomes identical with Fig. 4-21, merely a light controlled from two points by two 3-way switches. Now see Fig. 4-26, which is exactly the same as Fig. 4-25 except that

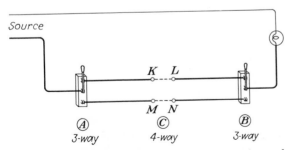

Fig. 4-25 The basic diagram for a 4-way switch, used with a pair of 3-way switches to control a light from three different points.

the handle of the 4-way switch *C* has been thrown to the opposite position.   Trace the circuit.   Chase the messengers any way at all; they
cannot get through and the light is off.   Draw a few diagrams similar to
Fig. 4-26, but with the handles of switches *A*, *B*, and *C* in different posi-

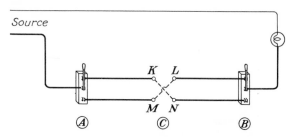

Fig. 4-26   The same as Fig. 4-25, but with the handle of
the 4-way switch thrown to the opposite position.

tions; the diagrams will show that the light can be controlled from any
one of the three switches.   To control a light from three positions, use
two 3-way switches and one 4-way switch.   The flush switch of Fig. 4-
7 in the 4-way type is identified by its *four* terminals and the fact that it
does *not* have the words ON and OFF on the handle (double-pole
switches also have four terminals but *do* have the words ON and OFF on
the handle).

Some manufacturers make their 4-way switches so that the internal
connections, when the handle is thrown, change as shown in Fig. 4-27.

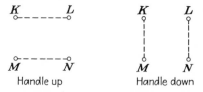

Fig. 4-27   On some brands of 4-way switches, the
connections inside the switch change as shown
above, when the handle is thrown.

In that case the diagram of Fig. 4-26 becomes that of Fig. 4-28—simply
cross two of the wires as shown.   As in the case of 3-way switches, no
harm can be done by wrong connections, except that the circuit will not
work, and if there is doubt as to the internal wiring of the switch, simply

proceed by trial and error, as far as the four terminals of the 4-way switch are concerned, until a combination that works is found.

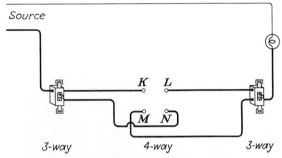

Fig. 4-28   With 4-way switches of the type shown in Fig. 4-27, use this diagram instead of the one in Fig. 4-25.

To control a light from four, five, or any number of points, use a 3-way switch at the point nearest the light, another at the point where the wires come from SOURCE, and 4-way switches at each of the other points; connect as shown in Fig. 4-29.

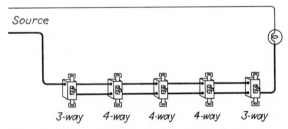

Fig. 4-29   When a light must be controlled from more than three points, use this diagram.

Miscellaneous Switches.   Switches are available in many types besides the conventional toggle type so far discussed; some of them will be mentioned here.   The lock type shown in Fig. 4-30 can be operated only by those having keys to fit.   The momentary-contact type of switch looks like the ordinary toggle type, but the handle is held in one position by a spring, returning to its original position when the operator releases the handle.   The surface type of Fig. 4-31 is used chiefly in exposed wiring, as in garages and farm buildings.

Ratings of Switches.   The maximum number of amperes that a switch is capable of handling, and the maximum voltage at which it may be used, are stamped into the metal mounting strap of the switch.   This

Fig. 4-30   This type of switch can be turned on and off only by using a special key.   (*Pass and Seymour, Inc.*)

Fig.   4-31   Surface-type switches are occasionally used.   (*General Electric Co.*)

might be a single rating such as "15A 277V" indicating that the switch may be used to control up to 15 amp at not over 277 volts.   Another common rating is "10A 125V—5A 250V" indicating that the switch may be used to control up to 10 amp at 125 volts, or up to 5 amp at not over 250 volts.   Of course switches with higher amperage and/or voltage ratings are available.

Kinds of Switches.   The Code and the Underwriters define two kinds of switches: (a) AC General-use (commonly referred to as "AC-only type), and (b) AC-DC General-use.   As their names imply, the first may be used only on AC circuits, and the other on either AC or DC circuits.   The AC-only type can be identified by the letters "AC" that appear at the end of the rating stamped on the strap of the switch; the AC-DC type however do *not* have the letters "AC-DC" on the strap; if the letters "AC" do not appear, the switch is automatically the AC-DC type.   Every student is urged to study the Code on the subject of switches: Art. 100 (Definitions) and Sec. 380-14 (Application and Use).

AC-only Switches.   These are the most common type being installed.   They cost a good deal less than the AC-DC type of equal rating.   They may be used anywhere to control any kind of load up to their full ampere and voltage rating, except that if used to control inductive[3] loads, they must be rated at least 125% of the amperage involved. AC-only switches have a minimum rating of 15 amp, some at 120 volts, others at any voltage up to 277. They are rather quiet in operation, and do not have the annoying click of the AC-DC type.

AC-DC General-use Switches. At one time this was the only kind of switch made. However, DC now is a genuine rarity, although it is still found in a few buildings in larger cities. Since AC-DC switches cost a good deal more to make, naturally the new AC-only type is used almost exclusively in new installations. However, many millions are still in use, having been installed in the past. As they fail, there is no reason why they cannot be replaced with the newer AC type, except in the rare cases where DC circuits are still in use.

Ordinary AC-DC switches are usually rated at 10 amp at not over 125 volts, or not over 5 amp at not over 250 volts. However, there are two subtypes: (a) those that are "T-rated" and (b) those not "T-rated." If they are T-rated, the letter "T" appears at the end of the amperage and voltage rating stamped on the mounting strap.

What is a T rating? A rather lengthy explanation is necessary. When an ordinary lamp is first turned on, for a tiny fraction of a second the amperage consumed by the lamp is from 8 to 12 times as much as when burning normally. A 100-watt lamp when first turned on consumes more nearly 1,000 watts for a small fraction of a second, then consumes its normal 100 watts; a 300-watt lamp momentarily consumes at least 3,000 watts. (The duration of this very high current is so short that it will not blow a fuse.)

This high momentary current is called the "cold inrush" of the lamp. If a switch is used to turn on a group of lamps totaling 1,000 watts, this inrush may be more than 10,000 watts (over 80 amp). That is a severe test for switches using AC-DC construction. AC-DC switches specially designed to handle such loads are designated as T-rated, the "T" standing for tungsten, the material in the filaments of the lamps.

AC-DC switches that are *not* T-rated, per Code Sec. 380-14, may be used to control lamps in private homes and living quarters of hotels and similar locations, but only if each switch is used to control lights in a single room, hall, attic, or basement. They may not be used to control

---

[3] An inductive load is a motor, a transformer, or other load containing windings of wire on a steel core.

lamps in other locations.   If used to control inductive loads, they must
be rated at least 200% of the amperage involved.

If T-rated they may be used in any location to control lamps or other
loads up to their full amperage ratings, unless the loads are inductive, in
which case they must still be rated at 200% of the amperage involved.

Type of Switch to Use.   In any location (unless the installation is one
of the rare ones supplied by DC) it makes sense to use only the AC-only
switches.   The quietness of their construction in itself commends them
for general use.   REA projects usually require T-rated switches, if the
old type is used.

Wall Plates.   Switches and receptacles cannot be mounted in walls
leaving untidy openings around them, nor can the terminals be left exposed, for that would not be safe.   Therefore they are covered with
"wall plates" or "face plates" after installation.   Figure 4-32 shows

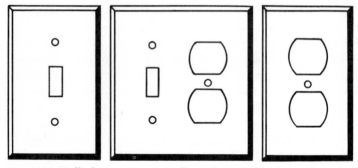

Fig. 4-32   Wall plates must cover all switches, receptacles, and similar
devices.

several plates.   The smaller ones are used for single devices.   Sometimes it is necessary to mount two or three or more devices side by side,
requiring wider plates known as "2-gang," "3-gang," or "4-gang"
plates, depending on how many devices the plate covers.   They are
available also in combinations so that switches, receptacles, and other
devices can be mounted side by side, as the same figure shows.

Wall plates are made of a great variety of materials, such as plastic in
brown or ivory, brass and other metals in natural finish or plated in
chromium, oxidized, and other styles to suit the user.   The nonmetal
plates are generally favored.

Sockets.   In the Code any device into which a lamp is inserted is

called a "lampholder." Practically everybody calls these devices
"sockets"; sometimes the term "receptacle" is used for certain types of
sockets, although this is not correct according to the Code, for a receptacle is a device where connection is made by plugging in an attachment plug.

Sockets are available in a very great variety of types. The type that
was shown in Fig. 4-2, commonly called a "cleat receptacle," is not actually used in house wiring. The most commonly known socket is the
brass-shell type; it may be either keyless or with a switching mechanism
to turn the lamp on or off. There are three common types of switching
mechanism: the key, the push-through, and the pull-chain type, all serving the same purpose. One of the pull-chain type is shown in Fig. 4-
33. The socket consists of the brass shell with an insulating paper liner
to insulate metal parts from the shell, the mechanism proper with two

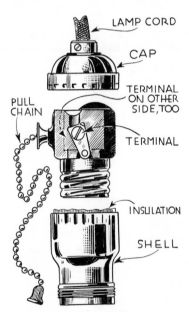

Fig. 4-33 A typical
brass-shell socket. Be-
sides the pull-chain type
shown, there are several
other types. (*General
Electric Co.*)

Fig. 4-34 Cross section of the
socket shown in Fig. 4-33.

terminals, and the cap. The cap may have a threaded hub, used when the socket is used on a floor lamp, fixture, or similar device, or it may have an insulating bushing when the socket is used on the end of a piece of cord. A cross-sectional view is shown in Fig. 4-34. Instead of brass for the outer shell, bakelite or porcelain can be used. Other sockets are of the type shown in Fig. 4-35, used on top of outlet boxes; the weather-

Fig. 4-35 The socket shown above fits directly on top of an outlet box. (*General Electric Co.*)

Fig. 4-36 The weatherproof socket shown is intended for outdoor use. (*General Electric Co.*)

proof type shown in Fig. 4-36, for outdoor use; and "sign receptacles" shown in Fig. 4-37, used chiefly in the manufacture of lighting fixtures.

Fig. 4-37 Sign receptacles are used mostly in the manufacture of lighting fixtures. (*General Electric Co.*)

Other Devices. There are dozens of other devices, and these will be described in later chapters of this book as their use is discussed.

New Work; Old Work. When a building is wired while it is under construction, the electrical work is known as "new work." If the building is completely finished before the wiring is started, it is known as "old work."

5

Overcurrent Devices

It is impossible for an electric current to flow through a wire without heating the wire. As the number of amperes increases, the temperature of the wire also increases. For any particular size of wire, the heat produced is proportional to the square of the amperage: Doubling the amperage increases the heat four times, tripling it increases it nine times, and so on.

Need for Protective Devices. As the temperature of a wire increases, the insulation may become damaged by the heat, leading to ultimate breakdown. With sufficient amperage the conductor itself may get hot enough to start a fire. It is therefore most necessary to limit carefully the amperage to a maximum value, one that is safe for any given size and type of wire. The maximum number of amperes that a wire can safely carry continuously is called the "ampacity" of the wire. As will be discussed later, the Code defines the ampacity of each kind and size of wire, under various conditions.

Any device that limits the current in a wire to a predetermined number of amperes is called an "overcurrent device" in the Code.   There are several kinds of overcurrent devices, and all of them may be considered the safety valves of electrical circuits.   The two types that will be discussed here are fuses and circuit breakers.

Besides being used to protect *wires* against too great amperage, overcurrent devices are frequently used also to protect electrical *equipment.* For example, an electric motor may require 15 amp to deliver the horsepower stamped on its name-plate.   Yet a good motor can safely deliver considerably in excess of that horsepower for short periods, but it will then draw a correspondingly greater amperage.   If the higher amperage is allowed to flow during a long period of overload, the motor probably will burn out.   Therefore an overcurrent device is provided to protect the motor.

Fuses.   The most common overcurrent device is a fuse.   A fuse is merely a short length of metal ribbon or wire, made of an alloy with a low melting point and of a size that will carry any given amperage indefinitely, but which will melt when a larger amperage flows.   When this wire inside the fuse melts, the fuse is said to "blow."   When it blows, the circuit is open, just as if a wire had been cut or a switch opened, at the fuse location.

Plug Fuses.   The common plug-type fuse is shown in Fig. 5-1.   The fusible link is enclosed in a sturdy housing which prevents the molten

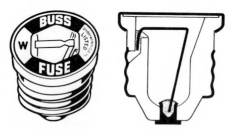

Fig. 5-1   Plug fuses are made only in ratings up to 30 amp.   (*Bussmann Mfg. Div.*)

metal from spattering when the fuse blows; there is a window through which you can see whether the fuse has blown; there are contacts for quick replacement when required.   The largest fuse in the plug type is the 30 amp; smaller standard sizes are 10, 15, 20, and 25 amp.   The Code requires that plug fuses rated at 15 amp or less be of hexagonal shape, or have a window or other prominent part of hexagonal form;

those rated at more than 15 amp are round.   This detail enables you to see at a glance whether a circuit in a home (where generally speaking a 15-amp fuse is the largest used) is overfused or not.

The Code limits the use of plug-type fuses to installations not exceeding 125 volts, although they are permitted on a 230-volt circuit, *if* that circuit consists of the two hot wires of a 115/230-volt installation.   On such a circuit, while the voltage between the two hot wires is 230, it is only 115 volts "to ground" (see Chap. 9).   That of course is the situation in ordinary residential and farm wiring; therefore plug fuses may be used on 230-volt motors, water heaters, and similar equipment, even if the neutral does not run to the equipment.

Time-delay Fuses.   Consider a circuit in a home, wired with the No. 14 wire generally used, having a maximum ampacity of 15 amp and protected by 15-amp fuses.   Most of the time the wire will be carrying considerably under 15 amp; the temperature of the wire and its insulation will be well within safe limits.   If the amperage is increased to 30 amp, the fuse will blow in a very few seconds.   On the other hand, 30 amp flowing for only a few seconds or even for half a minute would not heat the wire or its insulation to the danger point, especially if the amperage was very small before it was increased to 30.

Fig. 5-2   Cross section of a typical time-delay fuse, known as a "Fusetron."   Time-delay fuses carry *temporary* overloads safely without blowing.   (*Bussmann Mfg. Div.*)

In practice, there are often conditions just as described; perhaps 3 amp are flowing in the wire, representing about 300 watts of lights. Then a motor is turned on, for example, a washing machine.   The motor requires in the neighborhood of 30 amp for a few seconds while it is starting; after that it drops to a normal of around 6 amp.   Very frequently the fuse blows during this starting interval, although the wire and its insulation are in no danger whatever.

Accordingly time-delay or time-lag fuses have been developed which carry their rated amperage indefinitely and blow within a few minutes like ordinary fuses on an overload of, say, 50%, but which carry over-loads of 100% for about 30 sec and a 200% overload for about 5 sec. In other words, they *do not blow* like ordinary fuses on large but *temporary* overloads, but they do blow like ordinary fuses on *continuous* small overloads or on short circuits.   A plug fuse of this type is shown in Fig. 5-2.   The use of this type of fuse is very desirable, especially when motors are used.   Power suppliers, especially, find the use of time-delay or time-lag fuses by their customers most desirable, for it is well known that a large percentage of service calls are caused by nothing more serious than blown fuses—and usually such blown fuses could be

"Don't be silly – I've been putting
pennies in the fuse box
for years"

Fig. 5-3   Heed this message!

avoided by using the time-delay type, which carries short nondangerous overloads safely without blowing.

The Code in Sec. 240-4 requires that plug fuses in new construction in homes, if 20 amp or smaller, must be of the time-delay type.

"Type S" Nontamperable Fuses. Since each size of wire has a very definite ampacity, or maximum safe carrying capacity in amperes, the Code requires that the overcurrent device selected to protect the wire be of a rating no greater than that amperage. For example, the No. 14 wire used for ordinary residential wiring has an ampacity of 15 amp and accordingly should be protected by fuses no larger than 15 amp, yet all plug fuses up to 30 amp are interchangeable. Nothing prevents the homeowner from substituting a 30-amp for a 15-amp size. In doing this, he is of course defeating the purpose of the fuse, and in doing that he is stupid. See Fig. 5-3.

To prevent substituting oversize fuses, nontamperable fuses were developed, shown in Fig. 5-4; Fig. 5-5 shows a cross section of their con-

Fig. 5-4 A Fustat and its adapter. (*Bussmann Mfg. Div.*)

Fig. 5-5 Cross section of the fuse shown in Fig. 5-4. (*Bussmann Mfg. Div.*)

struction. The device consists of an adapter and the fuse proper. The adapters have amperage ratings just like the fuses; a 15-amp adapter will permit only 15-amp or smaller fuses to be inserted into it; a 25-amp adapter will permit only 25-amp or smaller fuses, and so on.

The adapters fit into ordinary fuseholders but are so designed that once inserted, they cannot be removed. Obviously then if a contractor when wiring a house with the usual No. 14 wire (which has an ampacity of 15 amp) installs 15-amp adapters, he makes it impossible to use fuses larger than 15-amp size, thus making overfusing impossible on the part of those who know no better, or are inclined to take chances. This eliminates one of the greatest causes of electrical fires and is obviously a sensible move.

In using nontamperable fuses of the type shown in Figs. 5-4 and 5-5, one caution is in order. When inserting the fuse into its adapter, turn it some more after it appears to be tightly in place. There is a spring under the shoulder of the fuse, and this spring must be flattened or the fuse will not "bottom" in the adapter. In that case no contact is made, and the circuit will appear to be open, just as if the fuse had blown.

The most common of these nontamperable fuses is known as a "Fustat" and, besides being of the nontamperable type, is also of the time-delay type, the advantages of which have already been discussed. Nontamperable fuses are called "Type S" in the Code. The Code in Sec. 240-21 requires that in all *new* construction, plug fuses must be Type S.

Cartridge Fuses. Although the plug type is the common fuse for homes, the cartridge type is by far more common for nonresidential purposes, and is the only kind that can be used anywhere if the rating must be more than 30 amp. There are two basic types of cartridge fuses: the ferrule-contact type shown in Fig. 5-6, and the knife-blade contact type shown in Fig. 5-7. The ferrule construction is used only on fuses rated 60 amp or less, the knife-blade construction is used on fuses rated over 60 amp.

Fig. 5-6 The ferrule-contact type of fuse is made only in sizes up to and including 60 amp. (*Bussmann Mfg. Div.*)

Fig. 5-7 The knife-blade-contact type of fuse is made only in sizes larger than 60 amp. (*Bussmann Mfg. Div.*)

Cartridge fuses are available in three types, depending on the maximum voltage rating of the circuit in which they are used: 250 volts, 300 volts, and 600 volts. Note the differing dimensions in the following table:

Fuse ratings, amperes	Dimensions in inches		
	250-volt type	300-volt Class G	600-volt type
10, 15	$\frac{9}{16} \times 2$	$\frac{13}{32} \times 1\frac{5}{16}$	$\frac{13}{16} \times 5$
20	$\frac{9}{16} \times 2$	$\frac{13}{32} \times 1\frac{13}{32}$	$\frac{13}{16} \times 5$
25, 30	$\frac{9}{16} \times 2$	$\frac{13}{32} \times 1\frac{5}{8}$	$\frac{13}{16} \times 5$
35 to 60	$\frac{13}{16} \times 3$	$\frac{13}{32} \times 2\frac{1}{4}$	$1\frac{1}{16} \times 5\frac{1}{2}$
70 to 100	$1 \times 5\frac{7}{8}$	Not made	$1\frac{1}{4} \times 7\frac{7}{8}$
110 to 200	$1\frac{1}{2} \times 7\frac{5}{8}$	Not made	$1\frac{3}{4} \times 9\frac{5}{8}$
225 to 400	$2 \times 8\frac{5}{8}$	Not made	$2\frac{1}{2} \times 11\frac{5}{8}$
450 to 600	$2\frac{1}{2} \times 10\frac{3}{8}$	Not made	$3 \times 13\frac{3}{8}$

The 250-volt and the 600-volt types have been in use for many years; the 300-volt type is a new kind called "Class G," introduced in 1963. But do note that the dimensions of the three types differ sufficiently so that it is impossible to use a type other than that originally installed.

From this table you can see that it is next to impossible to use a fuse of an amperage or voltage rating that differs widely from the amperage and voltage intended when the installation was first planned. This is an important safety feature.

Class G Fuses. This is a comparatively new type of fuse, called "Class G" by the Underwriters, and "Style SC" by the manufacturer. It was developed primarily for commercial and industrial buildings with lighting circuits operating at 277 volts (which will be discussed in Chap. 28). In such installations the overall size of the panelboards is very greatly reduced by using this type in place of the large 600-volt type that was formerly required. Of course the Class G may also be used for residential and farm installations. Do note it can be used only in panelboards or fuse cabinets designed for it. The Class G is of the time-delay type.

Panelboards and fuse cabinets for the Class G are furnished empty except for an assembly of terminals and bus bars. When installing, individual plug-in units of appropriate type and size are plugged in to complete the assembly. The plug-in units come in two types.

In the type shown in Fig. 5-8 each plug-in unit contains a switch for turning the circuit on and off; a fuse-holding block that can be removed for replacement of fuses only when the circuit has been turned off; and a

signal light that glows only when the fuse is blown.   It fits only panelboards designed for it.

In the less expensive type shown in Fig. 5-9, the switch is omitted.

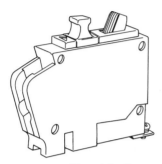

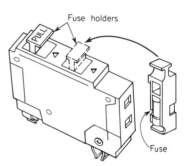

Fig. 5-8   A "Switchfuse" unit
for one circuit.   Only Class G
fuses can be used in it.
(*Square D Co.*)

Fig. 5-9   This unit is similar to that
shown in Fig. 5-8.   It has fuses for
two circuits, but no switches.
(*Square D Co.*)

Each plug-in unit contains two removable fuse-holding blocks for two
115-volt circuits (or in slightly different types, two fuses for one 230-
volt circuit).   Removing the fuse-holding block opens the circuit; inserting the block upside down turns off the circuit.   The signal light
glows only when a fuse is blown.   This type fits only fuse cabinets designed for it.

In both types the plug-in units carry ampere ratings, so that it is impossible to use fuses of a higher amperage rating than was originally
installed.   For example, if the original unit was for a 20-amp fuse, it
will accept only 15- or 20-amp fuses, but not larger.   The fuses can be
replaced only after removing the fuse-holding block from the plug-in
unit.

Fig. 5-10   A renewable type of fuse is easily disassembled for replacement of the fusible link.   (*Bussmann Mfg. Div.*)

Renewable Fuses. Cartridge fuses are divided into nonrenewable and renewable types. The nonrenewable, once blown, have no further value. Since only the fusible link is destroyed when the fuse blows, renewable fuses are available that permit the fusible link to be replaced after blowing. Figure 5-10 shows a clear view of the construction; it is a very simple matter to replace the fusible link. In external appearance there is no basic difference between nonrenewable and renewable except that the latter are so constructed that they can be taken apart. Class G fuses are not available in renewable type.

Cartridge fuses are available in ordinary type, as well as in the time-delay type shown in Fig. 5-11.

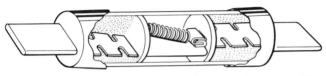

Fig. 5-11 Time-delay fuses are also made in cartridge type. They are especially useful in protecting motors, which usually require several times as many amperes while starting as while running. (*Bussmann Mfg. Div.*)

Circuit Breakers. The Code defines a circuit breaker as "a device designed to open and close a circuit by nonautomatic means, and to open the circuit automatically on a predetermined overload of current, without injury to itself when properly applied within its rating." Since all switches break circuits, they can in a sense be termed circuit breakers, and the definition merely states that as far as the Code is concerned, when it mentions circuit breakers, it refers only to the type that opens a circuit when an amperage greater than that for which it was designed flows through it.

A circuit breaker of the type used in homes looks like a somewhat overgrown toggle switch of the ordinary type used to turn lights on and off. One is shown in Fig. 5-12. Essentially it consists of a carefully calibrated bimetallic strip similar to that used in a thermostat. As the current flows through this strip, heat is created and the strip bends. If enough current flows through the strip, it bends enough to release a trip that opens the contacts, interrupting the circuit just as it is interrupted when a fuse blows or when a switch is opened. In addition to the bimetallic strip that operates on heat, most breakers have a magnetic arrangement that opens the breaker instantly in case of a short circuit. A circuit breaker, in fact, is a switch that opens itself in case of overload.

Fig. 5-12 A typical single-pole circuit breaker. (*Square D Co.*)

Circuit breakers are rated in amperes, just as fuses are rated. Breakers will carry their rated load indefinitely, will carry a 50% overload for perhaps a minute, a 100% overload for about 20 sec, and even a 200% overload for about 5 sec—long enough to carry the heavy current required to start a motor.

The trend is rapidly away from fuses to circuit breakers, for they have many advantages. When a fuse blows, spare fuses may or may not be on hand. When a circuit breaker trips, reset it like turning on a switch. The circuit breaker provides good protection and does not trip on large, but temporary, overloads. Modern homes are usually equipped with circuit breakers.

Determining Proper Rating of Overcurrent Device. The fuse must blow or the circuit breaker open when the amperage flowing through it exceeds the number of amperes that is safe for the wire in the circuit. The larger the wire, the greater the number of amperes it can safely carry.

The Code defines the number of amperes that may be carried by each size of wire. The word "ampacity" to designate "amperes carrying capacity" was first used in the 1965 Code. The ampacity of any size and kind of copper wire can be found in Code Tables 310-12 and 310-13 (see Appendix). There are some important exceptions, particularly in connection with motors, and these will be discussed in connection with related subjects in other chapters of this book.

It will be well to memorize carefully the ampacity of the smaller sizes of wire; these are also the maximum rating of the overcurrent device protecting that size of wire.   These ampacities are as follows:

No. 14	. . . .	15 amp	No. 4	. . . .	70 amp
No. 12	. . . .	20 amp	No. 2	. . . .	95 amp
No. 10	. . . .	30 amp	No. 1/0	. . . .	125 amp
No. 8	. . . .	40 amp	No. 2/0	. . . .	145 amp
No. 6	. . . .	55 amp	No. 3/0	. . . .	165 amp

Note carefully that these are the ampacities of the kinds of wire usually used in ordinary residential wiring.   Other types not generally used in residential work have higher ampacities, and are described in Chap. 27.

Joining Different Sizes of Wire.   If two different sizes of wire are joined as in Fig. 5-13, then the fuse (or other overcurrent protection)

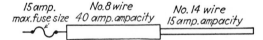

Fig. 5-13   When two different sizes of wire are connected in series, the largest fuse that may be used is one that protects the *smaller* wire.

may be no greater than that permitted with the smaller wire.   Since in this case the smaller wire has an ampacity of only 15 amp, this is the maximum-size fuse permitted.   In practice, a condition of this kind is sometimes met, especially in farm wiring, where a No. 8 wire is used for an overhead span to secure mechanical strength and to avoid voltage drop.   Although the wire has an ampacity of 40 amp, the circumstances may be such that more than 15 amp is never required, and the maximum fuse permitted, 15 amp, will not in any way prove inconvenient.

On the other hand, there may be a condition, as shown in Fig. 5-14, where under similar circumstances No. 8 is again used but where more than 15 amp is to be carried altogether; in this case a second fuse is used at the point where the wire size is reduced.   A 40-amp fuse may be used to protect the No. 8 wire, and a 15-amp fuse to protect the No. 14

wire.   Usually when these conditions are present, additional wires protected by individual fuses are used, as shown in the dotted lines.

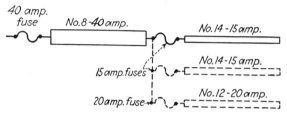

Fig. 5-14   If fuses are used where the wire size is reduced, select an amperage that protects the smaller wire.

There are a number of exceptions to these general requirements, the most important being the one permitted by Sec. 240-15 Exc. 6 of the Code.   No overcurrent protection is required at the point where the wire size is reduced if the smaller wire meets *all four* of the following conditions:

1. It must be not over 25 ft long.
2. It must be protected against mechanical injury.
3. It must have an ampacity at least one-third that of the larger wire.
4. It must end in a *single* overcurrent device of an amperage rating not greater than the ampacity of the *smaller* wire.

This condition is shown in Fig. 5-15, which portrays a combination of No. 8 with No. 14.   Beyond the final fuse or other overcurrent device at the end of the smaller wire, additional wires of any length or size (but

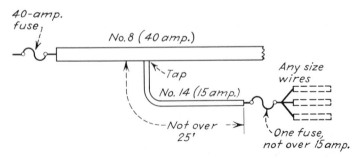

Fig. 5-15   Under certain conditions no fuse is required at the point where the wire size is reduced.

not smaller than No. 14) may be run; they are protected by the 15-amp protective device ahead of them.

Considering the ampacity of various sizes of wire and the requirements of the preceding paragraph, it will be evident that, if all those requirements are met and if ordinary Type T or Type R wire is used,

No. 14 may be tapped to No. 8 or lighter
No. 12 may be tapped to No. 6 or lighter
No. 10 may be tapped to No. 3 or lighter
No. 8 may be tapped to No. 1 or lighter

Another exception is permitted by Sec. 240-15 Exc. 5 of the Code. In the case of switchboards, panelboards, and similar devices, it is not unusual to have wires of considerable size run through such devices, with taps taken off to feed a smaller load.   In such cases no overcurrent protection is required where the smaller wire is tapped to the larger, provided it is not over 10 ft long, provided it has an ampacity equivalent to the total of the ampacities of all the circuits it feeds, and provided it does not extend beyond panelboards or other equipment in question.

An additional common-sense exception is contained in Code Sec. 210-19(c), which permits taps not over 18 in. long to be made from circuit wires of any size, to serve an individual socket, fixture, or outlet, provided only that the short wire is heavy enough to serve its specific load and never smaller than No. 14 (on 40- and 50-amp circuits, No. 12). Section 210-19(c) also permits wiring inside fixtures, also portable cords, to be smaller than the circuit wires serving them, provided they are heavy enough to carry their specific loads.   Section 240-5, in turn, authorizes the omission of overcurrent protection at such points where wire sizes are reduced.   These details will be discussed more fully in a later chapter.

6

Types and Sizes of Wires

Wires are used to conduct electric power from the point where it is generated to the point where it is used. Copper is the material used in practically all cases. The Code makes very little reference to "wire" but speaks frequently of the "conductor," which it defines as "wire or cable or other form of metal suitable for carrying current." All wires therefore are conductors, but not all conductors are wires. Copper bus bars, for example, are conductors but are not referred to as wires.

Previous chapters showed that all wire has resistance that prevents an unlimited flow of current and causes voltage drop. For any given load, you must select a size of wire that causes only a reasonable voltage drop.

Current flowing through a wire causes heat; the heat varies as the *square* of the amperage. There is a limit to the degree of heat that various types of insulation will safely withstand, and even a bare wire must not be allowed to reach a temperature that might cause fire. The Code carefully and in great detail specifies the ampacity or maximum amperage that is considered safe for wires of different sizes with different insulations and under different conditions. These ampacities will be given later.

Circular Mils. In order to discuss intelligently the different sizes of wire, you must understand something about the scheme used in numbering these sizes. The units used are "mils" and "circular mils." A mil is one one-thousandth (0.001) inch. A circular mil (abbreviated c.m.) is the area of a circle one mil in diameter. Thus a wire that is 0.001 in. or 1 mil in diameter is said to have a cross-sectional area of 1 circular mil. Since the areas of two circles are always proportional to the squares of their diameters, it follows that the cross-sectional area of a wire 0.003 in., or 3 mils, in diameter is 9 circular mils; that of one 0.010 in., or 10 mils, in diameter is 100 circular mils; that of one 0.100 in., or 100 mils, in diameter is 10,000 circular mils, etc. The cross-sectional area of any round wire in circular mils is equivalent to the diameter of the copper only, in mils or thousandths of an inch, squared or multiplied by itself.

Wire Sizes. Instead of referring to common sizes of wire by their areas, sizes or numbers have been assigned to them. The gauge commonly used is the American Wire Gauge, abbreviated AWG; it is the same as the Brown and Sharpe, or B&S gauge. This gauge is not the same as that used for steel wires used for nonelectrical purposes, for example fence wires.

Number 14 wire, which is a size most commonly used for ordinary house wiring, has a copper conductor 0.064 in., or 64 mils, in diameter. Wires smaller than this are Nos. 16, 18, 20, and so on. Number 40 has a diameter of approximately 0.003 in., as small as a hair; many still finer sizes are made. Sizes larger than No. 14 are Nos. 12, 10, 8, etc. Note that the bigger the number, the smaller the diameter of the wire.

In this way, sizes proceed until No. 0 is reached; the next sizes are No. 00, No. 000, and finally No. 0000, which is almost $\frac{1}{2}$ in. in diameter. Numbers 0, 00, 000, and 0000 are usually designated as $1/0$, $2/0$, $3/0$, and $4/0$ (one-naught, two-naught, etc.). As still heavier sizes are reached, they no longer are designated by a numerical size, but simply by their cross-sectional areas in circular mils, beginning with the 250,000 cm (250 MCM) up to the largest recognized standard size of 2,000,000 cm (2,000 MCM).

Figure 6-1 shows the approximate actual sizes of typical sizes of wire, without the insulation. The sizes from No. 50 (less than one-thousandth inch in diameter) to No. 20 are used mostly in manufacturing electrical equipment of all kinds.   Numbers 18 and 16 are used chiefly for flexible cords, for signal systems, and for similar purposes where relatively small amperages are involved.   Numbers 14 to 4/0 are used in ordinary residential and farm wiring and, of course, in indus-

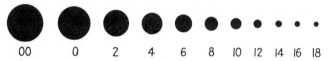

Fig. 6-1 Actual diameters of typical sizes of copper wire, without the insulation.

trial work, where the still heavier sizes are also used.   Number 14 is the lightest size permitted for ordinary wiring.   The even sizes of wire, such as Nos. 18, 16, 14, 12, 10, 8, etc., are commonly used; the odd sizes, as Nos. 15, 13, 11, 9 (with the exception of No. 3 and No. 1), are seldom used in wiring.   The odd sizes, however, are commonly used in the form of magnet wire for manufacturing motors, transformers, and so on, for which purposes even fractional sizes such as No. 15½ are not at all uncommon.

In Fig. 6-2 is shown the usual gauge used in measuring wire sizes. The wire is measured by the slot into which it will fit, not by the hole behind the slot.

Table 8 of the Code shows the commonly used sizes of wire, their areas in circular mils, their resistances in ohms per thousand feet, their dimensions in fractions of an inch, and their areas in fractions of a square inch.   For convenience this table is reproduced in the Appendix of this book.

You will find it useful to remember that any wire which is three sizes heavier than another will have a cross-sectional area exactly twice that of the other.   For example, No. 11 has an area exactly twice that of No. 14; No. 3 wire has an area exactly twice that of No. 6.   Any wire that is six sizes heavier than another has exactly twice the diameter, four times the area, of the other.   For example, No. 6 wire has exactly twice the diameter, and four times the area, of No. 12.

Stranded Wires.   When common sizes of wire are used for ordinary wiring purposes, there is usually no reason why the copper conductor should not be one single solid conductor.   Where considerable flexibility is needed, as in flexible cord, the conductor instead of being one solid

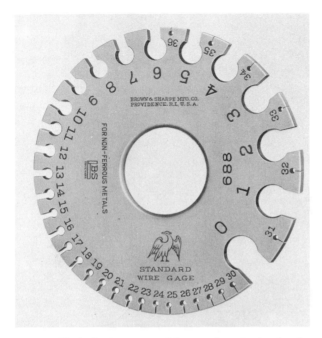

Fig. 6-2 Typical wire gauge. Measure the wire by the slot into which it fits. This illustration is actual size. (*Brown & Sharpe Mfg. Co.*)

wire consists of a great many strands of fine wire twisted together. The number assigned to such a conductor is determined by the total cross-sectional area of all these individual strands added together. For example, per Table 8 of the Code, the cross-sectional area of No. 16 wire is 2,583 circular mils. The total cross-sectional area of 65 strands of No. 34 wire is 2,585 circular mils; the total cross-sectional area of 26 strands of No. 30 wire is a trifle above this figure. Therefore wire made up of either of these two combinations, or any other combination totaling substantially 2,583 circular mils, is known simply as No. 16 wire. If it is necessary to describe such wire in more detail, the first-mentioned combination is described simply as No. 16, 65/34, and the second as No. 16, 26/30.

Building wires in No. 6 and heavier are stranded; solid wires are too stiff to be practical. The stranding of each size has been entirely standardized, so it is not necessary to specify the size of the individual strands. However, the number of strands and the size of each can be found for each size of wire in Table 8 of the Code (see Appendix).

Note that in Fig. 6-1, the dimensions shown for No. 6 and heavier are what they would be if the wires were solid. Actually they are stranded, and therefore somewhat larger than shown.

Colors of Wire. Building wires come in various colors, and there is, of course, a purpose in this. Only white wire may be used for the grounded neutral wire in wiring; this will be explained in more detail later. Other wires may not be white, or green. The color scheme established by the Code is as follows:

> Circuits of 2 wires . . . White, black
> Circuits of 3 wires . . . White, black, red
> Circuits of 4 wires . . . White, black, red, blue
> Circuits of 5 wires . . . White, black, red, blue, yellow

In cables and cords, the same color scheme is used. Sometimes there is an additional grounding conductor, which must be green, green with a yellow stripe, or in some cases may be bare, uninsulated.

In overhead outdoor runs, this color scheme is not required. As a matter of fact, the weatherproof wire used for the purpose, and described later in this chapter, is available only in black. The same is true of single-conductor underground cables.

Types of Wire. The Code recognizes many different types of wire that may be used in wiring buildings. The more ordinary ones will be described in this chapter; some others will be described in Chap. 27. Still other types less frequently used in the kinds of buildings discussed in this book, are described in Code Tables 310-2(a) in the Appendix of this book, and 310-2(b) which you will find in your copy of the Code.

Kinds of Locations. The Code limits some kinds of wire to use in dry locations, others to damp or dry locations, and still others to wet, damp, or dry locations. It is important to understand the definitions of these different locations. The Code defines them in Art. 100, as follows:

> DAMP LOCATION: A location subject to a moderate degree of moisture, such as some basements, some barns, some cold storage warehouses, and the like.
> DRY LOCATION: A location not normally subject to dampness or wetness. A location classified as dry may be temporarily subject

to dampness or wetness, as in the case of a building under construction.

WET LOCATION: A location subject to saturation with water or other liquids, such as locations exposed to weather, washrooms in garages, and like locations. Installations underground or in concrete slabs or masonry in direct contact with the earth shall be considered as wet locations.

Plastic-insulated Wires. Most kinds of wire used in wiring buildings have thermoplastic insulation. The copper conductor is sometimes tinned for easier soldering. The insulation consists of a layer of plastic insulating compound, the thickness of which depends on the size of the wire. The wire is clean and easy to handle, and strips easily. There are several types.

Types T, TW, THW. The most ordinary type of plastic-insulated wire is what the Code calls "Type T." It is shown in Fig. 6-3. It may be

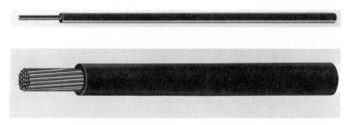

Fig. 6-3 Type T is used for general wiring. No. 6 and larger sizes are stranded. (*Crescent Insulated Wire & Cable Co.*)

used only in dry locations. Most manufacturers no longer make the ordinary Type T, instead produce Type TW, which is identical in appearance, but may be used in wet or any other locations. Also available is Type THW, which is similar to Type TW but withstands a greater degree of heat, consequently has a higher ampacity rating, in the larger sizes.

Types THHN, THWN. These are comparatively new types of wire, consisting essentially of the basic Types THH and THW but with less of the thermoplastic insulation, and with a final extruded jacket of nylon. Nylon has exceptional insulating qualities and great mechanical strength, all of which results in a wire which is much smaller in diameter than ordinary Types T, TW, and THW of corresponding size.

The Code limits the number of wires of any given size that may be installed in a particular size of conduit (as will be discussed in detail in Chap. 11).   In any given size of conduit, it permits a greater number of small-diameter wires (such as Types THHN or THWN) than it does for the more ordinary wires (such as Types T, TW, or THW).   Therefore using the more expensive Types THHN or THWN may result in lower total cost, because their use permits using smaller sizes of conduit. There are also other advantages.   All this will be discussed in more detail in Chap. 27.

Type XHHW Wire.   This is a type of wire first recognized in the 1968 Code.   It will be described in more detail in Chap. 27.   In appearance it resembles Types T, TW, or THW, but because of a somewhat thinner layer of insulation, the over-all diameter is smaller.   The insulation is "cross-linked thermosetting polyethylene," which has extraordinary properties as to insulating values, heat resistance, and moisture resistance.   It may be used in any location.   While at present it is an expensive wire, it would be no surprise if in due course of time, this one single type will replace all the many types and subtypes of Types T or R now recognized by the Code.

Rubber-covered Wire.   At the present time, most wire used is plastic-insulated.   Originally, all the wire used was what is commonly known as "rubber-covered wire," and it is still used in very considerable quantities.   According to the Code, Sec. 310-3(a), "rubber insulations include those made from natural and synthetic rubber, neoprene and other *vulcanizable* materials."   Its construction is shown in Fig. 6-4.

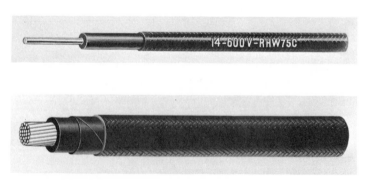

Fig. 6-4   Type R wire has rubber insulation, and a protective braid over the insulation.   This kind of wire is used less and less each year.   (*Crescent Insulated Wire & Cable Co.*)

It consists of the copper conductor, tinned to make it easier to remove the insulation, and for easy soldering. Over the copper is a layer of rubber, the thickness of which depends on the size of the wire. Then follows an outer fabric braid which is saturated with moisture- and fire-resistant compounds; if it is set on fire with a blowtorch, the flame dies out when the torch is removed. If the color is to be something other than black, the color is painted on during the manufacturing process. Over all is a wax finish to make the wire cleaner and easier to handle.

There are several subtypes, which will be discussed in detail in Chap. 27. Briefly however, the basic Type R, suitable for use only in dry locations, is no longer being made. The most ordinary kind now is Type RHW, which may be used in dry or wet locations. Type RHH has insulation which withstands more heat, therefore has a higher ampacity in the larger sizes. It may be used only in dry locations. However, throughout this book, to avoid awkward references to several types, the reference will be to the now nonexistent "Type R" and it must be understood that Types RHW or RHH are to be used even if "Type R" is mentioned.

Cable. A stranded wire heavier than No. 4/0 is called a "cable." Thus you would refer to a wire with a cross-sectional area of 1,000,000 circular mils as a "1,000,000 circular mil (or 1,000 MCM) cable." The word "cable" is also used equally often as outlined in the next paragraphs.

Cables. For many purposes, especially in residential and farm wiring, it is desirable to have two or more wires grouped together in the form of a cable. This makes a compact assembly which is easy to install, especially in wiring a building that was completed before the wiring is installed, for the cable lends itself well to being fished through hollow wall spaces.

A cable that contains two No. 14 wires is known as "14-2" (fourteen-two); if it contains three No. 12 wires it is known as "12-3"; if it contains only one No. 8 it is "8-1"; etc.

Nonmetallic-sheathed Cable. This cable consists of two or three Type T or Type R cables bundled together. It costs less than other types of cable, is light in weight and very simple to install; no special tools are needed. All that makes it very popular. The Code recommends it for use where an especially good ground is *not* found, which makes it the ideal cable for farms. There are two types which the Code calls Type NM and Type NMC.

Type NM is the ordinary kind that has been available for many years and may be used only in permanently dry locations. It is known by such names as Romex, Cresflex, Loomwire, etc. Figure 6-5 shows the

Fig. 6-5 Nonmetallic-sheathed cable is popular for ordinary wiring. This is Code Type NM and may be used only in dry locations.

type of construction if Type R conductors are used. Each wire is wrapped with a spiral layer of paper for additional protection. Over all comes an outer jacket of moisture- and fire-resistant material which may be either fibrous as shown, or made of plastic. The empty spaces between are filled with a moisture-resistant jute or similar cord. If the individual conductors are Type T there may or may not be a protective layer over each conductor.

This material is entirely suitable in permanently dry locations and once was also used in locations having high humidity, such as barns on farms. It was found that the fibrous materials used in its construction (the paper, the jute cords, the fabric outer braid) acted like wicks, pulling moisture into the inside of the cable. The result was rotting, of both the insulation and the other parts of the cable, which quickly led to dangerous conditions, with respect both to shock and to fire. For that reason it may now be used only in permanently dry locations, and a different kind of cable has been developed for wet locations.

Code Type NMC is shown in Fig. 6-6, and you will note that the individually insulated wires are embedded in solid plastic; no fibrous, wicklike materials are used. Therefore it may be used in any location,

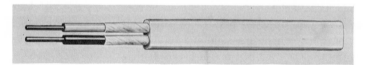

Fig. 6-6 Type NMC nonmetallic-sheathed cable may be used in wet or dry locations.

dry or damp, but not underground. This will be discussed in more detail in Chap. 11.

Most nonmetallic-sheathed cable comes with an additional bare, uninsulated wire in addition to the insulated wires. This is used for grounding purposes, and will be discussed later.

Armored Cable. Another common type of cable is armored cable, shown in Fig. 6-7. It is usually called by its trade name BX, or other names such as Flexsteel, etc. The spiral armor is made of galvanized steel.

Fig. 6-7 Armored cable has a steel armor for its final protection.

If the individual wires are Type R, the Code calls the cable Type AC. Over the group of wires there is a spirally wrapped layer of tough paper. If the individual wires are Type T, the Code calls it Type ACT. Each wire is separately wrapped in a layer of tough paper, but there is no paper over the group of wires. In both cases the paper serves as protection against abrasion by the steel armor, and serves another useful purpose that will be discussed later.

Inside the armor, outside of the paper, and parallel with the wires, there is a bare, uninsulated copper strip or wire, to provide better grounding conditions, and its use will be explained in a later chapter.

Service-entrance and Underground Cables. There are several kinds of special cables in use for bringing wires into a building in the service entrance, either overhead or underground. These will be described in Chap. 13.

Weatherproof Wire. When wires are run between buildings, there is no great likelihood that anyone will ever touch them. They are usually suspended a considerable distance above the ground and not close to each other. Therefore there is not the need for the same kind of insulation found on wires for interior use, where the wires usually lie next to each other inside conduit or cable. On the other hand, wire used out of doors must stand exposure to the weather, a factor that need not be considered for interior use.   A special wire known as "weatherproof wire" is used for outdoor work; it may never be used indoors.   The Code in no way prohibits the use of Type T or Type R wire out of doors, but it is common practice to use weatherproof wire for that purpose; as a matter of fact, it will last longer, and especially in the large sizes, it costs less.

Weatherproof wire was originally made as shown in Fig. 6-8, the

Fig. 6-8   Triple-braid weatherproof wire.   It may be used *only* outdoors.

copper conductor being covered with three separate cotton braids. They were saturated with weatherproofing moisture-resistant compounds usually of an asphaltic nature, followed by an application of flake mica for cleanliness.   This type has been almost completely superseded by a wire consisting merely of the copper conductor plus a neoprene jacket.   In appearance it is very similar to Type T wire.   You can tell it from Type T by the fact that the latter has the "Type T" printed on the wire at frequent intervals.

Weatherproof wire was formerly recognized in the Code as Type WP, but it was dropped in the 1959 Code.   The Code concerns itself primarily with interior wiring, and basically with *safety;* weatherproof wire is used only outdoors, and in such fashion that safety is practically automatic.   As a matter of fact, the Code does not even consider weatherproof wire an *insulated* wire; it considers it a *covered* wire, which in Art. 100 is defined as "having one or more layers of nonconducting materials that are not recognized as insulation under the Code."

Since weatherproof wire is installed only outdoors, any heat that develops is radiated into the air.   For this reason it is customary to assign higher ampacities to this type of wire than to ordinary Type T or Type R.   In practice the size of the wire used will depend on the necessary mechanical strength required for a span of any given length and on the voltage drop that can be tolerated.   Use the ampacities assigned by the Code to "Bare and Covered Conductors" (Table 310-13, see Appendix). In smaller sizes these are as follows:

No. 14	. . . .	30 amp	No. 2	. . . .	175 amp
No. 12	. . . .	40 amp	No. 1/0	. . . .	235 amp
No. 10	. . . .	55 amp	No. 2/0	. . . .	275 amp
No. 8	. . . .	70 amp	No. 3/0	. . . .	320 amp
No. 6	. . . .	100 amp	No. 4/0	. . . .	370 amp
No. 4	. . . .	130 amp			

Aluminum Wire.   There is a trend toward using aluminum in place of copper; the trend accelerates when there is a shortage of copper. Using aluminum however does introduce new problems.

First of all, aluminum has a higher resistance than copper; in other words aluminum does not conduct electricity as well as copper.   For that reason, for any given amperage, it is necessary to use an aluminum wire larger than is required when using copper.   For ordinary wiring, this usually means in practice that you must use aluminum two sizes larger than would be necessary using copper.   In other words, use No. 12 aluminum in place of No. 14 copper, No. 4 aluminum in place of No. 6 copper, and so on.   But this is not always correct.

In this book, whenever a reference is made to a wire size, the reference is to copper.

When using aluminum, instead of using the ampacities shown in Tables 310-12 and 310-13 of the Code, you must use the ampacities shown in Tables 310-14 and 310-15.   All four of these tables are shown in the Appendix of this book.

Aluminum is next to impossible to solder by methods available to most contractors.   In practice this has resulted in aluminum's being used mostly in heavy sizes and where long runs (few joints) are the rule. The most common usage is in weatherproof wire, and bare wires in transmission lines.

Formerly using aluminum wire in connection with copper or brass terminals led to electrolytic action, in turn leading to corroded high-resistance joints, and heating.   Moreover, aluminum tended to "flow" so that connections at terminals that were originally tight later became loose again leading to high resistance and heating.   Nevertheless much progress has been made in solving these problems so that now considerable quantities of aluminum are being used in ordinary construction.

Switches, receptacles, and similar devices if listed by Underwriters, and if they have terminal screws and terminals so constructed that the wire may be *wrapped* at least three-quarters of a turn *around the ter-*

minal screw, are suitable for aluminum. But note if the device has push-in terminals without screws (described in Chap. 8 in connection with Fig. 8-5) aluminum wire may *not* be used.

Solderless connectors of the type described in the next chapter may or may not be suitable for use with aluminum, as will be explained.

Flexible Cords. When wires are installed permanently, they need be only sufficiently flexible to permit reasonably easy installation. If the wires must be moved about, as on a floor lamp, a vacuum cleaner, or a portable motor, they must be very flexible. This is necessary first of all for convenience and second to prevent the conductors from breaking, which would be likely if they were solid copper of considerable diameter. Flexible wires of this type are called "flexible cords" in the Code. There are a great many different kinds, the more common of which will be described here. Others are described in Code Table 400-11.

Types SP and SPT. This is the cord commonly used on lamps, clocks, radios, and similar appliances. As shown in Fig. 6-9 the wires

Fig. 6-9 Type SPT lampcord is used mostly on floor and table lamps, clocks, and similar equipment.

are embedded directly in a solid mass of insulation. If the insulation is rubber, the Code calls the cord Type SP; if the insulation is plastic, it becomes Type SPT. The insulation is of a high quality, so that the cord requires no further protection such as an overbraid. Often the cord is made with a depression between the two conductors for ease in separating the conductors to make connections.

The Code further labels such cords with suffixes -1, -2, and -3. Types SP-1 and SPT-1 are available only in No. 18, and the insulation is $\frac{2}{64}$ in. thick. They are the types most commonly used. Types SP-2 and SPT-2 are available in Nos. 18 and 16 and have insulation $\frac{3}{64}$ in. thick. Type SPT-3 is available in Nos. 18 to 12 and has insulation $\frac{4}{64}$ in. thick.

Types S, SJ, SV. The cords described in the preceding paragraphs are designed for ordinary household devices, which generally speaking are moved very little once they are plugged into a receptacle. They will not stand a great amount of mechanical wear and tear. Neither are they particularly moisture-resistant.   A cord that has a sturdier construction is needed for vacuum cleaners, motors on washing machines, portable tools such as electric drills, and so on.

Such a cord is shown in Fig. 6-10.   It consists of two or more

Fig. 6-10   Type S cord is very tough and durable.

stranded conductors with a serving of cotton between the copper and the insulation to prevent the fine strands from sticking to the insulation. Jute or similar "fillers" are twisted together with the conductors to make a round assembly, which is held together by a fabric overbraid. Over all comes a jacket of high-quality rubber to complete the cord.

Cord as described is made with the outer rubber jacket in varying thicknesses, which determine the Code type.   Type S has the heaviest jacket, is available in Nos. 18 to 2, and is used for the hardest service, such as in industrial applications.   Type SJ has a lighter jacket, is available only in Nos. 18 and 16, and is used for household purposes and light industrial applications.   Type SV has a still lighter jacket, is available only in No. 18, and is used *only* on household vacuum cleaners.

Both Types S and SJ are available in two styles: stationary and constant service.   The stationary type has a conductor which is not so finely stranded as the constant-service type.   For example, in No. 18, the stationary type has a conductor consisting of 16 strands of No. 30, whereas the constant-service type has a conductor consisting of 41 strands of No. 34.   The constant-service type will last longer on tools and similar devices where the cord is flexed continuously in use.

Types ST, SJT, SVT.   If the outer jacket is made of plastic material instead of rubber, Types S, SJ, and SV become Types ST, SJT, and SVT.

Oil-resistant Cords.   When cords of the type just described are exposed to oil, the oil attacks the outer rubber or plastic jacket, which swells, deteriorates, and falls apart.   For that reason ordinary cords cannot be used where exposed to oil or gasoline, for example, in automobile service stations.   But such cords are also made with a special oil-resistant outer jacket, made of neoprene or similar material, in which case the letter "O" is added to the type designation: Type S, for example, becomes Type SO.

Heater Cords.   Cords that are used on flatirons, toasters, portable heaters, and similar appliances that develop a lot of heat are called "heater cords."   The construction shown in Fig. 6-11 was at one time the

Fig. 6-11   One type of heater cord.   It has a layer of asbestos over the insulation.   It is used on toasters, irons, and similar appliances.

only kind used, and is still in use.   The basic insulation is rubber, with a serving of cotton between the copper and rubber to keep the copper clean.   Over the rubber there is a serving of asbestos to withstand the heat of accidental contact with a hot surface.   Over all, there is a braid of cotton or rayon.   This describes the Type HPD illustrated.   If the outer jacket is rubber, it becomes Type HSJ.

More recently Type HPN has become the more popular.   There is neither rubber nor asbestos in its construction, but the conductors are embedded in solid neoprene.   There is no outer braid.   In appearance it is quite similar to Type SP lampcord shown in Fig. 6-9.

Ampacity of Cords.   The ampacity of the more ordinary cords is shown in the table below; for others see Code Sec. 400-9(b) in your

	A	B	C
No. 18	10 amp	7 amp	10 amp
No. 16	13 amp	10 amp	15 amp
No. 14	18 amp	15 amp	20 amp

copy of the Code.   Cords having two conductors carrying current (even if there is a third grounding conductor not normally carrying current) have the ampacity shown in Col. A.   If three conductors carry current (as in the case of 3-conductor cords to 3-phase equipment) the ampacity is slightly reduced as shown in Col. B.   Heater cords (any type with an "H" in the type designation) have a higher capacity, as shown in Col. C.

Cycles.   The smaller the size of the individual strands in a cord, the greater the flexibility of the cord.   Depending on the stranding, heater cords are known as "3,000 cycle" or "10,000 cycle," the latter being the more flexible.

Fixture Wire. For the internal wiring of lighting fixtures, special wire known as "fixture" wire is used. There are many types of fixture wire, and the particular type used depends to a great extent on the temperature that exists in the wire in use. Those with rubber or plastic insulation, as shown in Fig. 6-12, may be used only if the temperature of

Fig. 6-12 Fixture wire is used only in the internal wiring of fixtures. This shows one of several different constructions.

the wire *while carrying current* does not exceed 140°F (60°C). Type CF which has only cotton in its insulation may be used up to 194°F (90°C). Type AF with asbestos insulation may be used up to 302°F (150°C), and there is even Type SF with silicone insulation good for 392°F (200°C). Other types are described in Code Tables 310-2(a) and 402-6.

Fixture wires may be used only for the internal wiring of fixtures, never in a circuit leading up to a fixture.

Other Types. There are many other types of wires, cables, and cords. Some are rarely used in the type of wiring discussed in this book and will not be mentioned. Other types are used for specific purposes such as underground wiring and will be discussed in the chapters pertaining to that kind of wiring.

Low-voltage Wire. Certain types of wire are intended only for low-voltage work, usually under 30 volts, such as wires for doorbells, telephones, etc. Usually the source of current for operating such devices is very limited in capacity, so that ordinarily it is safe to assume that, even under short circuit, no danger of fire exists. Therefore the Underwriters do not concern themselves with wire for such purposes.

Fig. 6-13 Bell ("annunciator") wire has little insulation and is used only for low-voltage work.

Bell Wire. This wire is pictured in Fig. 6-13. It consists of a copper conductor over which are two layers of cotton, the two wrapped in opposite directions, then paraffined to give it some semblance of being moisture resistant.   It is commonly known as bell wire or annunciator wire.   Frequently two or more such wires are bundled together into one cable, which then receives a final outer braid of cotton, again paraffined.   An assembly of this kind appears in Fig. 6-14; it is commonly

Fig. 6-14   Thermostat cable consists of two or more separate bell wires in the form of a cable.

known as "thermostat cable" because it is most frequently used in connection with furnaces and thermostats.   Instead of fabric insulation, plastic material is becoming far more common.

7

Selection of Proper Wire Sizes

For any given combination of volts and amperes you must use a size of wire that is big enough to prevent the development of dangerous temperatures, and also big enough to avoid wasted power in the form of excessive voltage drop. Regardless of the size of wire selected, it is impossible to prevent all voltage drop; nevertheless, the drop must be held to nominal, practical proportions.

Advantage of Low Voltage Drop. Voltage drop is simply wasted electricity. If the drop is 5%, it means that 5% of the power is wasted as unwanted heat in the wires. Moreover, all electrical devices operate most efficiently on the voltage for which they are designed. If an electric motor is operated on a voltage 5% below its rated voltage, its power output drops almost 10%; if operated on a voltage 10% below normal, its power output drops 19%.

If a lamp is operated on a voltage 5% below its rated voltage, the amount of light it delivers drops about 16%; if the voltage is 10% below normal, its light output drops over 30%. So it is with most other electrical devices—the output drops off much faster than the reduction in voltage.   It should then be readily apparent that voltage drop must be limited to as small a figure as is practical.

Practical Voltage Drops.   The Code in Sec. 210-6(c) recommends that wire sizes be chosen so that the voltage drop will not exceed 3% in any branch circuit, measured at the most distant outlet.   It further recommends not over 5% drop for feeders and branch circuits combined.

In ordinary residential wiring there are no feeders; the branch circuits begin at the fuse or circuit breaker cabinet.   In farm wiring however, the wires between the meter on the pole, and the point where they enter a building, are feeders.   The drop in the feeders then should not exceed 2%, with an additional 3% in the branch circuits.   Do note that these are recommendations as to the *maximum* drop that should be permitted; good practice suggests that a lower figure be the goal.   A commonly accepted standard is 2% drop over all the wires from the point where they enter a building, to the farthest outlet.   On farms a compromise must be reached because of the feeders from pole to building.

This means that on a 115-volt circuit the voltage drop from entrance to most distant outlet should not exceed 2.3 volts; on a 230-volt circuit it should not exceed 4.6 volts.

In residential wiring, if No. 14 wire is used for the ordinary branch circuits, the voltage drop will usually not greatly exceed the 2% figure.   On the other hand, the lamps commonly used in floor lamps are getting bigger and bigger, appliances consuming 1,000 to 1,500 watts are becoming more and more common, and, all told, people are using more electric power every day, so that circuits are being loaded closer and closer to the limit of their carrying capacity.   Therefore there is good reason for the trend toward considering No. 12 the smallest size wire to be commonly used for residential wiring.   Some future Code may require No. 12 as the minimum size permitted for ordinary wiring, just as today No. 14 is the minimum.   Some local Codes already require No. 12 as the minimum.

Determining Minimum Wire Size.   First determine the maximum amperage that the wire will be called upon to carry.   Then refer to Table 310-12[1] of the Code (see Appendix) and determine the smallest wire that

---

[1] Use Table 310-12 for copper wire in conduit, cable, or buried in the ground, and Table 310-13 for open wiring with each wire exposed to free air.   For aluminum wire use Tables 310-14 and 310-15.   In any event, for purposes of illustration, it is assumed that ordinary Type T or Type R wire is used.   Other types will be discussed in Chap. 27.

may be used. For example, if 18 amp is to be carried, reference to this table will show that if Type T or Type R wire is to be used, No. 14 is too small, No. 12 is suitable. If, however, weatherproof wire is to be used, No. 14 is big enough. This table merely shows what the minimum size may be from a safety standpoint. The minimum size may be entirely too small when voltage drop is considered.

Calculating Voltage Drops by Ohm's Law. The actual voltage drop in any problem can be determined by the use of Ohm's law, which was discussed in Chap. 2:

E = IR or **voltage drop = amperes × ohms**

For example, assume that a 500-watt floodlight is to be operated at a point 500 ft from the meter; this requires 1,000 ft of wire. At 115 volts, 500 watts is equivalent to about 4.4 amp. Taking No. 14 wire as a random size, Table 8 in the Appendix shows that it has a resistance of 2.575 ohms per 1,000 ft. The voltage drop then is 4.4 × 2.575, or 11.33 volts, considerably over the limit of 2%, or 2.3 volts.

Trying other sizes, No. 6 with 0.410 ohms per 1,000 ft involves a drop of 1.804 volts; No. 8 with 0.641 ohms per 1,000 ft, 2.82 volts. Therefore, if the floodlight is to be used a great deal, use No. 6 wire; if it is to be used relatively little, No. 8 is acceptable; and if it is to be used only in emergencies, No. 10 with 4.479 volts drop (or even No. 12) will be entirely suitable, in that the amount of power wasted per year would not begin to pay for the extra cost of the heavier wire.

If in this example the distance had been 400 ft instead of 500 ft, the length of the wire would have been 800 ft instead of 1,000 ft. The voltage drop would then have been 800/1,000, or 80% of what it is for 1,000 ft.

Now assume that the same floodlight is to be operated at the same distance of 500 ft but at 230 volts instead of 115 volts. The amperage now becomes 2.2 instead of 4.4. Making the same calculations, No. 14 wire gives a drop of only 5.66 volts, still above the 4.6 volts (2% of 230 volts) considered permissible on a 230-volt circuit. Number 12 with a resistance of 1.619 ohms per 1,000 ft gives a drop of 2.2 × 1.619, or 3.563 volts, well under the 4.6-volt limit that has been set. This emphasizes the desirability of using higher voltages where a considerable distance is involved as well as where considerable power is involved.

Desirability of Higher Voltages. For any given *wattage* and any given *distance*, the voltage drop *measured in volts* on any given size of wire is exactly twice as great on 115 volts as it is on 230 volts. Doubling the voltage (regardless of what the actual voltages are) reduces the voltage drop *in volts* exactly 50% if the wattage, the distance, and the wire size remain the same.   It is the *wattage* and not the *amperage* that must remain unchanged for this statement to be correct.

When the voltage drop *in percentage* is considered, remember that in the case of the 230-volt circuit the initial voltage is twice as large but the actual voltage drop only half as large as in the case of the 115-volt circuit.   From this you can see that the voltage drop *in percentage* will be only one-fourth as great using 230 volts as it is using 115 volts—the wattage, the wire size, and the distance, of course, remaining unchanged during the discussion.   For example, in the first instance, 500 watts, 500-ft distance (1,000 ft of wire), No. 14 wire on the 115-volt circuit involved a drop of 11.33 volts, which is 9.8% of 115 volts; on the 230-volt circuit the drop was 5.5 volts, which is 2.45% of 230 volts; 2.45% is one-fourth of 9.8%.

All the foregoing can be simply restated: Any size of wire will carry any given wattage on 230 volts four times as far as on 115 volts with the same *percentage* of voltage drop.   This statement should not be confused with the statement made above, that any size wire will carry any given wattage twice as far with the same *number of volts* drop.

Another Method of Calculating Voltage Drop.   Another formula that is frequently used is

$$\text{Circular mils} = \frac{\text{distance in feet} \times \text{amperage} \times 22}{\text{volts drop}}$$

Applying this to the floodlight example, which involves a distance of 500 ft, 4.4 amp, and a voltage drop that is to be limited to 2.3 volts, the formula becomes

$$\text{Circular mils} = \frac{500 \times 4.4 \times 22}{2.3} = \frac{48,400}{2.3} = 21,043$$

In other words, to limit the voltage drop to exactly 2.3 volts, wire having a cross-sectional area of 21,043 circular mils must be used.   Reference to Table 8 of the Code (see Appendix) shows that there is no wire having exactly this cross-sectional area, which falls about halfway between No. 6 and No. 8.   Therefore compromise on No. 6 with a little under 2.3 volts drop or on No. 8 with a little over 2.3 volts drop.

If, instead of determining the size of wire that will produce a given voltage drop, the actual voltage drop with a given size wire is to be determined, merely transpose the formula to read

$$\text{Volts drop} = \frac{\text{distance} \times \text{amperes} \times 22}{\text{circular mils}}$$

To determine the number of feet any given size of wire will carry any given amperage, transpose the formula once more to read

$$\text{Distance} = \frac{\text{volts drop} \times \text{circular mils}}{\text{amperes} \times 22}$$

Three-phase Voltage Drop. The formulas given above are correct for direct current as well as for single-phase alternating current. In the case of 3-phase current a correction factor must be applied. Calculate the drop, using the formula above; then multiply the answer by 0.865.[2] A simple short cut is to deduct $\frac{1}{7}$ from the answer, whether it is in volts or circular mils. If preferred, the single-phase formula above can be used by substituting 19 for 22.

Voltage-drop Tables. For most purposes there is no need of going through tedious calculations to arrive at the right size wire to use. Suitable tables on pages 110 and 111, one for 115 volts, the other for 230 volts, each based on 2% drop. Under each wire size is shown the *one-way* distance which that size wire will carry the amperage shown in the left-hand column. To clarify, not the number of feet of wire in any problem, but the distance from the starting point to the load in question is given. The tables are correct only for *copper* wire. When the distance appears in **boldface** type, it indicates that Type T or Type R wires in open wiring, but not in conduit or cables, may be used. When the distance appears in *italics*, it indicates that only weatherproof wire will carry the corresponding amperage in the left-hand column; distances shown in ordinary type are applicable to all approved wires.

In the event that these tables are used for 3-phase runs, *do not* use the "volt-amperes" column; use only the "amperes" column. Then, increase all distances by $\frac{1}{6}$. In the formulas above, $\frac{1}{7}$ was deducted, which is the same as multiplying by $\frac{6}{7}$; for the tables, add $\frac{1}{6}$, which is the same as multiplying by $\frac{7}{6}$.

[2] 0.865 is $\frac{1}{2}\sqrt{3}$.

Wire Table—115 Volts—2 Per Cent Voltage Drop

Amperes	Volt-amperes* at 115 volts	No. 14	No. 12	No. 10	No. 8	No. 6	No. 4	No. 2	No. 1/0	No. 2/0	No. 3/0
1	115	450	700	1,100	1,800	2,800	4,500	7,000			
2	230	225	350	550	900	1,400	2,200	3,500			
3	345	150	240	350	600	900	1,500	2,300	3,750		
4	460	110	175	275	450	700	1,100	1,750	2,750	3,500	
5	575	90	140	220	360	560	880	1,400	2,250	2,800	
7½	860	60	95	150	240	375	600	950	1,500	1,900	2,400
10	1,150	45	70	110	180	280	450	700	1,100	1,400	1,800
15	1,725	30	45	70	120	180	300	475	750	950	1,200
20	2,300	**22**	35	55	90	140	225	350	550	700	900
25	2,875	*18*	**28**	45	70	110	180	280	450	560	720
30	3,450	*15*	25	35	60	90	150	235	340	470	600
35	4,025	...	20	**30**	50	80	125	200	320	400	500
40	4,600	...	*17*	**27**	45	70	110	175	280	350	440
45	5,175	...	...	25	**40**	60	100	155	250	310	400
50	5,750	...	...	*22*	**35**	55	90	140	225	280	360
60	6,900	...	...	...	*30*	45	75	120	185	240	300
70	8,050	...	...	...	*25*	**40**	65	100	160	200	260
80	9,200	...	...		...	**35**	55	85	140	180	220
90	10,350	...	...	...	...	*30*	**50**	75	125	160	200
100	11,500	...	...	...	...	28	**45**	**70**	115	140	**180**

* The figure in this column is also the wattage of the circuit if the power is direct current or if it is single-phase alternating current and the load has a power factor of 100 per cent, as is the case with lamp bulbs and most appliances.

In this table, the figures below each size wire represent the maximum distance which that size wire will carry the amperage in the left-hand column, with 2 per cent voltage drop.   All distances are one-way; in a circuit 100 ft long, of course 200 ft of wire is used, but look for the figure 100 above.

If a distance appears in **boldface** type, it indicates that the amperage in the left-hand column is too great for Type T or R wire in conduit but not too great for Type T or R wire in open air.

If a distance appears in *italics*, it indicates that the amperage in the left-hand column is too great for Type T or R wire under any circumstances but not too great for weatherproof wire.

Wire Table—230 Volts—2 Per Cent Voltage Drop

Am-peres	Volt-amperes° at 230 volts	No. 14	No. 12	No. 10	No. 8	No. 6	No. 4	No. 2	No. 1/0	No. 2/0	No. 3/0
1	230	900	1,400	2,200	3,600	5,600	9,000				
2	460	450	700	1,100	1,800	2,800	4,500	7,000			
3	690	300	480	700	1,200	1,800	3,000	4,600	7,500		
4	920	220	350	550	900	1,400	2,200	3,500	5,500	7,000	
5	1,150	180	280	440	720	1,020	1,750	2,800	4,500	5,600	
7½	1,720	120	190	300	480	750	1,200	1,900	3,000	3,800	4,800
10	2,300	90	140	220	360	560	900	1,400	2,200	2,800	3,600
15	3,450	60	90	140	240	360	600	950	1,500	1,900	2,400
20	4,600	45	70	110	180	280	450	700	1,100	1,400	1,800
25	5,750	35	55	90	140	220	360	560	900	1,100	1,440
30	6,900	30	50	70	120	180	300	470	680	940	1,200
35	8,050	...	40	60	110	160	250	400	640	800	1,000
40	9,200	...	35	55	90	140	220	350	560	700	880
45	10,350	...	...	50	80	120	200	310	500	620	800
50	11,500	...	...	45	70	110	180	280	450	560	720
60	13,800	...	...	...	60	90	150	240	370	480	600
70	16,100	...	...	...	50	80	130	200	320	400	520
80	18,400	...	...	...	...	70	110	170	280	360	440
90	20,700	...	...	...	...	60	100	150	250	320	400
100	23,000	...	...	...	...	55	90	140	230	280	360
125	28,750	...	...	...	...	...	75	110	180	220	290
150	34,500	...	...	...	...	...	...	95	150	190	240
175	40,250	...	...	...	...	...	...	80	130	165	210
200	46,000	...	...	...	...	...	...	...	115	140	180

° The footnotes on page 110 apply to this table also.

Note that these tables are based on an assumed voltage drop of 2%. If, under certain circumstances, other voltage drops are to be permitted, the tables are easily converted, as follows:

> For a voltage drop of 1% decrease all distances by 50% (to 50%)
> For a voltage drop of 2½% increase all distances by 25% (to 125%)
> For a voltage drop of 3% increase all distances by 50% (to 150%)
> For a voltage drop of 4% increase all distances by 100% (to 200%)
> For a voltage drop of 5% increase all distances by 150% (to 250%)

Suppose you use any wire size for a circuit longer (or shorter) than the distance shown for a 2% drop for that size of wire; what will the drop be? It is easy to determine. For example, the table for 115 volts shows that using No. 10 wire with a load of 25 amp, the drop will be 2% on a circuit 90 ft long. But you are going to use it on a circuit 120 ft long. What will the drop be? Since 120 ft is 133% of 90 ft (120/90 = 133%), the drop will be 133% of 2% or 2.66%.

If you use any size of wire for a load *smaller* than shown in the left-hand column, the drop will be correspondingly smaller. The same No. 10 wire in the example above will carry 25 amp 90 ft with 2% drop. But if instead of 25 amp the wire is to carry only 22 amp, the drop will be 88% of 2% (22/25 = 88%) or 1.76%

Cost of Voltage Drop. As already discussed, voltage drop represents wasted power, power used to heat the wires in carrying the power to the point where it is to be used. The smaller the wire and the greater the distance, the greater the loss will be. Voltage drop cannot be eliminated, but it can be and must be kept to a reasonable percentage.

Assume you want to operate a 115-volt motor consuming 20 amp at the end of a 100-ft circuit. At 115 volts, 20 amp is 2,300 watts. Per the table on page 110, No. 8 wire will result in 2% drop if the circuit is 90 ft long; at 100 ft the drop will be about 2.2%. Suppose the circuit is in use 3 hr per day, about 1,000 hr per year. In 1,000 hr that circuit will consume about 2,300 kwhr, which at 3 cents per kilowatthour will cost you $69.00; of that, 2.2% or about $1.52 is wasted. That does not seem an unreasonable total. But suppose again, that in order to reduce your initial investment, you had elected to use No. 12 wire. Carrying 20 amp, the same table tells you that the drop will be 2% if the circuit is 35 ft long, or about 6% for a 100-ft circuit. Then 6% of your $69.00, or $4.14, is wasted. That is a difference of $2.62, which in 5 years becomes $13.10, which would have paid the difference between the small and the larger wire.   More important: Your motor would perform more efficiently, deliver more horsepower, with the larger wire.

Now, instead of operating the motor at 115 volts, you decide to operate it at 230 volts.   As pointed out in earlier paragraphs, when a fixed wattage load is changed from 115 to 230 volts, *without changing the wire size*, the voltage drop changes to one-quarter of the percentage of drop at 115 volts.   In other words, using No. 8 wire, the drop changes from 2.2 to 0.55%, and using No. 12 wire, changes from 6 to 1.5%. You can use the smaller wire, and still have less voltage drop.

In debating whether to use the smaller or the larger of two sizes of wire, remember that the labor for installing the larger sizes is very little if any more than for the smaller size.   The difference in cost is basically the difference in the cost of the wire only.   Consider also that with a larger wire, with less drop, your motor (or other load) will operate more efficiently, than when using smaller wire.

Outdoor Wiring.   In outdoor wiring there is one additional factor to be considered: mechanical strength.   The wires must be heavy enough to support not only their own weight but also the strain imposed by winds, ice loads, etc.   In many areas it is not unusual to see a layer of ice an inch thick around outdoor wires after a severe sleet storm.   For this reason it is best to use nothing lighter than No. 12 wire for spans up to 25 ft, No. 10 up to 50 ft, No. 8 up to 100 ft, and No. 6 over 100 ft.   If distances over 150 ft are involved, use heavier wires or install extra poles for supports.

In installing outdoor overhead wires in northern areas, take into consideration the expansion and contraction that take place with changes in the temperature.   A 100-ft span of copper wire will be almost 2 in. shorter when it is 30° below zero than on a hot summer day when the thermometer stands at 100°.   Therefore if the wires are installed on a cold winter day, they may be pulled as tight as practical.   If installed on a hot day, allow considerable sag in the span so that when the wires contract in the winter, no damage will be done.

8

Wire Connections and Joints

Wires must be connected to switches, receptacles, and other devices. This is very simply done, yet many times such connections are poorly made. Study carefully the following points.

Removing Insulation. Cut through the insulation down to the copper conductor, holding the knife not at a right angle but at about 60 deg. This precaution reduces the danger of nicking the conductor, which weakens it and sometimes leads to breaks. After the insulation has been cut all around, pull it off, leaving the conductor sticking out far enough to suit the purpose (see Fig. 8-1). For a good electrical connection you must clean off all traces of rubber or other insulation very carefully.

Terminals. Wire is fastened to devices by means of terminals designed for the purpose. If the device is intended for wire No. 8 or lighter, it is usually provided with ordinary screw terminals of the type shown in Fig. 8-2. The end of the terminal is bent upward to prevent the wire from slipping from under the terminal screw. This screw is often "upset" so that it cannot be removed entirely and lost.

RIGHT   WRONG

Fig. 8-1 In removing insulation from wire which is to be attached to a terminal, hold knife at an angle of about 60 deg.

Bend the end of the wire into a loop to fit and insert it under the terminal screw in such a way that tightening the screw tends to close

Fig. 8-2 Terminals for connecting small wires.

rather than to open the loop.   Figure 8-3 should make this clear.   It is best to close the loop completely with long-nosed pliers after it is in-

Screw closes loop   *Screw opens loop*
RIGHT   WRONG

Fig. 8-3 Insert the loop under the screw of the terminal so that tightening the screw tends to *close* the loop.

serted under the screw.   Cut off any excess length of wire so that the insulation comes up close to the terminal (see Fig. 8-4).

RIGHT   WRONG

Fig. 8-4 Do not leave exposed conductor next to a terminal.

Many switches and receptacles are now made so that the straight, stripped end of the wire is merely pushed into holes in the device; auto-

matically a good connection is made.   See Fig. 8-5.   Strip the insulation off the wire as far as indicated by the marker on the device.   If it is
necessary to remove a wire, push a small screwdriver into a slot on the

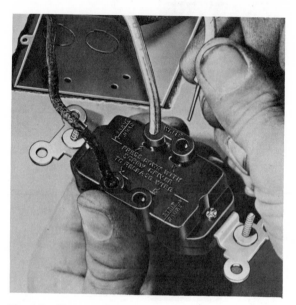

Fig. 8-5   Many modern devices do not have terminal
screws.   Push the bare wire into openings in the device
for a good permanent connection.   (*General Electric
Co.*)

device to release the wire.   Some of such devices have no terminal
screws; others do have the screws, so that the user can use either
method for connections.

Fig.   8-6 A   typical   solderless
connector.   (*Ilsco Corp.*)

Fig. 8-7   A typical solder-
type lug.   (*Ilsco Corp.*)

For No. 6 or heavier wire, solderless terminals of the type shown in Fig. 8-6 are commonly used. Simply insert the stripped and cleaned end of the conductor into the terminal or connector, drive home the nut or screw, and the connection is completed.

Soldering lugs of the type shown in Fig. 8-7 are still occasionally used. In such cases it is necessary to solder the wire into the lug; the way to do this will be explained later in this chapter.

Joints. In many cases joints between two different pieces of wire are prohibited by the Code; these will be mentioned as the related work is discussed in this book. At other times joints are necessary; consequently it is most important that they be made properly. The requirements for joints are simple. Mechanically the joint must be as strong as a continuous length of the wire. Electrically it must be as good a conductor as a continuous piece of the same wire. After the joint is completed, insulation equivalent to the original insulation on a continuous piece of the wire must be placed over it.

To accomplish these three things, it is necessary to remove the insulation where the wires are to be joined, make the mechanical joint, using solder or solderless connectors, and replace the insulation by means of tapes made for the purpose.

Removing Insulation. In removing insulation for a joint, hold the knife at an angle as in sharpening a pencil, rather than at a 60-deg angle as in preparing wire for a terminal. When stripping stranded wires, be careful not to damage any of the individual strands. After the insula-

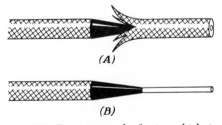

Fig. 8-8 Preparing end of wire which is to be joined to another, and later soldered.

tion has been cut all the way around the wire, it can often be removed to the end of the wire with one pull. In the case of wire with a fabric cover it is necessary to remove also the outer braid only, for some distance back, depending on the size of the wire. Be careful to cut only

through the outer braid, not into the insulation proper.   All this is
shown in steps in Fig. 8-8.

   Making the Splice.   The simplest and most common method of joining two solid wires is shown in successive steps in Fig. 8-9 and hardly
needs additional explanation.

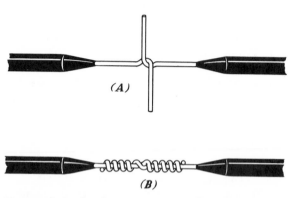

Fig. 8-9   A simple splice; it is strong mechanically.

   Small stranded wires can be spliced in the same way.   If the
stranded wire is of considerable size, it will be better to make the type of
splice shown in successive steps in Fig. 8-10.   This may appear a bit
difficult but with practice becomes relatively easy.   Spread the individual strands evenly, as in A.   Place the two wires end to end with the
several strands intersecting, as in B.   Then wrap one of the strands
around the assembly and repeat with one of the strands of the opposite
piece, wrapping it in the opposite direction; this is shown in C.   Follow
through with alternate strands of each piece and continue in this fashion
until each strand has been wrapped, giving the final appearance of D in
Fig. 8-10.   Naturally considerable practice is required to learn just
how long to leave the strands for different wire sizes.

   Taps.   The simplest and most commonly used form of tap is shown in
Fig. 8-11.   It is so simple that the picture should be self-explanatory.

   With stranded wire the same kind of tap can be used, but, especially
in heavier sizes, it is better to proceed as shown in successive steps in
Fig. 8-12.   First separate the strands in the main wire into two groups,
as in A, so that the tap wire can be inserted into the opening.   Divide

the strands of the tap wire into approximately equal groups, as in *B*. Next wrap each group around the main wire, one in the clockwise and

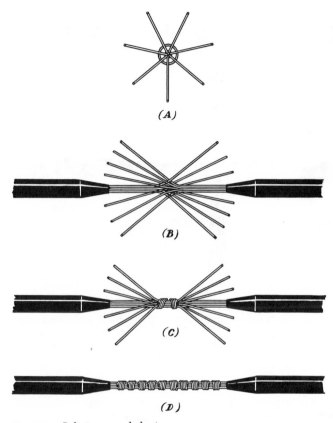

(A)

(B)

(C)

(D)

Fig. 8-10   Splicing stranded wires.

the other in the counterclockwise direction, working toward opposite ends, until the tap is completed as in *C* of Fig. 8-12.

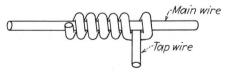

Main wire

Tap wire

Fig. 8-11   A simple tap.

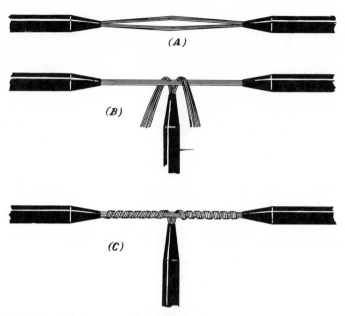

Fig. 8-12 Making a tap with stranded wires.

Fixture Joints.   Where there is no mechanical strain whatever on the wire, usually the two ends are merely twisted together, as shown in Fig. 8-13, then soldered and taped, or connected by solderless connector.

Fig. 8-13 Where there is no strain on wires, they may be spliced by merely twisting.

Splices in Lamp Cords.   In splicing cords, splice each conductor separately, but stagger the two joints so that, when the splice is completed, they will not lie next to each other in the finished job (see Fig. 8-14). This makes a much less bulky joint in the finished job.   A much better practice is never to make such a splice at all—follow the requirement of Code Sec. 400-5 and use only continuous lengths of cord.

Soldering.   The joint having been made mechanically, the next step is to solder it.   It is best to practice first on solid wires rather than stranded.   Good soldering is an art; to become proficient in it requires considerable practice.   The requirements for good soldering are:

1. Absolutely clean conductors.
2. Careful use of flux.
3. Clean soldering copper.
4. Correct temperature of soldering copper.
5. Proper solder.

Fig. 8-14   It is best never to splice a cord.   If a splice must be made, stagger the joints in the two conductors.   This makes a smaller and a safer splice.

Clean the Conductor.   The insulation having been removed, the next step is to clean the conductor carefully.   If this is not done, it will be utterly impossible to do a good soldering job.   The cleaning is usually done by scraping with a knife and is no great task.   Most building wires have tinned conductors which make it easier to strip the insulation.   If you use care in scraping off the remaining traces of insulation, the tinned surface remains intact, making soldering much easier.

The cleaning of small stranded conductors does call for considerable patience, for it is definitely necessary to clean each strand separately. Do not skimp on this detail.

Flux.   In soldering, a flux of some type is necessary to permit solder to form a solid bond with the copper.   For some kinds of nonelectrical soldering, acid is used as a flux, but under no circumstances may it be used for electrical soldering.   The acid reacts with the copper to form a new compound, usually of an insulating nature, and, especially on fine stranded wires, often eats through the copper, destroying the conductor.   Use any kind of noncorrosive paste, of which several brands are on the market.   Use it sparingly.   Rosin makes a good flux, and rosin-core solder is entirely practical.

Keep Soldering Copper Clean and at Right Temperature.   If a soldering copper is used, whether it is electrically heated or heated by a blowtorch, keep it clean.   Applying the hot copper to a cake of sal ammoniac will clean it.   The right temperature can best be learned by experience, too hot a copper being just as impractical as one that is not hot enough.

Apply the soldering copper directly to the conductor, as shown in Fig. 8-15. You must heat *the conductor* to a temperature high enough

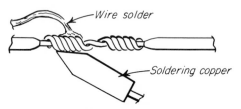

Fig. 8-15 In soldering, do not drop hot solder on cold wire. Heat the wire until it is hot enough to melt the solder.

so that when the solder is touched to the conductor (not the soldering copper), it will melt completely and flow into every little space. If the soldering copper is not hot enough, it will *slowly* increase the temperature of the conductor, but much of the heat will follow the conductor away from the spot where the heat is being applied, and flow along the conductor so that it will be heated not only where you want it hot, but also for some distance inside the insulation, probably damaging the insulation. Have the soldering copper hot enough so that it will heat the conductor *quickly* at the joint, but not inside the insulation.

While the actual solder that will remain in the finished joint is applied to the hot conductor, you will find that a drop of solder on the soldering copper itself, at the point where you apply it to the conductor, will help in transferring heat from the soldering copper to the conductor; the unsoldered joint will heat faster. If you have done no work of this kind, experiment on scrap pieces of wire to gain experience.

In the case of very fine wires, the drop of solder on the soldering copper will probably be enough to solder the joint. This then becomes an exception to the rule of applying the solder to the conductor and letting the heat in the conductor melt the solder.

If you use a blowtorch for soldering, remember that the tip of the flame is the hottest spot; there is no heat in the inner cone of the flame. When using a torch it is easy to heat the conductor too much, damaging the insulation. Practice on scrap till you get the knack of it.

Figure 8-16 shows the difference between dropping hot solder on a cold conductor and heating the conductor sufficiently to melt the solder, which then flows into the smallest crevice.

Fig. 8-16   When hot solder is applied to cold wire as at left above, a poor joint results.   When hot wires melt the solder, it flows into every crevice and makes a good joint, as shown at right above.

Reinsulating the Joint.   After a joint has been soldered, you must replace the insulation originally on the wire.   The insulation on the finished joint must be electrically and mechanically as good as that on a continuous piece of wire.   Formerly this required two kinds of tape, but today a single variety of tape is used.   Both methods will be described.

Plastic Tape.   Plastic electrical tape has largely replaced the two kinds of tape formerly used.   It is tough mechanically, and has very high insulating value (per mil or thousandth of an inch of thickness) so that a comparatively thin layer of it is sufficient, thus doing away with clumsy, bulky joints.   This leads to a less crowded condition inside boxes, especially when there must be many joints within the same box.

In applying this tape on a joint, start at one end, laying the end of the tape over the tapered end of the original insulation, then winding it diagonally toward the opposite end, letting the successive turns slightly overlap (see A of Fig. 8-17).   Keep the tape stretched so that, wherever the turns overlap, they will fuse to each other.   From the opposite end, work back toward the starting point in exactly the same manner.   The individual turns of tape in successive layers will be almost at right angles to each other (see B of the same picture).   Work back and forth in this fashion until the thickness of the tape you have applied is as thick as the original insulation on the wire.   The job is finished.

Do note that Fig. 8-17 at C shows in exaggerated fashion a cross section of the finished joint; the small openings between turns and layers will in actual practice be nonexistent, for the pressure created by keeping the tape tight during application will result in one solid mass of insulation.

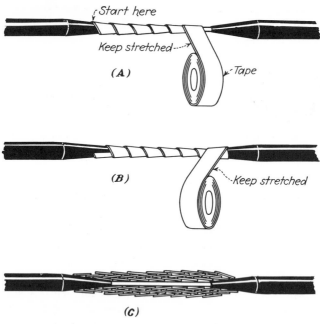

Fig. 8-17   Applying tape to a splice.

Splicing Compound and Friction Tape.   In the older method, two kinds of tape are used: splicing compound (commonly called rubber tape), which replaces the original insulation, and friction tape to protect the splicing compound mechanically.

Splicing compound is a very high grade of rubber put up in rolls. The rubber is unvulcanized and under slight pressure vulcanizes with another layer of the same material.   A layer of cloth between layers of rubber in the roll prevents the layers from sticking to each other; the cloth is thrown away as the tape is used.   Wrap several layers of this tape around a pencil, one layer on top of the other, stretching the tape tightly as it is applied.   Cut through the mass of rubber with a knife and you will find a solid mass of rubber instead of the several layers you have applied; the layers have vulcanized into a solid mass.

Apply this tape to a joint as already described for plastic tape and as shown in Fig. 8-17.   Then follow with several layers of friction tape, and the joint is finished.

Weatherproof Wire.   If the joint is in weatherproof wire (which has a relatively poor insulation), use either plastic tape alone or friction tape alone.   Rubber tape is not required.

Soldering Lugs.   A typical soldering lug was shown in Fig. 8-7. The conductor is soldered into this lug, which then is connected to the device with bolts provided for the purpose.   This sounds simple, yet it is safe to say that a goodly percentage of such soldering jobs are very poorly done, resulting in high-resistance joints which lead to arcing, overheating, and sometimes damage from the resulting heat.   Figure 8-18 shows a cross section of a poorly soldered lug.   Note how most of the

Fig. 8-18   Cross section of a poorly soldered lug.   Not enough solder was used, probably not enough flux, and certainly not enough heat.

individual strands of wire have no trace of solder.   Careful study of this picture should indicate the points to watch in soldering a wire into a lug.

To solder a wire into a lug is undoubtedly more difficult than soldering, for example, a simple splice.   The copper conductor first must be properly cleaned.   If the wire is stranded, the individual strands must not be spread out, which would prevent a good mechanical fit in the lug.   Tin the exposed end of the conductor; apply a reasonable quantity of flux; dip the end into a ladle of melted solder, making sure that it is hot enough so that it will flow freely into every crevice, every space, between individual strands.   When it cools, there will be a single solid mass of metal: copper and solder.   If a blowtorch is used instead of a ladle, heat the end of the conductor sufficiently so that, when solder in wire form is applied, it will melt and flow down into the space between the strands.   Do not overheat the wire lest the insulation be damaged.

Tin the inside of the soldering lug in the same way; then melt sufficient solder into it so that, when the conductor is inserted, the lug, the conductor, and the solder will form a single mass of metal.   Of course, heat must be applied to the lug, by means of either a soldering copper or a blowtorch, to melt the solder previously inserted in order to permit complete fusion of the three elements.

Because it is recognized that a large percentage of soldered joints involving tubular lugs and heavy sizes of wire are poorly made in the field, the Code in Sec. 230-72 prohibits their use in connection with service equipment and in Sec. 250-113 prohibits their use in connection with grounding conductors and clamps. Connections must be made by means of "pressure connectors," as heavy-duty solderless lugs are termed in the Code.

Solderless Connectors. When there is no strain on the joint, as, for example, in an outlet box, solderless connectors ("wire nuts") of the type shown in Fig. 8-19 may be used.

Fig. 8-19 Two types of solderless connectors. (*Ideal Industries, Inc.*)

One type has a brass insert with a set screw. The wires are clamped into this insert; then the insulating shell is screwed on over the insert. The one-piece type is screwed directly onto the stripped wires. If one of the wires is lighter than the others, let it project a bit beyond the heavier wires before installing the connector. Connectors of this type are made in several sizes; use a size suitable for the number and size of the wires being connected. Figure 8-20 shows several connectors in use.

Regardless of which of these two types you use, strip the wires just far enough so that no bare conductor will be exposed beyond the end of the insulating shell. No tape is required.

The spring-type connectors shown in Fig. 8-21 are also popular. They are available in two types—uninsulated and preinsulated—and in several sizes. Lay the wires to be joined so that the stripped ends are parallel. Using the uninsulated type, screw the connector over the wires with the projecting end of the connector serving as a handle. The twisting action tends to uncoil the spring during the screwing-on process, but it promptly contracts and puts very heavy pressure on the

wires when the work is completed.   Then twist the "handle" at right
angles, and it will break off, leaving the completed splice.   Tape, of
course, must be applied over it.

Fig. 8-20   This shows the use of solderless connectors.
(*Ideal Industries, Inc.*)

The preinsulated type is far more popular than the other, and there is
no handle to break off.   The insulation increases the diameter, and is
provided with "wings," thus providing a good grip.   No tape is necessary if you have been careful in stripping the insulation off the wires so
that no bare wire is exposed.

A still different type of connector which is very popular, especially in
farm wiring, is shown in Fig. 8-22.   The body has a U cross section, and
the nut slips over the threaded legs of the U, making it very handy for
tapping one wire to another continuous wire.   When there is no strain
on the finished joint and heavy sizes of wire are involved, the simple
type of connector shown in Fig. 8-23 is commonly used.   Being made
of metal, the connectors must be taped after the joint is made, although
snap-on molded plastic insulating covers are available in some
brands.

Solderless Connectors for Aluminum.   Connectors as described may or may not be suitable for use with aluminum wire.   On the smaller

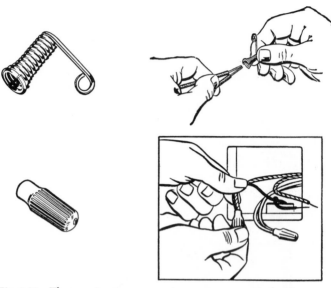

Fig. 8-21   These spring-type connectors are very practical.   Screw them over the wires to be joined.   If the preinsulated type is used, taping is not necessary.   (*3-M Co.*)

sizes, only the statement on the carton holding the connectors will provide the necessary information.   If in doubt, assume the connectors are suitable only for copper.

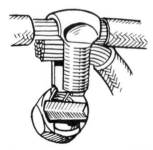

Fig. 8-22   A connector of this type permits one wire to be tapped to another *continuous* wire.

Fig. 8-23   If no strain is involved in the joint, simple connectors of the type shown here may be used.

In the larger sizes however each connector is marked either CU[1] or AL[1] or CU/AL. If unmarked, or marked CU, the connector is suitable only for copper. If marked AL it is suitable only for aluminum. If marked CU/AL it may be used with either copper or aluminum, but never for a combination of the two in the same connector. A few connectors however are made which may be used with copper and aluminum in the same connector; the connector includes a divider so that the two metals do not touch each other. They are identified as such on the cartons.

Connectors on heavy service switches, circuit breakers, and similar equipment are similarly marked AL or CU.

[1] CU is the chemical abbreviation for copper, AL for aluminum.

9

Grounding: Theory and Importance

In all discussions concerning electrical wiring, you will regularly meet the terms ground, grounded, grounding. They all refer to deliberately connecting parts of a wiring installation to ground, usually to a buried pipe of a water system, or where such pipe is not available, to a ground rod driven into the earth. Grounding falls into two categories: (a) system grounding, or grounding one of the current-carrying wires of the installation, and (b) equipment grounding, or grounding non-current-carrying parts of the installation, such as the service switch[1] cabinet, the frames of motors or electric ranges, the metal conduit or armor of armored cable, and so on.

The purpose of grounding is *safety*. If an installation is not properly grounded, it can be exceedingly dangerous as to shocks, fires, and dam-

[1] In any installation there is either a main service switch, or a circuit breaker cabinet serving the same purpose. When in this chapter the reference is to "service switch" it will refer to either of the two, whichever is installed.

age to appliances and motors.   Proper grounding reduces such dangers, and also minimizes damage from lightning, especially on farms. Grounding is a most important subject; it is so important that many points will be repeated throughout this book, for emphasis.   Study the subject thoroughly; *understand it.*

The Code rules for grounding are quite complicated, and at times appear to be ambiguous.   However, for installations in homes and farm buildings they are relatively simple.   In this chapter will be discussed only the basic principles of grounding; other chapters will discuss sizes of the ground wire, ground clamps, ground rods, and similar details.

Terminology.   In this book the term "ground*ed* wire" means the "neutral" wire of a circuit, normally carrying current, but also connected to ground.

The term "ground wire" will refer only to the wire which runs from the service switch to ground.   In the service switch it is connected to the neutral wire mentioned above, thus grounding it, and also to the metal cabinet itself, thus grounding the cabinet and also the conduit or the metal armor of armored cable, which are anchored to the cabinet.

The term "ground*ing* wire" refers to a wire which does not carry current at all during normal operation; it is connected to parts of the installation such as frames of motors or clothes washers, the outlet boxes on which switches or receptacles are installed, and so on.   It runs with the current-carrying wires.   In other words it is connected only to components that normally do not carry current, but which do carry current in case of damage to or defect in the wiring system, or the appliances connected to it.   In the case of wiring with conduit, or cable with a metal armor, a grounding wire as such is not installed separately, for the conduit or the armor of the cable serve as the ground*ing* conductor.

Grounded Neutral Wire.   In ordinary residential or farm wiring, the power comes into your premises from the power supplier's line over three wires.   See Fig. 9-1 in which the wires are marked *N*, *A*, and *B*. Wire *N* is grounded and is called the neutral wire, and wires *A* and *B* are called "hot" wires.   Note that the voltage between the neutral *N* and either *A* or *B* is 115 volts; between the two hot wires *A* and *B* it is 230 volts.   The neutral is usually insulated just like the hot wires, in which case it must be white; in some cases a bare, uninsulated wire is permitted, as will be explained elsewhere.   The two hot wires may be any color except white or green, and are usually black and red, or sometimes both black.

If you touch both *A* and *B* you will receive a 230-volt shock.   If you touch either *A* or *B*, and also *N*, you will receive a 115-volt shock.   But note that *N* is connected to the ground; so to receive a 115-volt shock you don't have to actually touch *N*.   Touching *A* or *B* while standing on the ground is the same as touching *A* or *B* and also *N*—and you will receive the 115-volt shock.

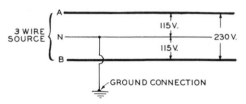

Fig. 9-1 Using only three wires, two different voltages are available. The neutral wire *N* is grounded.

White Neutral Wire.   As already mentioned, the white wire is known as the neutral wire.   The Code in Sec. 200-6 requires that the neutral wire must be "identified" which is further defined as having a white or natural gray color.   In this book it will be referred to as the white or neutral wire.   It is a grounded wire.   Likewise the Code with one exception (which will be explained later) requires that white wire may never be used for a purpose other than the grounded neutral.   However, No. 4 and heavier wires are rarely stocked except in the black color.   For this reason the Code permits black wire in these heavier sizes to be used throughout an installation, provided that in each case where a white wire should be used, the ends of the black wire are painted white.   Painting the ends makes a white wire out of a black one.

For outdoor use it is not necessary to use white wire for the grounded neutral.   Weatherproof wire, which is the type ordinarily used for the purpose, is available only in a black color.

The grounded neutral wire is never interrupted by a fuse, circuit breaker, switch, or other device unless the device used is so designed that in opening the grounded wire it simultaneously opens also all the ungrounded wires, but such devices are not used in *residential* work. The white wire, with the exception noted, always runs directly from the point where it enters the building up to the device where the current is finally consumed. This simple fundamental requirement of wiring must at all times be kept firmly in mind.

The neutral must run without interruption to all equipment operating at 115 volts, but not to anything operating at 230 volts. Only hot wires run to 230-volt loads.[2] A separate grounding wire runs to 230 loads (unless the conduit or the armor serve as a grounding conductor). There are a few exceptions where the grounded neutral of a cable runs to a 230-volt appliance, thus serving also as a grounding wire, and these cases will be explained later. Most important of all: The neutral wire is grounded to the water pipe or ground rod, at the service switch. The power supplier also grounds that wire at the transformer serving the premises.

If the neutral wire is *properly* grounded both at the transformer and at the building, it follows that if you touch an exposed neutral wire at a terminal or splice, no harm follows, no shock, any more than if you touch a water pipe or a faucet, for the neutral and the piping are connected to each other. Any time you touch a pipe, you are in effect also touching the neutral.

Since the neutral wire is actually grounded and there is no possible danger in touching it, why put insulation on that wire? Many engineers argue in favor of using uninsulated neutral wires, but this scheme is not permitted by Code; the few exceptions permitted today will be covered separately in other chapters. Unless otherwise stated, it is necessary to use the same kind of insulation, the same care to avoid accidental grounds, and the same careful splices for the neutral wire as for the hot wires.

How Grounding Promotes Safety. See Fig. 9-2 which shows a 115-volt motor with the neutral grounded, and a fuse in the hot wire (actually the fuse and the ground connection would be at a considerable distance from the motor, at the fuse cabinet, not close to the motor as shown, although there might be an additional fuse near the motor). Assume the fuse blows (which is equivalent to cutting the wire at the fuse location); the motor stops. You as the owner may inspect the motor, may accidentally touch one of the wires at the terminals of the motor. What happens? Nothing! The circuit is hot only up to the

---

[2] Anything that is connected to a circuit and consumes power constitutes a "load" on the circuit. The load might be a motor, a toaster, a lamp—anything consuming power. Switches and receptacles do not *consume* power, therefore are not loads.

fuse.   Between the fuse and the motor the wire is dead, just as if the wire had been cut at the fuse location.   The other wire to the motor is the grounded neutral, so is harmless.   You are protected, provided the motor is not defective.   But do note that if the fuse is *not* blown and you touch the hot wire, you will receive a 115-volt shock, through your body, to the earth, and through the earth back to the other or neutral wire.

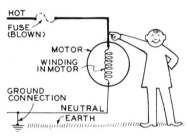

Fig. 9-2 A 115-volt motor properly installed   except   for   a   grounding wire.   It is a safe installation so long as the motor remains in perfect condition.

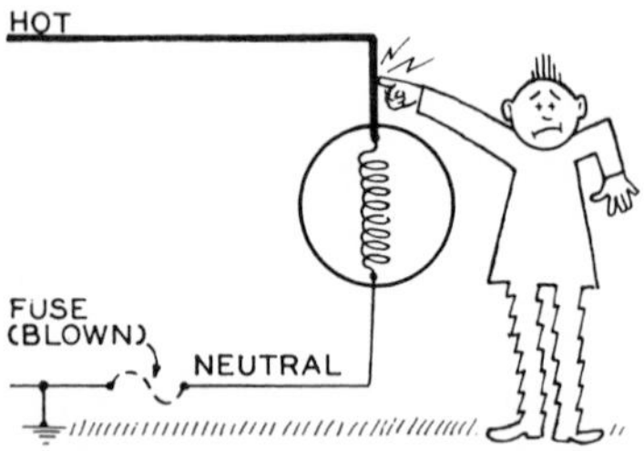

Fig.   9-3 A   115-volt   motor   installed   with a fuse in the neutral. It is a dangerous installation.

Now see Fig 9-3 which shows the same circuit except with the fuse wrongly placed in the neutral instead of the hot wire.   The motor will operate properly.   The fuse blows, the motor stops.   But the circuit is still hot, through the motor, up to the blown fuse.   You touch one of the wires   at   the   motor.   What   happens?   You   complete   the   circuit through your body, through the earth, to the neutral wire ahead of the fuse.   You are directly connected across 115 volts, and as a minimum you will receive a shock, and at worst will be killed.   The degree of shock and danger will depend on the surface on which you are standing.   If you are on an absolutely dry surface you will note little shock; if you are on a damp surface as in a basement, you will experience a severe shock; if you are standing in water you will undergo maximum shock, often leading to death.

Now see Fig. 9-4, which again shows the same 115-volt motor as in Fig. 9-2.   But suppose the motor is defective, so that at the point marked G the winding inside the motor accidentally comes into electrical contact with the frame of the motor; the winding "grounds" [3] to the frame.   That does not prevent the motor from operating.   But suppose you choose to inspect the motor, touch just the frame of the motor. What happens?   Depending on whether the internal ground between winding and frame is at a point nearest the neutral, or nearest the hot wire, you will receive a shock up to 115 volts, for you will be completing the circuit through your body back to the grounded neutral.   It is a potentially dangerous situation; shocks of a whole lot less than 115 volts can be fatal.

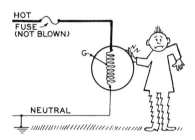

Fig. 9-4   The motor of Fig. 9-2, but the motor is defective.   Its windings are accidentally grounded to the frame.   It is a dangerous installation.

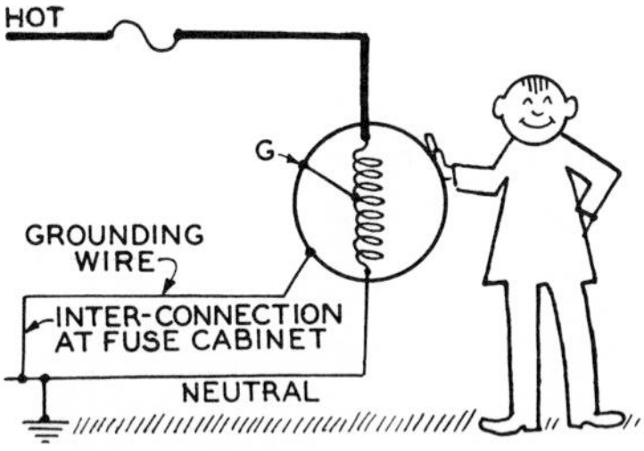

Fig. 9-5   The motor of Fig. 9-4, but a grounding wire has been added from the frame of the motor, to ground.   It is a safe installation.

Now see Fig. 9-5 which shows the same motor, with the same accidental ground between winding (or cord) and frame, but protected by a ground*ing* wire, connected to the frame of the motor, and running back to the ground connection at the service switch.   When the internal ground occurs, current will flow over the grounding wire.   It will sometimes but not always blow the fuse.   But even if the fuse does not blow, the grounding wire will protect you.   The grounding wire re-

[3] Breakdowns in the internal insulation of a motor, so that an electrical connection develops between the winding and the frame of the motor, are not uncommon.   The entire frame of the motor becomes hot.   The same situation arises if the motor is fed by a cord which becomes defective where it enters the junction box of the motor, so that one of the bare wires in the cord touches the frame.   If there is no cord but the motor is fed by the circuit wires, a sloppy splice between the circuit wires and the wires in the junction box on the motor can lead to the same result: The frame of the motor becomes hot.

duces the voltage between the frame of the motor, to substantially zero as compared with the ground on which you are standing; you will not receive a shock *provided* that a really good job of grounding was done at the service.   If there is a poor ground you will still receive a shock.

Green Wire.   The ground*ing* wire from the frame of the motor (or from any other normally non-current-carrying component) may be green, or green with a yellow stripe,[4] or in many cases bare, uninsulated.   If conduit, or cable with armor, is used, the conduit or armor serves as the grounding wire.   Green wire may not be used for any purpose other than the ground*ing* wire.   Other chapters will discuss just when a separate grounding wire must be installed.

Now refer to Fig. 9-6, which shows a 230-volt motor installed with a fuse in each hot wire (all hot wires must be fused, or protected by circuit breaker).   Remember that in such 230-volt installations, the white

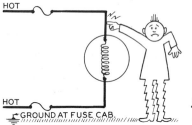

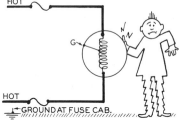

Fig. 9-6   A 230-volt motor properly installed except for a grounding wire.   It is a safe installation so long as the motor remains in perfect condition.

Fig. 9-7   The motor of Fig. 9-6 but the motor is defective; its windings are accidentally grounded to the frame.   It is a dangerous installation.

neutral wire does not run to the motor, but the neutral is nevertheless grounded at the service switch.   If you touch both hot wires, you will be completing the circuit from one hot wire to the other, and you will receive a 230-volt shock.   But if you touch only one of the wires, you will be completing the circuit through your body, through the earth, back to the grounded neutral, and you will receive a shock of only 115 volts: the same as touching the neutral and one of the hot wires of Fig.

---

[4] Green with yellow stripe is basically the European color scheme.   Using green with yellow stripe permits American appliances to be exported without modification for the European market.

9-1.   The difference between 115-volt and 230-volt shocks may be the difference between life and death.   This illustrates one of the benefits of proper grounding.

But assume that the motor becomes defective, that the winding of the motor becomes accidentally grounded to the frame, as shown in Fig. 9-7.   You will recognize this as the same as Fig. 9-4, except that the motor is operating as 230 volts instead of 115 volts.   Touching the frame will produce a 115-volt shock.   But if the frame has been properly grounded as shown in Fig. 9-8, one of the fuses will probably blow, but even if it does not blow, touching the frame will not result in shock because the frame is grounded, again assuming that a really good ground was installed at the service.

Advantages of Conduit or Armored Cable.   Now let us consider a different aspect of the situation.   In any of the illustrations of Figs. 9-2 to 9-8, if there is an accidental contact between any two wires of the cir-

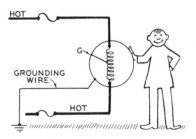

Fig. 9-8 The motor of Fig. 9-7, but a grounding wire has been installed from the frame of the motor, to ground.   It is a safe installation.

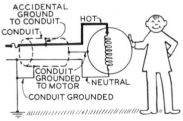

Fig. 9-9 A 115-volt motor properly installed, now using wires in steel conduit or steel armor.   The conduit is grounded at the fuse cabinet: it is also grounded to the frame of the motor.   It is a safe installation.

cuit, this of course constitutes a short circuit and a fuse will blow, regardless of whether the short is between one of the hot wires and the neutral, or between the two hot wires.   For all this to happen, there must be bare places on two different wires touching each other, which does not happen very often in a carefully installed job.   But consider a wiring system in which all the wires are installed in a metal "raceway": iron pipe called conduit, or in flexible metal conduit which will be discussed in a later chapter, or using cable with a metal armor as in the cable commonly called "BX."   The raceway or armor is grounded at the service switch.   It is also connected to the motor itself (assuming that the motor is not connected by a flexible cord and plug).   No separate grounding wire is then required: If there is now an accidental ground from winding to frame, or short circuit (not from one of the hot wires to the neutral, but from a hot wire to the raceway or armor), it has the same effect as a short between one of the hot wires and the neutral, because the neutral and the raceway or armor are connected to each other *and the ground* at the service switch.   A fuse will immediately blow, whether the motor operates at 115 or 230 volts.   A considerable advantage has been gained, for accidental grounds that otherwise might remain undiscovered are automatically disclosed.   All this is shown in Fig. 9-9.

In spite of the advantages of a metal raceway system, in some locations (especially farms) there are conditions that will be explained later, which make the use of a cable *without* metal armor more desirable. Then nonmetallic-sheathed cable is used, which in addition to the usual insulated wire has a bare, uninsulated grounding wire which runs throughout the wiring system.

Continuous Grounds.   As will be explained in Chap. 10, at every location where there is an electrical connection, a metal "outlet box" or "switch box" is used.   When a grounded-neutral wiring system is used (which is 100% of the time for residential and farm wiring, and most of the time in other work), and you use metal conduit or cable with a metal armor, you must ground not only the neutral wire but also the conduit or the cable armor.   The neutral wire is grounded only at the service switch, but the conduit or armor must be securely grounded to every box or cabinet.   When fixtures are installed on outlet boxes they become automatically grounded.   When conduit or armor is connected to motors or appliances (without cords and plugs) they become automatically grounded.

When using conduit or armored cable, it is of the utmost importance that the locknuts on the conduit (or on the connectors of the cable) be driven down solidly, so that the teeth of the locknut bite into the metal of the box.   In that way you provide a continuous metallic circuit, entirely independent of any of the wires installed, that is such an important part of grounding.

If you are using nonmetallic-sheathed cable, it will contain an extra wire, a bare, uninsulated grounding wire, which must be carried from outlet to outlet, as will be explained in another chapter, providing a continuous ground. Regardless of the wiring method used, a *continuous* ground is of the utmost importance.

Other Advantages of Grounding. Suppose a 2,300-volt line accidentally falls across your 115/230-volt service, during a storm, as shown in Fig. 9-10. If the system is not grounded, you can easily be subject to

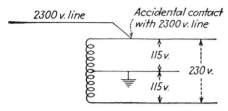

Fig. 9-10 In an ungrounded system, if a 2,300-volt line falls across the 115/230-volt wires, it raises their voltage to 2,300; in a grounded system the voltage will be considerably above 115/230, but much less than 2,300 volts.

2,300-volt shocks, and wiring and appliances will be ruined. If the system is *properly* grounded, the highest voltage of a shock will be much more than 230 volts, but very much less than 2,300 volts.

Lightning striking on or even near a high-voltage line can cause great damage to your wiring and your appliances, and can cause fire and injuries. Proper grounding throughout the system largely eliminates the danger.

Receptacles. Most people are used to the original receptacle having two parallel openings for the plug, as shown in Fig. 9-11. Anything plugged into it duplicates the condition of Fig. 9-2. If the appliance is defective, and the owner handles it, he could receive a shock as shown in Fig. 9-4. This led to the development of what are called grounding receptacles, shown in Fig. 9-12. Note that such a receptacle has the usual two parallel slots for two blades of a plug, plus a third round or U-shaped opening for a third prong on the corresponding plug. In use the third prong of the plug is connected to a third or grounding wire in the cord, running to and connected to the frame of the motor or other appliance. (This third wire in the cord is green, sometimes green with a yellow stripe.) On the receptacle the round or U-shaped opening leads to a special *green* terminal. In turn, when installing the receptacle, a wire must be connected from this green terminal, to ground. If conduit or cable with armor is used, that wire runs to the outlet or switch box and the conduit or armor becomes the grounding conductor.   In the case of cable without armor, it will contain an extra wire, bare, uninsulated, in addition to the insulated wires, and this must be connected to the green terminal of the receptacle, and also grounded to the boxes. In this way, the frame of the motor or appliance is effectively grounded, leading to extra safety as shown and discussed in connection with Fig. 9-5.   The details of how to connect the grounding wire from the green terminal will be discussed in later chapters.

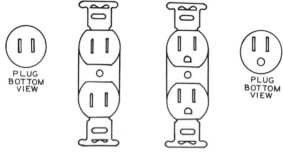

Fig. 9-11 An ordinary receptacle and plug. Only plugs with two blades will fit.

Fig. 9-12 A grounding receptacle and plug. Plugs with either two or three blades will fit.

The Code requires the use of 3-wire cord and 3-prong plugs on specific appliances such as clothes washers, clothes dryers, dishwashers, and hand-held appliances such as drills,[5] sanders, hedge clippers, and some others.   The 1968 Code added refrigerators, freezers, and air conditioners; manufacturers of course are allowed plenty of time for the changeover.   It is not required on ordinary household appliances such

[5] It should be noted that in the case of ordinary portable tools, the Code does not require grounding if the tool is constructed using what is called *double insulation*.   Ordinary tools have only what is known as functional insulation, that is, only as much as is necessary for the proper functioning of the tool; this includes the insulation on the wires in the winding of the motor, an insulated switch, and so on.   Double insulation goes much farther: Usually the case of the tool is made of insulating material rather than metal; the shaft is insulated from the steel laminations of the rotor of the motor; switches have handles of an especially tough insulating material; many other points of extra and specially durable insulation.

as toasters, irons, radios, TV, razors, lamps, and similar items.   This might lead you to think you must install two kinds of receptacles, one for use with 2-prong plugs, and others for use with 3-prong plugs.   Not so: *the grounding receptacles are designed so that either plug will fit.*

The Code requires that in all new construction, only this grounding receptacle may be installed.   If in an existing installation, you replace a defective receptacle the new one must be of the grounding type, provided you can effectively ground it.   If that is difficult or impossible, then the replacement receptacle may *not* be the grounding type.

How Dangerous Are Shocks?   Most people think it is a high voltage that causes fatal shocks.   This is not necessarily so.   The amount of current flowing through the body determines the effect of a shock.   A milliampere is one thousandth of an ampere.   A current of one milliampere through the body is just barely perceptible.   One to eight milliamperes causes mild to strong surprise.   Currents from 8 to 15 milliamperes are unpleasant, but usually the victim is able to free himself, to let go of the object that is causing the shock.   Currents over 15 milliamperes are likely to lead to "muscular freeze" which prevents the victim from letting go.   Currents over 75 milliamperes are likely to be fatal; much depends on the individual involved.

Of course the higher the voltage, the higher the number of milliamperes that would flow through the body, under any given set of circumstances.   We must distinguish between shocks resulting from touching *two* hot wires, and those resulting from touching *one* hot wire. In the latter case, a shock from a relatively high voltage while the victim is standing on a completely dry surface will result in fewer milliamperes than a shock from a much lower voltage while he is standing in water.   Many deaths have been caused by shock on circuits considerably below 115 volts; many have survived shock from circuits of 600 volts and more.

Another determining factor in danger is the number of milliamperes a source of power can produce.   When you walk on the dry carpeting of a house in the winter when the humidity is low, you pick up a static charge, so that you feel a mild shock when you touch a grounded object.   The voltage is high, but the current is infinitesimal.   If you touch the spark plug of a car while the motor is running, you receive a shock of at least 20,000 volts, but the current is exceedingly small, so no harm is done.   The same principle applies on "electric fences" used on farms: a voltage high enough to be uncomfortable, but a very small current that does no harm.

It should be noted that farm animals are much less able to withstand shocks than are human beings.   Many cattle have been killed by shocks that would be only uncomfortable to a man.

Voltage to Ground.   This term is often used in the Code.   If one of the current-carrying wires in the circuit is grounded, the voltage *to ground* is the maximum voltage that exists between that grounded neutral wire, and any hot wire in the circuit.   If no wire is grounded, then the voltage to ground is the maximum voltage that exists between *any* two wires in the circuit (Code, Art. 100).

Polarizing.   The process of maintaining a grounded wire throughout a wiring system, always identified by its white color, is known as polarizing.   The hot wires may be of other colors, but never white or green, as already explained.   If a cord or cable contains an insulated ground*ing* wire (not the white ground*ed* wire), it must be green, or green with yellow stripe.

Colors of Terminals.   The color of the terminals or lugs on switches, receptacles, and so on, identifies the kind of wire that may be connected to each one.   Natural copper or brass terminals are for hot wires. Terminals of a whitish color such as nickel, tin, or zinc plated are for a ground*ed* wire.   Terminals of a green color are for a ground*ing* wire.

Methods of Grounding.   The details of how to ground a wiring installation will be discussed in Chap. 17.

10

Outlet and Switch Boxes

In the very early days of electrical wiring, it was the general practice to run wires on the surface of walls or inside them, without further protection, up to the devices to be connected. Fixtures were mounted directly on the ceiling, switches on the plaster walls, without any further ado. All this has been changed in the interest of safety from both fire and shock, and now outlet boxes or equivalent devices must be used at every point where wires are spliced, or connected to terminals of electrical equipment.

Purposes of Boxes. Outlet boxes house the ends or splices in wires at all points where the original insulation has been removed. There is little danger in a continuous piece of wire, but a poorly made joint may lead to short circuits, grounds, or overheating at that point. Inside walls there are naturally loose dust, cobwebs, and other easily ignitible materials. Therefore there is some danger of fire at joints, but when the joint is enclosed in a metal outlet box, this danger is practically eliminated. Moreover, outlet boxes provide a continuity of ground, as explained in the previous chapter.   The Code requires that boxes be supported in walls and ceilings according to definite standards which provide mechanical strength for supporting fixtures, switches, and other devices, eliminating the danger of mechanical breakdown.

Common Types of Outlet Boxes.   Figure 10-1 shows a 4-in. octagon box, one of the most common boxes in use.   Around the sides and in the bottom are found "knockouts"—sections of metal that can be easily knocked out to form openings for wire to enter.   The metal is com-

Fig. 10-1 Typical octagon outlet box. (*All-Steel Equipment, Inc.*)

Fig. 10-2 Square outlet box. This box is 4 in. square and much roomier than an octagon box. (*All-Steel Equipment, Inc.*)

pletely severed around these sections except at one small point which serves to anchor the metal until it is to be removed.   It is a simple matter to remove these knockouts—usually a stiff blow with a pair of pliers on the end of a heavy screwdriver held against the knockout will start it, and with a pair of pliers the metal disk is then easily removed.   If the knockout is near the edge of a box, a pair of pliers is the only tool needed.

On some brands of boxes the pry-out type of knockout is furnished. The pry-out is simply a small slot near or in the knockout, into which it is necessary only to insert screwdriver to pry out the metal disk, which prepares the knockout for use.

Remove only as many knockouts as are actually needed for proper installation.   If accidentally you remove a knockout leaving an opening that won't be used, you must close the opening using a closure washer.

The outlet boxes are provided with ears and screws to make it easy to mount covers, switches, or other devices used on the boxes.

There is also in common use a box similar to the one shown in Fig. 10-1 but only $3\frac{1}{4}$ in. in size. The two are more or less interchangeable; the larger size is by far the more common because it is easier to do a good job using it, especially if a large number of wires enter the box.

Another common box is the 4-in. square box shown in Fig. 10-2, more or less interchangeable with the octagon but decidedly roomier and handier to use. Especially when conduit rather than cable is used, this box is used almost exclusively for reasons that will be explained later. There is another box identical in appearance but larger, $4\frac{11}{16}$ in. square, used mostly for commercial work as distinguished from residential.

Depth of Outlet Boxes. The Code in Sec. 370-14 requires that boxes of all descriptions be at least $1\frac{1}{2}$ in. deep, except when the use of a box of this depth "will result in injury to the building structure or is impracticable," in which case a box not less than $\frac{1}{2}$ in. deep may be used.

Switch Boxes. For mounting switches, receptacles, and similar devices flush in the wall, switch boxes of the type shown in Fig. 10-3 are

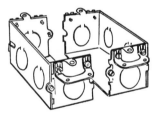

Fig. 10-3 A typical switch box. The sides of such boxes are removable.

Fig. 10-4 Two single boxes may be ganged to form one larger box of double size. Any number of boxes may be ganged to form a box of any necessary size.

used. The sides are removable; this makes it easy to make a double-size, or "2-gang," box out of two single ones by simply throwing away one side on each box and joining together the two boxes. No extra parts are needed. This is shown in Fig. 10-4. In similar fashion it is possible to make boxes of any required size, to mount three or more devices side by side.

Depth of Switch Boxes.   Switch boxes range in depth from 1½ in. to a maximum of 3 in.   The 2½-in. depth is the most popular, for it provides generous room for connectors, wire, etc., between the switch or other device and the bottom of the box.   Use the 1½-in. depth only when two boxes in two different rooms happen to come back to back in the wall separating the two rooms; usually the wall is not thick enough to permit two deeper boxes to be used.

Material and Finish of Boxes.   Boxes are usually made of steel, with a choice of galvanized or black-enamel finish.   The galvanized is by far the more durable of the two, and the black enamel is little used today.   In fact it is usually difficult to locate boxes with black-enamel finish, and

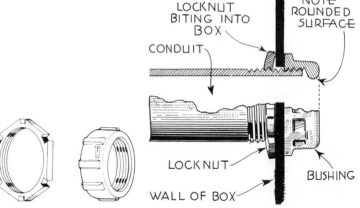

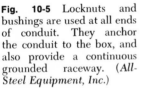

Fig.   10-5 Locknuts   and bushings are used at all ends of conduit. They anchor the conduit to the box, and also provide a continuous grounded   raceway.   (*All-Steel Equipment, Inc.*)

Fig. 10-6 Cross section showing how locknut and bushing are used at outlet boxes or other boxes.

many local ordinances prohibit them.   For certain purposes boxes made of an insulating material such as plastic or porcelain are used; these will be discussed in Chap. 24, which covers farm wiring.

Securing Conduit and Cable to Boxes.   To provide a good, safe continuous ground throughout an installation, it is absolutely necessary that every length of conduit or cable entering a box be firmly and solidly anchored to the box.

In the case of conduit this is very simply done by means of a locknut and bushing, both of which are shown in Fig. 10-5. Note that the locknut is not a flat piece of metal, but is dished or bent, so that the lugs around the circumference become teeth on one side. The smallest internal diameter of the bushing is slightly less than the internal diameter of the conduit. This causes the wire where it emerges from the conduit, to rest on the rounded surface of the bushing; this prevents damage to the insulation of the wire that might occur if the wires were to rest against the sharp edge of the cut end of the conduit.

Slip a locknut on the threaded end of the conduit, with the teeth facing the box. Slip the conduit into the knockout. Then install the bushing on the conduit inside the box, screwing it on as far as it will go. Only then tighten up the locknut on the outside of the box, driving it home solidly so that the teeth will bite into the metal of the box, thus making for a good sound ground. This construction is shown in Fig. 10-6. Detailed instructions for cutting and using conduit will be found in the next chapter.

In the case of cable, connectors of the type shown in Fig. 10-7 are

Fig. 10-7 Cables are anchored to boxes with connectors of this type.

used. After the connector is fastened to the cable, slip the connector into a knockout, install the locknut on the inside of the box, and run the locknut home tightly, as in the case of the conduit, so that the teeth of the locknut bite into the metal of the box.

Clamps. The use of boxes having built-in clamps which eliminate the need for special connectors for cable is common. Typical boxes of this type are shown in Fig. 10-8, and the picture should be self-explanatory.

Round Boxes. Instead of being octagonal or square, boxes may be round, in which case locknuts and bushings may not be used on the rounded wall of the boxes but only on the bottom. Round boxes usually have clamps for cable.   They are used mostly in "old work," in wiring a building after it is completed.

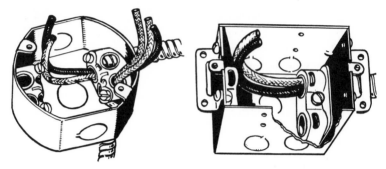

Fig. 10-8 Often boxes are provided with clamps, serving the same purpose as separate connectors.

Supporting Outlet Boxes.   The usual method of supporting an outlet box in a new building is by means of a hanger, shown in Fig. 10-9. Remove the center knockout in the bottom of the box, slip the box over

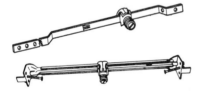

Fig. 10-9 Outlet boxes are usually supported in buildings using hangers of this type. (*All-Steel Equipment, Inc.*)

Fig. 10-10   Factory-assembled combinations of box and hanger are convenient and save time. (*All-Steel Equipment, Inc.*)

the fixture stud that is part of the hanger, and drive the locknut home on the fixture stud inside the box.   The fixture stud later may be used to support a fixture installed on the box.   There are also available factory-assembled combinations of the type shown in Fig. 10-10.

In any case the assembly is mounted between joists, as shown in Fig. 10-11.   The hanger described is of the "shallow" type, and its depth is such that with boxes 1½ in. deep the front of the box will be flush with the plaster.   The plaster must come right up to the box; there may be no open space around the box.

This shallow-type hanger is used with all wiring systems with the exception of conduit. When conduit is to be used, "deep" hangers are used instead of "shallow." The only difference is that the offset is about ½ in. deeper, which brings the front edge of the outlet box about ½ in. below the surface of the plaster but leaves sufficient room behind

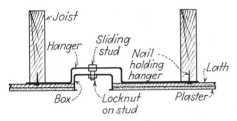

Fig. 10-11 The mounting of boxes, using hangers, is very simple.

the plaster for the conduit, the locknut, etc. Since the Code in Sec. 370-10 requires that the front edge of the box must be not more than ¼ in. below the finished surface of the wall or ceiling (if the surface is combustible, it must be flush), some expedient must be adopted to overcome the fact that the front edge is over ¼ in. below the surface. Ac-

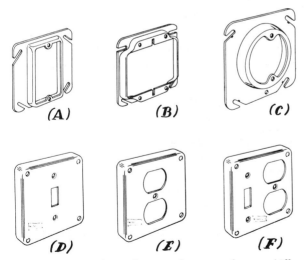

Fig. 10-12 Typical raised covers for square boxes. (*All-Steel Equipment, Inc.*)

cordingly, it is customary to use covers, the front edges of which are flush with plaster.   Figure 10-12 shows an assortment of covers fitting the 4-in. square box which is usually used with conduit.   Each cover serves a specific purpose, that at *A* accommodating one switch or receptacle, that at *B* two such devices.   At *C* is shown a "plaster ring" which has an opening of the same size as a 3¼-in. box.   When this is mounted on top of the 4-in. square box, it permits devices designed for the 3¼-in. box to be used.   At *D*, *E*, and *F* are shown covers that are used only when the boxes are mounted on the surface, as in basements, factory walls, etc.   They accommodate, respectively, one toggle switch, one

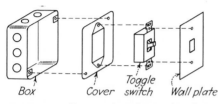

Fig. 10-13   All parts of an electrical outlet are standardized in size so as to fit properly and easily.

receptacle, and a switch and receptacle.   Many other types are available.   Of all these covers the types at *A*, *B*, and *C* are the most common.   Figure 10-13 shows a completed installation.

Instead of requiring hangers to support boxes, the Code generally permits any other type of support which is sturdy and which becomes part of the building structure.   Wooden strips may be used provided

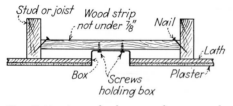

Fig. 10-14   An outlet box may be mounted on a wooden strip, instead of using a hanger.

they are at least ⅞ in. thick.   This method is shown in Fig. 10-14 but is little used.

Boxes in a run of conduit must still be independently supported to or by a structural member of the building.   The support provided by being anchored to a run of conduit, even if the conduit is rigidly supported, is not sufficient.

In old work (buildings wired after completion), as compared with new work (buildings wired while under construction), different methods of support are used, and these will be discussed in the chapter on old work.

Supporting Switch Boxes.   Instead of switch boxes, 4-in. square boxes with covers are frequently used, as previously described.   Usually, however, switch boxes are used, and since they are available in depths up to 3 in., there is no problem in connection with the use of conduit as there is with 1½-in.-deep outlet boxes.

Perhaps the most common switch box for new work is the bracket type of box, of which several are shown in Fig. 10-15.   The bracket is

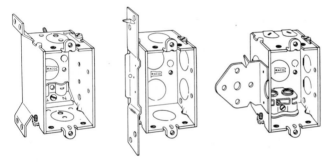

Fig. 10-15   Switch boxes with mounting brackets save time during installation.   (*All-Steel Equipment, Inc.*)

merely nailed to the studding of the building.   Some boxes have a trough which holds up the ends of the lath that end at the box and that would otherwise be left unsupported.   The brackets themselves have a number of projections that form a good support and anchor for the plaster when it is applied.

Also available and generally used are mounting strips which permit any number of switch boxes to be mounted at any point between two studs.   These are shown in Fig. 10-16, and the picture will be self-explanatory.

Switch boxes may be mounted on wooden strips like outlet boxes, and a finished installation is shown in Fig. 10-17.   The strips must be at least ⅞ in. thick and so mounted that the front edge of the boxes will be flush with the plaster.

Ordinary switch boxes of the types shown in Figs. 10-3 and 10-8 have an external bracket at each end, held to the box by two machine screws. The part of the bracket next to the box has two slotted openings, so that the brackets can be adjusted up and down, as required to compen-

Fig. 10-16 Switch boxes may be mounted between studs, using special steel mounting strips.

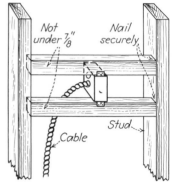

Fig. 10-17 Switch boxes may also be mounted on wooden strips.

sate for varying thicknesses of plaster, so that the front of the box will be flush with the plaster.   The brackets can be reversed for use with plasterboard or similar construction, as will be explained in more detail in Chap. 22, Fig. 22-7.

Outlet Box Covers.   An outlet box may never be left uncovered. When a fixture is mounted on top of the box, no further cover is necessary.   An outlet box cover must be used in every other case.   Figure 10-18 shows an assortment for the conventional $3\frac{1}{4}$- or 4-in. boxes; there are corresponding ones for other styles of boxes.   At A is shown a blank cover used to cover the box when it serves merely to house joints in wire.   The box then is known as a "pull" box or "junction" box. Such boxes may be located only where permanently accessible without damaging the structure of the building.   At B is shown a drop-cord cover used with drop cords; the hole is bushed to do away with sharp edges which might otherwise injure the insulation of the cord.   At C is shown a "spider cover" on which surface style switches are mounted. At D is shown a cover with a duplex receptacle, used mostly in basements, workshops, and similar locations. E shows an outlet cover and F a similar cover provided with pull-chain control. These covers are widely used in closets, attics, basements, farm buildings, and similar locations.

The covers illustrated are the ones most commonly used; most jobs can be completed using only the ones shown.   There are, however, dozens of other types, each serving a specialized purpose.   For example, there are blank covers similar to that shown in A of Fig. 10-18, but

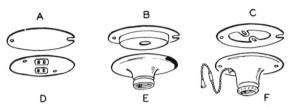

Fig. 10-18   Outlet box covers are made in dozens of different types.   The six shown here are typical.

with a knockout in the center, permitting an armored-cable connector to be used when running the flexible cable to a stationary device like a motor.   There are covers with openings to accommodate sign receptacles of the type that were shown in Fig. 4-37.   For square boxes there are covers with many other combinations of openings in addition to the ones shown in Fig. 10-12.

Surface Boxes.   When the wiring is on the surface of a wall so that it will be permanently exposed, ordinary boxes are impractical because of the sharp corners, both on the boxes and on the covers for switches or similar devices.   For that reason a special kind of box known as a utility box or handy box was developed, together with various kinds of covers,

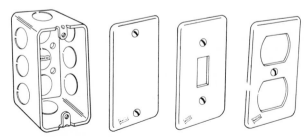

Fig. 10-19   The "handy box" and covers shown here are used for surface wiring.

both shown in Fig. 10-19.   Square boxes and covers shown in Fig. 10-12 are also often used for surface wiring.

Nonmetallic Boxes.   Instead of steel boxes, others made of plastic material are in common use.   While originally used only on farms, their use in other locations is becoming very common when nonmetallic-sheathed cable is used.   More information about such boxes will be found in Chap. 24 concerning farm wiring.

Number of Wires in Box.   If you run too many wires into a box, it becomes very crowded, making it difficult to do good work.   Moreover, the crowded condition leads to short circuits and grounds.   Therefore the Code in Secs. 370-6(a-1) and 370-6(a-2) limits the number of wires permitted in a box, depending on the cubical content of the box, and the size of the wires, as in Table 370-6(a-1).

For shallow boxes, the figures are as in Table 370-6(a-2).

TABLE 370-6(a-1)   Deep Boxes

Box dimensions, inches trade size	Cubic inch cap.	Maximum number of conductors			
		No. 14	No. 12	No. 10	No. 8
$3\frac{1}{4} \times 1\frac{1}{2}$ octagonal	10.9	5	4	4	3
$3\frac{1}{2} \times 1\frac{1}{2}$	11.9	5	5	4	3
$4 \times 1\frac{1}{2}$	17.1	8	7	6	5
$4 \times 2\frac{1}{8}$	23.6	11	10	9	7
$4 \times 1\frac{1}{2}$ square	22.6	11	10	9	7
$4 \times 2\frac{1}{8}$	31.9	15	14	12	10
$4\frac{11}{16} \times 1\frac{1}{2}$ square	32.2	16	14	12	10
$4\frac{11}{16} \times 2\frac{1}{8}$	46.4	23	20	18	15
$3 \times 2 \times 1\frac{1}{2}$ device	7.9	3	3	3	2
$3 \times 2 \times 2$	10.7	5	4	4	3
$3 \times 2 \times 2\frac{1}{4}$	11.3	5	5	4	3
$3 \times 2 \times 2\frac{1}{2}$	13	6	5	5	4
$3 \times 2 \times 2\frac{3}{4}$	14.6	7	6	5	4
$3 \times 2 \times 3\frac{1}{2}$	18.3	9	8	7	6
$4 \times 2\frac{1}{8} \times 1\frac{1}{2}$	11.1	5	4	4	3
$4 \times 2\frac{1}{8} \times 1\frac{7}{8}$	13.9	6	6	5	4
$4 \times 2\frac{1}{8} \times 2\frac{1}{8}$	15.6	7	6	6	5

See Section 370-18 where boxes are used as pull and junction boxes.

Table 370-6(a-2)   Shallow Boxes

Box dimensions, inches trade size	Maximum number of conductors		
	No. 14	No. 12	No. 10
3¼	4	4	3
4	6	6	4
1¼ × 4 square	9	7	6
4¹¹⁄₁₆	8	6	6

Any box less than 1½ inch deep is considered to be a shallow box.

In Table 370-6(a-1), boxes called "device" are what are ordinarily called just switch boxes.   The last three are the "handy boxes" shown in Fig. 10-19.

The tables must be correctly interpreted depending on many factors, as follows:

1. The wires from a fixture to wires in the box are not counted.

2. Deduct one from the numbers in the tables if the box contains a fixture stud, a hickey, or one or more cable clamps (as in Fig. 10-8). The total deduction is one regardless of the number of clamps.   Connectors for cable of the type shown in Fig. 10-7 are *not* cable clamps, for this purpose.

3. Deduct one from the numbers in the tables for each switch, receptacle, or similar device, or a combination of them if mounted on a single strap.

4. A wire originating in a box and ending in the same box (for example the wire from the green terminal of a receptacle, grounded to the box) is not counted.

5. The bare grounding wire of nonmetallic-sheathed cable is counted.

6. A wire running through a box without interruption (without splice or tap), as is often the case in conduit wiring, is counted as only one wire.

For combinations of several sizes of wire entering the same box, follow Code Sec. 370-6(b) in your copy of the Code.

11

Different Wiring Methods

The Code recognizes many different wiring systems. Many of them are used only in rather large buildings and will not be discussed here. In ordinary residential and farm wiring, six different systems all using similar basic materials, are in common use. These systems are:

1. Rigid conduit.
2. Thin-wall conduit.
3. Flexible conduit.
4. Nonmetallic-sheathed cable.
5. Armored cable.
6. Knob-and-tube.

The knob-and-tube is so little used today that it no longer warrants space in this book. Basic principles of each system will be discussed in this chapter; detailed information concerning each system will be found in a later chapter.

Rigid Conduit. In this method of wiring, all wires are enclosed in steel pipe known as conduit. Conduit differs from ordinary water pipe in that it is specially annealed to soften it, for easy bending. The inside surface is carefully prepared so that the wires can be pulled into it with a minimum of effort and without damage to the insulation.   It has a corrosion-resistant finish inside and outside, so that the installation may be permanent.

It comes in 10-ft lengths, each with an Underwriters' label, as shown in Fig. 11-1.   It is made with either a black enamel or galvanized fin-

Fig. 11-1   Rigid conduit looks like water pipe, but differs in many important ways.

ish.   The black enamel is rarely stocked, and may be used only indoors.   The galvanized may be used indoors or outdoors, above ground or underground.   However, per Code Sec. 346-3, it may not be used in or under cinder fill where subject to permanent moisture, unless buried a minimum of 18 in. under the fill, or if encased in ordinary concrete at

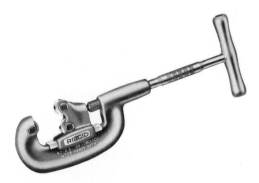

Fig. 11-2   A typical pipe cutter.   (*Ridge Tool Co.*)

least 2 in. thick.   Your local inspector is the final judge, if there is any question.

The ½-in. size is the smallest used in ordinary wiring.   All sizes are identical in dimensions with the corresponding sizes of water pipe, and therefore, as in the case of water pipe, the nominal size in no way denotes the actual physical dimensions.   See table on page 159.

To thread conduit, use the same tools as used in threading water pipe, except with dies that cut a thread with a taper of ¾ in. to the foot.

Cutting Conduit.   Conduit can be cut with an ordinary pipe cutter, as used for water pipe, of the type shown in Fig. 11-2.   This unfortunately leaves a sharp edge at the cut, as shown in Fig. 11-3.   This sharp edge might seriously damage the insulation of the wire as it is pulled into the conduit; therefore common sense suggests, and the Code requires, that each cut be reamed smooth.   A reamer of the type shown in Fig. 11-4 serves the purpose.   Some prefer to use a hack saw with 18 teeth to the inch, claiming that this leaves less burr to be reamed than when a pipe cutter is used.

Fig. 11-3   When pipe is cut, a sharp edge is left; this must be removed before conduit is installed.

Fig. 11-4   A reamer of this type is used to remove sharp burrs from cut end of pipe.   (*Ridge Tool Co.*)

Bending Conduit.   Because the wires are pulled into conduit after it is installed, it is important that all bending be carefully done so that the internal diameter is not substantially decreased in the process.   Make the bends uniform and gradual.   The Code in Sec. 346-10 specifies the minimum radius of the bend, which varies from six to eight times the nominal inside diameter of the conduit (unless lead-sheathed cable is to be installed in the conduit, in which case it varies from ten to twelve times the inside diameter).   See table on page 159.

Naturally great care must be used so that the conduit is not reduced in diameter at the bend, and will fit properly as intended.   Use a bender of the type shown in Fig. 11-5.   Assume a right-angle bend is to be made, and that the conduit is to have a "rise" of 13 in. above the

Dimensions of Rigid Conduit

Trade size, inches	Internal diameter, inches	Internal area, square inches	External diameter, inches
½	0.622	0.30	0.840
¾	0.824	0.53	1.050
1	1.049	0.86	1.315
1¼	1.380	1.50	1.660
1½	1.610	2.04	1.900
2	2.067	3.36	2.375
2½	2.469	4.79	2.875
3	3.068	7.38	3.500
3½	3.548	9.90	4.000
4	4.026	12.72	4.500
4½	4.506	15.94	5.000
5	5.047	20.00	5.630
6	6.065	28.89	6.620

Minimum Radius of Bends in Conduit

Size of conduit, inches	Minimum radius of bend, inches	
	Conductors without lead sheath	Conductors with lead sheath
½	4	6
¾	5	8
1	6	11
1¼	8	14
1½	10	16
2	12	21
2½	15	25
3	18	31
3½	21	36
4	24	40
4½	27	45
5	30	50
6	36	61

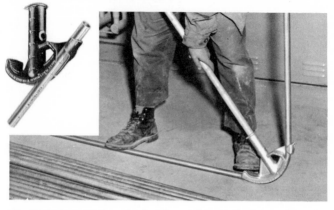

Fig. 11-5 A typical conduit bender, and method of use. (*Republic Steel Co.*)

floor.   How far from the end must the bend begin?   From the rise, subtract 5 in. for the ½-in.; 6 in. for the ¾-in., and 8 in. for the 1-in.

In the example of Fig. 11-6, assuming ½-in. conduit, subtracting 5 in. from the rise of 13 in. leaves 8 in.   Hook the bender over the conduit

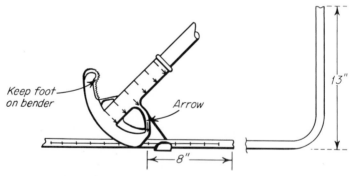

Fig. 11-6 This shows how to figure right-angle bend.

so that the arrow on the bender points to a spot 8 in. from the end of the conduit.   Full instructions for many types of bends usually come with the bender when purchased.   Factory-made elbows are usually used in larger sizes, although power benders for larger sizes may be used on the job.

Number of Wires in Conduit.   The number permitted depends on the size of the conduit, and the size *and type* of wire. This subject is covered by Code Table 1, which you will find in the Appendix of this book. A part of the table is shown below, covering the types and sizes of wire ordinarily used in residential and farm wiring.

Size of wire	Maximum number of wires								
	½-in.	¾-in.	1-in.	1¼-in.	1½-in.	2-in.	2½-in.	3-in.	3½-in.
14	4	6	10	18	25	41	58	90	121
12	3	5	8	15	21	34	50	76	103
10	1	4	7	13	17	29	41	64	86
8	1	3	4	7	10	17	25	38	52
6	1	1	3	4	6	10	15	23	32
4	1	1	1	3	5	8	12	18	24
2	...	1	1	3	3	6	9	14	19
1	...	1	1	1	3	4	7	11	15
1/0	...	...	1	1	2	4	6	9	12
2/0	...	...	1	1	1	3	5	8	11
3/0	...	...	1	1	1	3	4	7	9
4/0	...	...	...	1	1	2	3	6	8

If your installation involves sizes larger than shown, see Code Table 1. If you will have more than three wires per conduit, their ampacity must be derated, as explained in Chap. 27. If you are using certain types of more expensive wires (which are smaller in diameter than ordinary wires), more wires are permitted in any size of conduit, as will be explained in Chap. 27. If you are replacing wires in an existing conduit, a greater "fill" is permitted; again this will be explained in Chap. 27.

If your installation involves lead-sheathed cable, inside conduit, the following applies for one cable per conduit:

Use ¾-in. conduit for 14-2, 12-2, 10-2, 14-3 cable.
Use 1-in. conduit for 8-2, 12-3, 10-3, 8-3 cable.
Use 1¼-in. conduit for 6-2, 4-2, 2-2, 6-3 cable.
Use 1½-in. conduit for 1-2, 4-3, 2-3 cable.

For larger sizes of cable, or more than one cable per conduit, see Table 2 in your copy of the Code.

For combinations of conductors (plain or lead-sheathed) not covered by Tables 1 and 2, see Tables 3 to 7 in your copy of the Code.   Note that if one of the wires is bare, uninsulated, you may per Sec. 346-6 use the dimensions of the bare wire as given in Table 8 of the Code.

Number of Bends.   The Code in Sec. 346-11 prohibits more than four quarter-bends or their equivalent in one "run" of conduit, or the distance between outlet boxes or other openings.   The fewer the bends, the easier it is to pull wire into the conduit.

Splices in Wire.   Wires must be continuous, without splice, throughout all conduit (Sec. 300-13).   Splices are permitted only at outlet boxes, pull or junction boxes, or where the splice is otherwise permanently accessible.

Pulling Wires into Conduit.   If the run is very short and the wires occupy a relatively small portion of the area of the conduit, they can sometimes be pushed in at one outlet and through the conduit up to the next outlet.   If the run is of considerable length, especially if it contains bends, then "fish tape" is used.   For occasional work, a length of ordinary galvanized steel wire may serve the purpose, but a special fish tape made of stiff but flexible steel is more frequently used.   In size it is usually about ⅛ by 0.060 in.   Often it is put up in special reels (see Fig. 11-7).   Bend a small loop or hook on the end of the tape; this will per-

Fig. 11-7   Fish tape is necessary to pull wires into conduit.   The tape is made of springy steel. (*Ideal Industries, Inc.*)

mit it to go easily around bends as it is pushed into conduit.   Since this tape is highly tempered, it will break if bent sharply unless the temper is taken out; this can be done by heating it to a red heat with a blowtorch, then letting it cool.

Push the tape into the conduit through which the wires are to be pulled; when the end emerges, attach all the wires which are to be pulled into the conduit to the fish tape, taking care to leave no sharp ends which might catch at the joints in the conduit.   Then pull the wires into position; this usually requires one man pulling at one end and another feeding the wires into the opening at the other end, to make sure there will be no snarls and in general to ease the wire on its way. Powdered soapstone may be used as a lubricant to make it easier to pull the wires.   The Code in Sec. 300-14 requires a minimum of at least 6 in. of wire projecting at each outlet box where a connection is to be made; it is easy to cut off a few inches later if there is too much, but it is hard to do good work if the ends are too short.   Don't be too enthusiastic in cutting off every last possible fraction of an inch.   Leave a little extra.   Remember, if a switch or receptacle must be later replaced, the replacement may have terminals located somewhat differently than the original device, and may require a little longer length of wire.

If a wire running *through* a box is to be electrically continuous, but with its insulation stripped off for a short distance for connection to a receptacle or other device (as will be shown in B and C of Fig. 18-6 in Chap. 18), you need to provide only enough slack to permit easy connection to the device.   Naturally if a wire is to run through a box on to another box, with its insulation intact, no slack need be provided at the box through which it runs.

Fig. 11-8   Either one-hole or two-hole straps may be used for supporting conduit.

Supporting Rigid Conduit.   Conduit must be supported within 3 ft of every box, fitting, or cabinet.   Use straps shown in Fig. 11-8.   Per Sec. 346-12, it must be further supported at intervals not greater than shown in the following table:

½- and ¾-in. . . . . 10 ft
1-in.   . . . . 12 ft
1¼- and 1½-in.   . . . . 14 ft
2- and 2½-in.   . . . . 16 ft
3-in. and larger   . . . . 20 ft

Electrical Metallic Tubing.   The material which the Code has labeled with this rather unwieldy name is commonly known by its abbreviation "EMT" or by its descriptive common name of "thin-wall conduit."   It is shown in Fig. 11-9, and as in rigid conduit, each 10-ft length bears the

Fig. 11-9   Thin-wall conduit ("EMT") is never threaded. It is lighter and easier to use than rigid conduit.

Underwriters' label.   It is available only in sizes up to and including 4-in.   It must be supported within 3 ft of every box, fitting, or cabinet, and must be further supported at least every 10 ft, regardless of size. For residential and farm wiring it may be used interchangeably with rigid conduit.

The *internal* diameter, size for size, is the same as for rigid conduit, but, as the name implies, the walls are thinner.   Because of this the material is never threaded, but all joints and connections are made using special threadless fittings which hold the material through pressure.   Figure 11-10 shows both a coupling and a connector.   Each

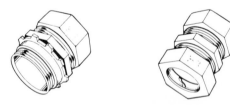

Fig. 11-10   A connector and coupling used with thin-wall conduit.   (*All-Steel Equipment, Inc.*)

consists of a body plus a split ring through which tremendous pressure is exerted on the conduit when the nut is forced home tightly.   Another type of fitting is shown in Fig. 11-11, and the indenter tool in Fig. 11-12.   In use, slip the fitting over the conduit; then use the indenting tool to deeply indent both fitting and conduit. This makes a joint that is mechanically secure, and at the same time has low electrical resistance, thus providing a good continuous ground that is so necessary.

Thin-wall conduit may be cut either with a hack saw (use a blade with 32 teeth to the inch) or with a special tool designed for the pur-

Fig. 11-11 The telescopic type of fittings shown here may also be used with thin-wall conduit. (*All-Steel Equipment, Inc.*)

Fig. 11-12 Use this indenting tool to indent conduit and the fittings of Fig. 11-11, at the same time. (*All-Steel Equipment, Inc.*)

pose. After the cut, the end must be reamed to remove burrs or sharp edges. Bends are made in the same way and under the same conditions as for rigid conduit.

Occasionally it will be necessary to join a length of thin-wall conduit to a length of rigid conduit or to a fitting with threaded hubs for rigid conduit. The simplest method is to use a connector of the type shown in Fig. 11-10. The threaded portion of the thin-wall connector will always fit the internal thread of any fitting designed for the corresponding size of rigid conduit. The adapter shown in Fig. 11-13 can also be used.

Fig. 11-13 This adapter makes it possible to use thin-wall conduit in fittings designed for rigid conduit. (*Killark Elec. Mfg. Co.*)

Nonmetallic-sheathed Cable. The Code in Sec. 300-1(c) recommends this cable for locations where *poor* grounds are the rule rather than the exception: farms and other locations where a continuous underground water system is not available for grounding.   As you have already learned in Chap. 6, there are two kinds of nonmetallic-sheathed cable.   Code Type NM has wires enclosed in an outer fabric braid or plastic sheath, and Type NMC has the wires embedded in a solid plastic mass.   The two types are shown in Figs. 11-14 and 11-15.   Statements

Fig. 11-14   Type NM nonmetallic-sheathed cable may be used only in dry locations.

made in this chapter will refer to either type unless otherwie mentioned.

Type NM may be used only indoors and only in permanently dry locations.   It may not be embedded in plaster.   It may not be used in farm buildings where the humidity is high, specifically barns and similar buildings.   The Type NMC on the other hand may be used wherever

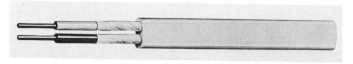

Fig. 11-15   Type NMC nonmetallic-sheathed cable may be used in dry or wet locations.

Type NM is acceptable, but may also be used in moist or damp locations, but not in permanently wet locations, or underground.   It may be embedded in plaster but must then be protected by a strip of steel $\frac{1}{16}$ in. thick and $\frac{3}{4}$ in. wide to guard against nails.   The Type NMC is the successor to the so-called "barn cable" that was in use many years ago.

The facts that nonmetallic-sheathed cable is very easy to install, is light and comparatively inexpensive all are partly responsible for the growing popularity of this cable.   If in doubt as to whether its use is permissible locally, consult local codes, your power supplier, or your electrical inspector.

Removing the Outer Cover.   The outer cover must be removed at the ends for a distance of about 8 in.　On the Type NM this can be done by slitting the braid with a knife; usually the slit is merely started, then by pulling on the jute or fiber filler with a pair of pliers, you can rip the opening as far as you wish.　Cut off the dangling cover with a jackknife.　Be very careful not to damage the individual conductors.　The use of a cable ripper, shown in Fig. 11-16, will save time and avoid damage to the insulation.

Fig. 11-16 This cable ripper saves much time when installing Type NM cable. (*Ideal Industries, Inc.*)

For the Type NMC a knife is the only tool needed.　Again be especially careful not to damage the insulation.

Anchoring Cable to Boxes.　The cable is anchored to switch and outlet boxes by the use of connectors, several types of which are shown in Fig. 11-17.　The connector is first solidly fastened to the cable; the

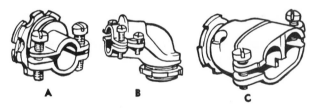

A **B** **C**

Fig. 11-17　An assortment of connectors used in anchoring cable to outlet and switch boxes.

locknut then removed, the connector slipped into the knockout of the box, and the locknut of the connector then driven solidly home on the inside of the box.　As mentioned in Chap. 10, many boxes have built-in clamps that serve the purpose of connectors.

Joints.　Joints and splices are never permitted in nonmetallic-sheathed cable except in outlet boxes, where they are made as in any other type of wiring.

Mechanical Installation.　If installed while a building is under construction, nonmetallic-sheathed cable is installed inside the walls.

The Code requires that it be anchored at least every 4½ ft and in any case within 12 in. of every outlet box. Straps of the types shown in Fig. 11-18 are used for the purpose. It is best not to use staples of any kind, because damage to cable can result from staples driven too hard.

Fig. 11-18 A typical strap for supporting nonmetallic-sheathed cable.

In old work where the cable is fished through the walls, this requirement is waived. All bends in cable must be gradual so as not to injure the cable; the Code requirement of Sec. 336-10 is that, if a bend were continued so as to form a complete circle, the diameter of the circle would be at least ten times the diameter of the cable.

Where the cable is run exposed, as in basements, attics, barns, etc., it must be given reasonable protection against mechanical injury. This protection can be provided in a variety of ways (see Fig. 11-19). If the cable is run along the side of a joist, rafter, or stud as at *A* or along the

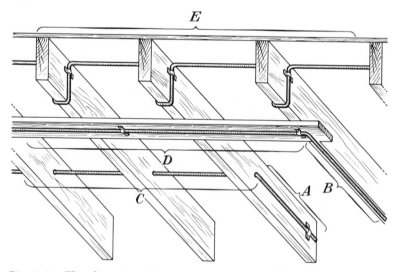

Fig. 11-19 This shows five different ways in which cable may be run on an open ceiling. Various methods are used to protect the cable, depending on the method of installation.

bottom edge of a timber as at *B,* no further protection is required. If it is run at an angle to the timbers, the cable may be run through bored holes as at *C.* No additional protection is required; neither are porcelain tubes, loom, or any similar materials needed where the cable goes through the bored holes. The holes should be bored in the approximate center of the timbers. If the cable is not run through bored holes but instead is secured to the bottoms of the joists, then it must be run on substantial running boards, as shown at *D;* this requirement is waived if the cable is size 6-2, 8-3, or heavier. A final method is to let the cable follow the structure of the building as shown at *E,* when again no further protection is required. This method, however, is very wasteful of material, leads to unnecessarily long lengths of cable with consequently large voltage drops, and is therefore to be discouraged. Whichever method is used, the cable must be supported at least every 4½ ft and also within 12 in. of every outlet box.

In accessible attics, if the cable is run at angles across the top of floor joists, the cable must be protected by guard strips at least as high as the cable, as shown in Fig. 11-20. If run at right angles to studs or rafters,

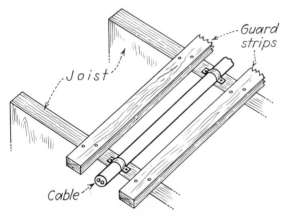

Fig. 11-20 When cable runs crosswise of floor joists in attics, guard strips must be used.

it must be protected in the same way at all points where it is within 7 ft of the floor joists. No protection is required under other conditions. If the attic is not accessible by means of permanent stairs or ladder, this protection is required only for a distance of 6 ft around the opening to the attic.

In any event, the cable must always follow the approximate contour of the building—never any short cuts across open space.

Cable with Grounding Wire.   Nonmetallic-sheathed cable is also available with a bare, uninsulated grounding wire in addition to the insulated wires; see Fig. 11-21.   Cable with grounding wire was first re-

Fig. 11-21   Nonmetallic-sheathed cable in both Types NM and NMC is available with an extra, bare, grounding wire for grounding purposes.

quired by the 1962 Code, in runs to boxes containing plug-in receptacles.   It is now required also to any box that is in contact with metal lath or any other grounded object, and to any box located where it (or the device that it contains) can be reached by a person standing on the ground,[1] or where he can touch a grounded object such as plumbing. In general this means all reachable boxes in bathrooms, kitchens, basements, and all farm buildings.

The grounding wire contributes to safety.   It does complicate the wiring a bit, but it is not difficult to learn how to do it.   How to connect this bare grounding wire will be explained in detail later in this chapter.

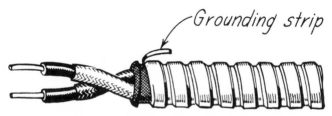

Fig. 11-22   Armored cable.   A steel armor protects the wires.   A bare grounding strip runs inside the armor, for better grounding.

Armored Cable.   Today's armored cable, as pictured in Fig. 11-22, is known as the "ABC type"—armored bushed cable.   It consists of two

---

[1] For this purpose a concrete floor is considered to be the ground, even if it is tiled as in basement recreation rooms.

or more Type T or Type R wires, wrapped with a spiral layer of kraft paper, and a spiral outer steel armor.   Under the armor there is a narrow, uninsulated strip of copper, which reduces the resistance of the armor itself, thus providing better continuity of ground.   This copper strip thus becomes a safety device in case of short circuits or accidental grounds.

Cutting Cable.   A hack saw is generally used to cut armored cable. Do not hold it at a right angle to the cable, but rather at a right angle to the strip of armor as it runs around the cable, as shown in Fig. 11-23.

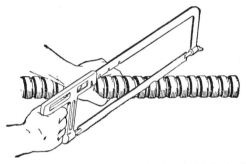

Fig. 11-23   In sawing armored cable, hold the hack saw as shown.

After the cut is made, grasp the two ends of armor, give a twist as shown in Fig. 11-24, and the two ends will separate; it is then a simple matter to cut through the wires.

To use the cut end, first remove about 8 in. of the armor.   Proceed as before, holding the hack saw in the same position, being extremely careful to saw through only the armor, not touching the insulation, or the

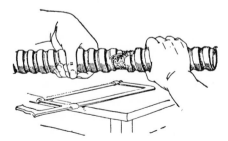

Fig. 11-24   After sawing, twist the cut ends to separate them.

grounding strip.   The paper inside the armor provides some little spacing, making this possible.   Nevertheless, considerable practice is necessary to get the knack of sawing through the armor without damaging the insulation.   When the armor is severed, a twist will remove the short end.

Fiber Bushings.   Careful examination of the cut end of the armor will show that the hack saw has left sharp teeth on the armor, sometimes quite long.   These teeth point inward toward the wires and might damage the insulation, causing a short or ground.   Therefore a bushing of thin but tough fiber which has a high insulating value is inserted between the armor and the wires.   Such a bushing is shown in Fig. 11-25.   Since there is little room between the paper and the armor, space

Fig. 11-25   A bushing of tough fiber is inserted between the armor and the wires, to protect them against danger of grounds from cut ends of armor.

is provided by removing the paper.   The steps shown in Figs. 11-26 and 11-27 demonstrate how to unwrap the paper beneath the armor,

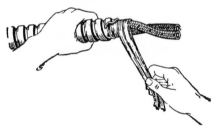

Fig. 11-26   Unwrap the paper over the wires as far as you can; do not tear off.

and then with a sudden yank to tear it off some distance inside the armor, thus leaving room for the bushing, which is inserted as shown in Fig. 11-28.   Figure 11-29 shows the final assembly.

Connectors.   The connectors used with armored cable are practically identical with those shown in Fig. 11-17 for nonmetallic cable, except that the end of the connector which goes into the box has openings or peepholes through which the red color of the antishort bushing can be seen by the inspector. Because of these peepholes, the connectors are known as the "visible type."

Fig. 11-27 Yank the paper wrapper, removing it for some distance *inside* the armor.

To install the connector properly, first you must bend the bare copper grounding strap that runs under the armor back over the outside of the armor. Then insert the fiber bushing inside the armor, and slip the connector over the cable. Push the cable into the connector as far as it

Fig. 11-28 Insert the fiber bushing.

Fig. 11-29 Cross section of cable, showing paper removed inside the armor, and fiber bushing in place.

will go, so that the fiber bushing cannot slip out of place, and so that it can be seen through the peepholes on the connector. Then tighten the screw on the connector to anchor it solidly to the cable. Be sure that the bare grounding strip is placed in the connector so that it will be solidly squeezed by the connector, rather than lying loosely in the connector. Remove the locknut, slip the connector into the knockout of the box, and drive the locknut solidly home inside the box, as with other types of cable (see Figs. 11-30 and 11-31). If this is carefully done, with the teeth of the locknut biting into the metal of the box, all the outlets are then tied together through the armor of the cable, and the grounding strip under the armor, providing the continuity of ground discussed in Chap. 9.

Where Used.   Like Type NM nonmetallic-sheathed cable, armored cable may be used only in permanently dry locations.   For residential

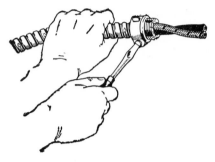

Fig. 11-30   Installing connector on cable.

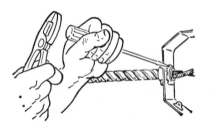

Fig. 11-31   Drive the locknut of the connector down tightly, so that teeth of locknut bite into metal of box.

wiring it is more or less interchangeable with nonmetallic-sheathed cable, but it must not be used on farms or other locations where poor grounds are the rule.   Some local codes prohibit it for new work.

Supporting.   Armored cable is supported and protected exactly as is nonmetallic-sheathed cable.   Staples of the type shown in Fig. 11-32 are perhaps a bit more convenient than ordinary straps.   These staples,

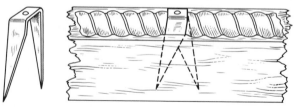

Fig. 11-32   Staples of this type are used to support armored cable on wooden surfaces.

while not prohibited by the Code, should not be used with nonmetallic-sheathed cable because there is danger that they might be driven home so solidly that they would actually damage the cable. This danger does not exist with armored cable, which has steel armor for protection.

Flexible Conduit. This material is generally called "greenfield." It is shown in Fig. 11-33 and is substantially the empty armor of armored

Fig. 11-33 Flexible conduit. The wires are pulled into place after the conduit is installed.

cable, without the wires, and of course in a larger diameter. The smallest size used in ordinary wiring is the ½-in., which is actually over ¾ in. in outside diameter. Except in a few scattered areas, this material is little used for general wiring. Where it is used, it is used like armored cable, except that the flexible conduit is first installed, and the wires are pulled into it later just as in the case of rigid conduit or EMT. Connectors are used as in armored cable, except that the peepholes are not required, since fiber bushings are not used. It must be supported at the same intervals as armored cable.

Flexible conduit is sometimes used in runs of rigid conduit, where the rigid variety would involve very difficult or awkward bends. If this is done in a service conduit, the flexible conduit must be "jumpered" with a length of wire of the same size as the ground wire. This can easily be done using a pair of ground clamps, as in jumpering a water meter, which will be explained in Chap. 17.

Another application is in connection with installations where a certain amount of permanent flexibility is required, for example when a motor is installed with a sliding base, to permit adjustment of belt tension. Obviously the wiring to a motor so installed can't be in rigid conduit, for then the motor would be immovable. Section 350-4 of the Code limits the length of the *unsupported* portion of the flexible conduit to 36 in.

One point that can be debated is the use of a grounding wire. You will recall that armored cable contains a bare, ungrounded grounding conductor, because of the relatively high resistance of the armor of the cable.   The grounding strip is used to provide a better grounding path, in case of a breakdown in the wiring.   But the resistance of the armor of flexible conduit is higher than the resistance of the armor of armored cable, so it would be logical to expect the Code to require a separate grounding wire inside of the flexible conduit.   Beginning with the 1968 Code, it *is* required, although a casual reading of Secs. 250-57(a) and 250-91(b) might indicate the contrary.   Reading those sections carefully you will note that a separate grounding wire is not required *if* the flexible conduit and the fittings used with it are "both approved for grounding purposes."   The catch is that so far the Underwriters have listed neither the flexible conduit nor the fittings specifically suitable for this purpose.   Therefore a grounding wire must be run in each length of flexible conduit.   The wire may be bare or insulated, but if insulated must be green, or green with a yellow stripe.   It should be run inside the conduit; at each end ground it as you would the grounding strip of armored cable.   The size required depends on the rating of the over-current unit in the circuit involved.   With 15-amp protection, use No. 14 wire; with 20-amp use No. 12; with 30- to 60-amp, use No. 10; with 100-amp use No. 8; with 200-amp use No. 6.   These values are from Code Table 250-95.

Grounding Receptacles.   In Chap. 9 we discussed the importance and advantages for safety in using receptacles of the grounding type, with two parallel slots, plus one U-shaped opening for the U-shaped prong on a 3-wire plug.   The grounding wire of a cord is connected to this third prong on the plug.   If you will take apart one of these grounding-type receptacles, you will find that its U-shaped opening is grounded to the mounting strap of the receptacle, and is also connected to the special grounding screw which is identified by its *green* color. In every case, the U-shaped opening must be *effectively* grounded to the wiring system.

Conduit or Armored Cable.   Since the conduit or the armor of armored cable is already grounded (through the locknuts and bushings, or connectors, anchored to outlet and switch boxes all the way back to the service switch), it is only necessary to ground the U-shaped opening to the box.   It could be argued that since the strap of the receptacle is installed in the box using steel screws, grounding will be automatic. That is theoretically a correct conclusion, were it not for the fact that the metal mounting strap is rarely in direct, solid contact with the box. In most cases the plaster ears of the receptacle strap hold the strap some distance away from the box, so that whatever ground there is depends on two small mounting screws; these more often than not make very poor contact with the mounting strap. In other words, whatever ground exists automatically is a very poor and undependable ground.

For that reason, the Code in Sec. 250-74 requires that a grounding wire (bare or green) must run from the green terminal of the receptacle, to the box. Anchor it to the box using a special metal clip shown in Fig. 11-34, or install a small screw through an unused mounting hole

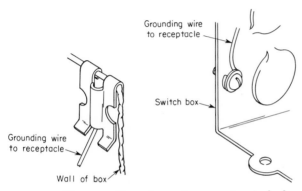

Fig. 11-34 The grounding wire to the green terminal of a receptacle may be grounded to the box using either method shown here.

of the box, as shown in the same illustration. Most outlet and switch boxes have an extra hole tapped for a 10-32 screw for the purpose. *This screw may not be used for any other purpose.*

Type NM or NMC Cable. If you are using metal boxes, you must use cable with the extra bare grounding wire described earlier in this chapter. At the starting point, connect the grounding wire to the neutral in your fuse cabinet or circuit breaker cabinet. At each box, connect the ends of all the grounding wires entering the box together; a connector of the type that was shown in Fig. 8-19 will prove convenient. From the junction you have just made, run a short piece of wire to the box itself as described in preceding paragraphs. If the box contains a receptacle, run another short piece of wire from the junction to the green terminal of the receptacle. All this is shown in Fig. 11-35. If construction as described has been used, it will be possible to remove a receptacle, and still preserve the continuity of ground (while the box is empty) that is so important, and as required by Code.

However, if the box is one that does not need to be grounded (as explained earlier in this chapter), then the length of cable entering the box

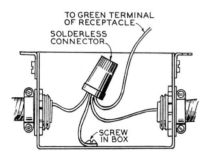

Fig. 11-35 How to connect bare grounding wire of nonmetallic-sheathed cable.

does not need to contain the grounding wire, unless it runs on to another box that must be grounded.

If you are using nonmetallic boxes, the procedure is as already outlined, except that the grounding wire does not need to be connected to the box itself.

Exceptions.   If surface wiring is installed, the grounding wire from receptacle to box is not required, for then the mounting strap of the receptacle is in direct, solid contact with the box.   The Code also permits

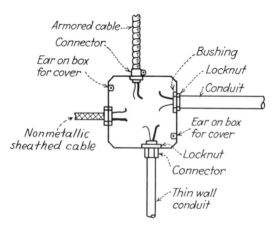

Fig. 11-36   How to change from one wiring system to another.   A cover must be installed on the box.

omitting the wire from receptacle to box, *if* the receptacle has specially designed mounting straps or other suitable means for grounding; unfortunately such receptacles are not available; so the Code provision merely leaves the door open for something suitable to be designed. However, there is available a special mounting screw, to be used in place of the screws that normally come with receptacles, that may be used in place of wire from receptacle to box.

Switches with Grounding Screw.   There are on the market toggle switches with a green grounding screw, just as on receptacles.   In case of breakdown in the switch this provides additional safety.   It would be wise to use such switches, for example, in bathrooms, and especially in wiring farm buildings if switches are installed on *metal* covers on *nonmetallic* boxes.

Changing from Conduit to Cable.   At times it will be necessary to change from conduit wiring to cable wiring, but this presents no problem.   Make the change at an outlet box as shown in Fig. 11-36.   Connect all black wires together, also all the white.   The box must be permanently accessible.   Cover with a blank box cover.

12

Adequate Wiring

An adequately wired home is one that has been wired so that it is completely safe, and so that the occupant will get the maximum of convenience and utility from the use of electric power, with a minimum of inconvenience.

He must have light available where needed, in the amount needed, from permanently installed fixtures or portable lamps. He must be able to plug in lamps, radio, and TV where he pleases, without resorting to extension cords, even after the furniture is moved around. He must be able to turn lights on and off in any room without stumbling through darkness to find a switch, to move from basement to attic with plenty of light but without leaving an unneeded light turned on behind him.

He must be able to plug in needed appliances without first unplugging others. He must get full utility out of the appliances by having them heat quickly, and without lights dimming as the appliances are turned on. Circuit breakers must trip rarely.

Many people feel that if a house is wired "according to the Code," it will automatically be adequately wired. Far from it—despite the real estate advertisements that so imply. The introduction to the Code contains this statement: "This Code contains basic minimum provisions considered necessary *for safety.* Compliance therewith and proper maintenance will result in an installation essentially free from hazard, but not necessarily efficient, convenient, or *adequate for good service* or future expansion of electrical use."

A house will be adequately wired only if you carefully plan it that way. Many houses are being wired with entirely too little thought about adequacy, even though today electric power is expected to provide ten to fifty times as much light per room as in the early days of electricity and is expected to run radios and television, vacuum cleaner and toaster, ranges and ovens, washing machine and clothes dryer, air conditioners and fans, and dozens of appliances not even thought of then. There is no reason to think that we are now at a time when additional new uses for electric power will not come along in the next 10 or 20 years. These things must be planned for.

An inadequately wired home is like the automobile of 60 years ago, which furnished transportation but did not have such conveniences as spare tires, electric starters, an enclosed body, or a lighting system, not to speak of such refinements as heaters or radios. Today these things are considered essentials. Plan the house to include even those things that today you may not consider essential but within 10 or 15 years will be.

Factors in Adequate Wiring. In order that a home may be adequately wired, careful attention must be paid to the following details:

1. Service entrance of sufficient capacity.
2. Wires of sufficient capacity throughout the home.
3. Sufficient number of circuits.
4. Receptacle (plug-in) outlets in sufficient number.
5. Lighting outlets in sufficient number.
6. Lighting fixtures of scientific design.
7. Wall switches in sufficient number for complete flexibility.
8. Miscellaneous outlets and devices for signaling, radio, and so on, in proportion to the size and pretentiousness of the house.

Service Entrance. This general term includes all wires and equipment from the outside of the building up to and including the meter and the overcurrent protection (circuit breakers or fuses). It must be of sufficient size so that the maximum load in use at one time will neither overload the entrance wires, causing excessive voltage drop and wasted electricity, nor trip breakers or blow fuses.   Provision must be made for future equipment that the owner will no doubt install later.

Wire Sizes.   Remember that the Code specifies only minimum sizes.   While No. 14 wire may generally speaking be used throughout the average installation, the trend is toward No. 12 as a minimum.

Circuits.   If all the lights in a house were protected by a single circuit breaker or fuse, the entire house would be in darkness when the breaker tripped or the fuse blew.   To avoid this the outlets are subdivided into groups or circuits each protected by an individual breaker or fuse.   The greater the number of circuits, the greater the flexibility, the less danger of tripping breakers or blowing fuses (because it reduces the likelihood of overloading any one circuit), and the less the voltage drop, thus making for brighter lights.

Receptacle Outlets.   Sufficient receptacle or plug-in outlets do away with the need of extension cords, which are unsightly, inconvenient, and dangerous, both from the standpoint of possible injury caused by tripping over them and also from the standpoint of electrical and fire hazards caused by fraying and short circuits.   The Code in Sec. 210–22(b) requires that

> In every kitchen, family room, dining room, breakfast room, living room, parlor, library, den, sunroom, recreation room, and bedroom, receptacle outlets shall be installed so that no point along the floor line in any wall space is more than 6 ft, measured horizontally, from an outlet in that space, including any wall space 2 ft wide or greater and the wall space occupied by sliding panels in exterior walls.   The receptacle outlets shall, in so far as practicable, be spaced equal distances apart.

This is a Code minimum, but remembering that floor lamps, radios, and other electrical devices are seldom equipped with cords 6 ft long, you may want to reduce the 6 ft to 5 ft for a really adequate installation.

Lighting Outlets.   Usually each room with the exception of the living room requires a ceiling outlet for general lighting.   Additional lighting, of course, is provided by floor or table lamps.

Careful attention should be paid to lighting outlets in miscellaneous locations.   It costs very little to install lights in clothes closets, hall, porches, and attics; this subject will be treated at greater length later in this chapter.

Lighting Fixtures.   The selection of lighting fixtures will be discussed in Chap. 14.

Wall Switches.   Lights that are controlled by a pull chain, or a similar switch on the fixture itself, are inexcusable today, except perhaps in closets or other rooms so small that it is impossible to miss the cord or pull chain.   Outside this one exception, every light should be controlled by a wall switch.

If there is only a single door leading into a room, the logical location for the switch is near the door.   If, however, there are two entrances, it is equally logical that there should be a switch at each so that the light can be controlled from either point; in other words, use a pair of 3-way switches.   Should there be three entrances, a switch at each of the three entrances is a touch of luxury that the owner will appreciate.   In a house that has been really adequately wired, you can enter by any entrance and move from basement to attic without ever being in darkness, yet without ever having to retrace your steps to turn off lights.

Miscellaneous Outlets.   Every house will have a minimum of at least a doorbell system which will permit signaling from either front or back door.   Pilot lights will be used at switches which control lights that cannot be seen from the location of the switch, to indicate whether the lights are on or off; common uses are in connection with basement, attic, or garage lights.   Careful consideration of these details will make a home much more livable.   Other suggestions will be found in Chap. 20.

Adequacy by Rooms.   Some rooms require much more light than others.   Likewise the need for receptacle outlets is greater in some rooms than others.   Consider what is good practice in each room.

Living Rooms.   At one time lamps larger than 100 watts were not often used in floor lamps.   Today 300-watt lamps are common.   This trend toward larger lamps was responsible for a tendency to eliminate lighting fixtures entirely from living rooms.   With white ceilings, high-wattage floor lamps do produce very excellent lighting, but since these lamps must necessarily be used mostly in corners or at least in locations along the wall, they frequently leave dark spaces in the middle or the farther end of the room.   They sometimes do not provide enough general illumination.   For this reason some people want a ceiling fixture in the living room.   Provide one ceiling outlet; in a very large room provide two.

In the living room more than in any other room, be generous with receptacle outlets.   The Code ruling that no point along the wall may be more than 6 ft from an outlet is a *minimum* requirement; for a living room 5 ft would be more nearly adequate.   Place outlets in such a way that a floor lamp can be placed anywhere without using an extension cord.   Also remember to locate them in such a way that after the furniture is located in the room, at least one outlet will always be accessible for the vacuum cleaner.

The ordinary duplex receptacle is so constructed that both halves are either on or off.   There is also available a type so constructed that one of the two outlets in each device is permanently live for clock or radio but the other is controlled by a wall switch.   Instead of each floor lamp having to be turned off separately, the entire group of lamps can be controlled by a single wall switch (see Chap. 20).

In planning the switches for the living room be sure to use 3- and 4-way switches so that lights can be turned on or off from all entrances (and if there is an upstairs, from that location too).

Sunroom, Den.   Provide a ceiling outlet for a lighting fixture, with a wall switch, and a generous number of receptacle outlets for floor lamps, radio, or similar devices.

Dining Room.   Be sure to provide a ceiling outlet for a lighting fixture, controlled by 3-way switches located at both entrances to the room.   Visualize the arrangement of the furniture, and locate the ceiling light so that it will be over the center of the dining table rather than in the center of the room.   Wall brackets are little used; they are more decorative than useful.   Lots of receptacle outlets should be provided, taking into special consideration the probable location of the furniture. Too many dining-room outlets are located where it is impossible to get at them easily for vacuum cleaner, fan, and table appliances.

Kitchen.   There are kitchens—and kitchens.   One will be the modest kitchen in a small five-room house, with relatively few appliances; another will be the de luxe kitchen in a larger house, with several thousand dollars' worth of equipment.   What is adequate in the one will be too little in the other.   But whatever the nature of a particular kitchen, more of the housewife's working hours are probably spent there than in any other room.   Therefore it is but logical that special attention should be paid to adequacy of wiring in that room.

For general lighting, there should be a ceiling outlet, controlled by switches at each entrance to the kitchen.   A light over the sink and another over the stove are essential, for without these the housewife will be standing in her own shadow when she works at these points.   These lights should preferably be controlled by wall switches. If, however, they are to be controlled by a pull chain, be certain that there is an insulating link in each chain as a safety measure.

The kitchen more than any other room needs lots of receptacle outlets. The minimum number of receptacles specified by Code is quite acceptable for other rooms, for floor and table lamps, vacuum cleaner, radio and clocks, and so on, each consuming a small amount of power. In the kitchen however, many of the appliances consume 1,000 watts or more, and several are often in use at the same time. The Code recognizes this and requires two special 20-amp circuits for appliances; no lighting outlets may be connected to these circuits. The details of these circuits will be discussed in the next chapter under the heading of "Special Appliance Circuits."

Install a clock-hanger type of the kind shown in Fig. 12-1, at the loca-

Fig. 12-1 This outlet supports the clock. The receptacle is in a "well" so that the cord and plug of the clock are completely concealed. (*Pass & Seymour, Inc.*)

Fig. 12-2 A door switch turns the closet light on when the door is opened, off when it is closed. (*Pass & Seymour, Inc.*)

tion where a clock is to be placed. The receptacle is located at the bottom of a well on the device; the outlet supports the clock. Cut the cord on the clock to a few inches in length; the cord and plug will be completely concealed behind the clock. This clock-hanger outlet may be installed on a lighting circuit, for a clock is not considered an appliance so far as the special 20-amp appliance circuits are concerned.

Even if you do not intend to install an electric range immediately, it is wise to provide a special range receptacle in your initial installation, for it is quite likely that sooner or later an electric range will be installed.

Should your kitchen be of the type that includes utilities that are more often placed in the basement (automatic washer, clothes dryer, water heater, and similar appliances), be sure to provide the extra individual outlets required for that purpose.   To add such receptacles later as an afterthought usually is quite expensive, much more so than when included in the beginning.

Breakfast Room.   A ceiling fixture controlled by a wall switch is necessary.   At least one receptacle outlet at table height is an essential for operating a toaster or coffee maker.

Bedrooms.   Every bedroom should be provided with a ceiling light controlled by a wall switch.   In addition, provide a minimum of three receptacle outlets, preferably four.   One at least should be located where readily accessible for the vacuum cleaner.   The other two should be located on opposite sides of the room and will serve bed lamps, electric heating pad, radio, and so on.   Again special attention should be paid to the location of these outlets with regard to the probable location of the furniture so that they may be readily accessible and yet not leave the cords to the lamps unduly prominent.

It would be well to remember that sooner or later you will probably install a room-size air-conditioning unit.   While the smaller models can be plugged into an ordinary circuit, a larger one might require a circuit of its own.   One receptacle on its own circuit would be a wise investment for the future.

Clothes Closets.   It is most exasperating to grope around trying to find something in a dark closet.   Provide a ceiling light; usually closets are so small that it is fairly easy to find the pull chain or a cord on the fixture.   A de luxe installation will include an automatic door switch of the type shown in Fig. 12-2, which automatically turns on the light in a closet or other room as the door is opened, and turns it off as the door is closed.   It requires a special narrow outlet box.   In some brands the box comes with the switch as shown in the illustration; in other brands it must be obtained separately.

The Code in Sec. 410-8 totally prohibits drop cords with a lamp in a socket at the bottom, so far as clothes closets are concerned. The fixture may be located on the wall *above the door,* or on the ceiling, provided there is clear space all the way to the floor. In every case, the fixture must be so located that there is at least 18 in. of clearance between the lamp in the fixture, and any combustible item in the closet. These are wise requirements. Many fires have started from clothing in direct contact with lamps in drop cords, or fixtures that were installed ignoring the very considerable heat given off by lamps, and the fact that the glass bulb of a lamp may reach a temperature above 400°.

Bathrooms. Provide a ceiling light for general illumination. However, such a light does not give enough light for shaving or make-up; for that reason provide additional light near the mirror, either one light above the mirror (which, however, casts too much shadow for easy shaving or make-up) or preferably two lights, one on each side of the mirror. Provide an outlet near the mirror for using an electric razor.

In general the use of portable appliances in bathrooms is to be emphatically discouraged. In bathrooms, the occupant of a tub or shower is in direct contact with ground, the ideal condition for a fatal shock. In case of a defective appliance, the person already in contact with ground, touching such an appliance, can easily receive a fatal shock. As a matter of fact there are on record dozens of fatal accidents each year caused by people in bathrooms, especially while in tubs, touching a defective appliance or letting some appliance such as a heater or radio (even if not defective) drop into the water. The same fatal result can be brought about by touching defective cords, switches, or fixtures while at the same time touching a faucet or other grounded object.

There is real need in the bathroom for a quick-action electric heater, but it should be one built into the wall and controlled by a wall switch, not one controlled by being plugged into an outlet.

Porches. If the porch is a simple stoop, a ceiling or wall light illuminating the floor and steps is sufficient. Illuminated house numbers are a touch an owner and his friends will appreciate. If the porch is larger, so that it is used in summer as an outdoor living room, provide a number of receptacle outlets for radio or lamps.

Basements. First of all there should be a light that illuminates *the stairs,* controlled by a switch at the head of the stairs. If the switch is in the kitchen or some other point from which the basement light cannot be seen, install at the switch a pilot light which will always be on when the basement light is on.   Beyond this, the requirements vary, depending entirely on how elaborate the basement is.

If there is an all-purpose room, which might be anything from a children's playroom to a second living room, provide as good lighting as in the living room.   Since the ceilings will probably be relatively low, flush fixtures of the type shown in Fig. 12-3 may be considered.   Natu-

Fig. 12-3 Flush ceiling fixtures are convenient when the ceiling is low.

rally they will not provide illumination over so wide an area as the more conventional fixtures but do give most excellent light directly below, especially convenient for cards, ping-pong, and other games.   Provide receptacle outlets generously.   Let the lights be controlled by wall switches.

The 1968 Code brought a new requirement for the laundry space. Section 220-3(b) demands at least one 20-amp outlet in the laundry; it must be located within 6 ft of the intended location of an appliance such as an automatic clothes washer.   This receptacle may *not* be connected to one of the two special appliance circuits otherwise required, but must be on a separate 20-amp circuit.   (Do note however that this is not adequate for a clothes dryer, which must have its own circuit.)   It will be a considerable convenience to have the laundry outlet controlled by a wall switch.   Many housewives like to iron in the basement; a good ceiling light, preferably with reflector, is essential.

At other points in the basement, install ceiling lights as required.   If there is a storage room, it needs a ceiling light; surely one is needed near the furnace.   Nearly every basement has at least a corner that becomes a workshop, and a good ceiling light plus at least one receptacle is essential.

On all basement lights, especially if the ceiling is not plastered, it is wise to install reflectors of the general type shown in Fig. 12-4. Use of

Fig. 12-4 A reflector behind a lamp greatly increases the quantity of *useful* light obtained from a lamp.

such reflectors will greatly increase the amount of useful light. Dark ceilings absorb light; reflectors throw the light downward where it is needed.

Halls. A ceiling light controlled by a wall switch is a necessity. In long narrow halls, two ceiling lights controlled by wall switches at both ends are most desirable. Don't overlook a receptacle outlet for vacuum cleaner; it will save many steps in cleaning.

Attics. If the attic is used mostly for storage, a single light placed so that *the stairway* itself is well lighted, may be sufficient. It must be controlled by a switch at the bottom of the stairs. If it is a large attic, provide additional lights as required; they can all be controlled by the same switch. If the attic is really an additional floor, unfinished but capable of being finished later into completed rooms, do not make the mistake of providing merely a single *outlet,* from which cable will later radiate in octopus fashion to a large number of additional outlets. Bear in mind future construction, and provide at least one extra *circuit,* terminating it in some convenient location.

Stairs. Thousands of people every year are involved in serious accidents due to falls on stairs. They stumble because they cannot see some object that is lying on one of the steps, or fall because they think they have reached the bottom of the stairs when, in fact, there is one more step. A fixture that lights up the general area of the stairs but does not light each individual step is an invitation to such an accident. If necessary, provide two lights, one at the foot of the stairs and one at the head. Stairway lights should be controlled by 3-way switches so that they can be controlled from either level.

13

Service Entrance and Branch Circuits

A single set of wires brings electric power from the power supplier's lines into the house. Inside the house wires are run to each outlet where power is to be used. Each group of outlets protected by a single circuit breaker or fuse is called a branch circuit. The Code defines a branch circuit as "that portion of a wiring system extending beyond the final overcurrent device protecting the circuit."

The wires *between* the *main* fuse of an installation and the final fuse *protecting the circuit* beyond that fuse are feeders. (In an ordinary residential installation, where the branch-circuit fuses and the main fuses are usually in the same cabinet, there are no feeders.) Note carefully the words "protecting the circuit." An overcurrent device installed just ahead of a motor, or built into the motor, is there to protect the motor and not the circuit. Therefore you must disregard it when determining where a circuit begins.

This chapter will discuss how to decide how many circuits to install, the sizes of wire in each circuit, size of the incoming wire, and similar details.   Chapter 17 will discuss the actual selection and installation of the particular components involved.

In this chapter there will be many references to overcurrent devices, which, as you have already learned, may be either fuses or circuit breakers, which are rapidly replacing fuses in new installations.   It would be tedious to use repeatedly the long phrases "circuit breakers or fuses" or "the circuit breaker trips or the fuse blows."   Likewise it is not practical to duplicate the illustrations showing circuit breakers in one and fuses in its counterpart.   Therefore the references will be to fuses, but it must be understood that the reference applies equally to circuit breakers and fuses.

Advantages of Numerous Branch Circuits.   Having separate circuits for groups of outlets leads to the practical result that an entire building is never in complete darkness on account of a blown fuse except on the rare occasions when a main fuse blows.   There is added safety in having a considerable number of separate branch circuits.   Most of the time each fuse carries but a portion of its maximum carrying capacity.   However, there are times during each day when a considerable number of lights are turned on.   A washing machine may be running; perhaps a flatiron is being used; other devices may also be put into service at the same time.   With a sufficient number of circuits in a properly designed installation, the load will be fairly well divided among the various fuses with the result that none is overloaded and none blows.   Where there are only a few circuits, each fuse will carry a heavier load and fuses will blow far more frequently.   Fuses that blow often tempt the owner to use an amperage larger than permitted with the usual No. 14 wire used in the average installation, or perhaps even to resort to substitutes which defeat the purposes of fuses entirely.   Naturally, this introduces distinct danger of fire.   Another practical consideration is that the greater the number of circuits, the less will be the voltage drop in each circuit, with less wasted power and higher efficiency of lamps and appliances.

Area Determines Number of Circuits.   The Code in Secs. 220-2 and 220-3 has definite requirements concerning the *minimum* number of circuits that may be installed.   The starting point is floor area.   The Code requires that this shall be computed "from the outside dimensions of the building, apartment, or area involved, and the number of floors; not including open porches, garages in connection with dwelling occupancies, nor unfinished spaces and unused spaces in dwellings *unless adaptable for future use."*

Those last five words "unless adaptable for future use" are important.   Many houses are being built that have unfinished spaces not at first used for living purposes but intended to be completed later by the owner, when his need for added living space or his financial ability suggests that this be done.   Too often this future space is disregarded in planning the original electrical installation.   One outlet may possibly be installed in such a space, and when the space is later finished, many more outlets are added, branching off from the one and only outlet originally installed.   That overloads the existing circuit.   Run a separate circuit to the unfinished space.

The question of basements is not too clear in the Code.   If the basement space is to be used for ordinary basement purposes as in older and less pretentious houses, it can be safely disregarded in your calculations so far as *lighting* circuits are concerned.   But if any part of the basement can be finished off into an amusement room or similar area, add its area to the total otherwise determined.

Houses and Apartments.   The Code in Sec. 220-2 says you must allow a minimum of 3 watts for every square foot of floor area to determine the number of circuits for *lighting only.*   Lighting circuits, of course, include not only outlets for permanently installed fixtures, but also receptacles into which you plug floor and table lamps.   These lighting circuits will also take care of minor devices such as clocks, radio, TV, and vacuum cleaner, all of which consume comparatively small amounts of power, but positively will not handle larger appliances such as kitchen appliances, room coolers, and so on.

A house that is 25 by 36 ft has an area of 900 sq ft per floor or 1,800 sq ft for two floors.   Assume that it has space for a finished recreation room in the basement, 12½ by 16 ft, or an area of 200 sq ft.   This makes a total area of 2,000 sq ft and will require for *lighting* a minimum of $3 \times 2,000$, or 6,000 watts.

The average circuit is wired with No. 14 wire, which has an ampacity of 15 amp and at 115 volts is equivalent to 15×115, or 1,725 watts.   For 6,000 watts, 6,000/1,725 or 3.4 circuits will be required.   Since there cannot be a fraction[1] of a circuit, three circuits must be installed to meet the Code minimum for lighting only.

---

[1] If an answer involves a fraction smaller than one half, drop the fraction. If the fraction is half or more, use the next higher number.

You can reach the same answer another way. Since each circuit can carry 1,725 watts, and since 3 watts is required for each square foot, each circuit can serve 1,725/3, or 575 sq ft. For 2,000 sq ft there will then be required 2,000/575, or 3.4, which means three circuits. The answer is the same whichever method is used.

Now bear in mind that the Code is concerned primarily with *safety*. As the Code itself points out, an installation made strictly in accordance with Code requirements will be safe, but it may not be practical, convenient, or adequate. Few people will be satisfied with a house wired using the minimum number of circuits required by the Code.

The Code in Sec. 220-3(a) recommends, but does not require, one circuit for every 500 sq ft of floor area for lighting. In the case of a 2,000-sq-ft house, following the recommendation will result in four lighting circuits instead of the minimum of three required by the Code. Actually, providing one circuit for every 400 sq ft would be even more modern and practical and would result in five lighting circuits, affording more flexibility and more provision for future needs.

Special Appliance Circuits. Once upon a time the kitchen appliances consisted of an electric iron and a toaster, each one consuming about 600 watts. These were plugged into ordinary circuits and worked fairly well; rarely were both used at one time. A modern kitchen on the other hand is equipped with many appliances in addition to toaster and iron: coffee maker, deep-fat fryer, frying pan, roaster, mixer, to mention some of the more common. Individual appliances, such as toaster and iron, consume 1,000 watts, and an electric roaster even more. Often several appliances are used at the same time. The ordinary circuits can no longer handle such loads.

The Code now requires special circuits for such appliances. Code Sec. 220-3(b) reads as follows:

(b) Small Appliance Branch Circuits, Dwelling Occupancies. For the small appliance load in kitchen, pantry, family room, dining room, and breakfast room of dwelling occupancies, two or more 20-ampere appliance branch circuits in addition to the branch circuits specified in Section 220-3(a) shall be provided for all receptacle outlets in these rooms, and such circuits shall have no other outlets.

Receptacle outlets supplied by at least two appliance receptacle branch circuits shall be installed in the kitchen.

At least one 20-ampere branch circuit shall be provided for laundry receptacle(s) required in Section 210-22(b).

Receptacle outlets installed solely for the support of and the power supply for electric clocks may be installed on lighting branch circuits.

A three-wire 115/230 volt branch circuit is the equivalent of two 115 volt receptacle branch circuits.

No lighting outlets may be connected to these circuits. They must be wired with No. 12 wire and protected by 20-amp breakers or fuses. Each then has a capacity of 20×115 or 2,300 watts, permitting two appliances, in most cases, to be used on each of these circuits.

Both these circuits must run to the kitchen; each of the other rooms may be served by either of these circuits, or both if you prefer.

If there is to be a built-in counter, locate the receptacles about 6 or 8 in. above the counter top, preferably not over 36 in. apart, so that appliances may be placed wherever wanted, and still within easy reach of a receptacle. The use of plug-in strip described in Chap. 20 will be found to be a good investment. Whatever method is used, be sure to wire the receptacles so that one will be on the first circuit, the next on the second, then alternately first and second circuit. This will reduce the likelihood of overloading one of the two circuits when several appliances are being used at the same time.

Instead of providing two 2-wire circuits, it is wise to consider, instead, one 3-wire circuit providing equal capacity and less voltage drop. The 3-wire circuit is described in Chap. 20.

Types of Branch Circuits.   The Code recognizes two *types* of branch circuits. The first type is the circuit serving a single current-consuming device such as a range, a water heater, or similar load. The second type is the ordinary circuit serving two or more outlets. The load may consist of devices plugged into receptacles, or permanently connected lighting fixtures, or appliances, or a combination of them.

Branch Circuits Serving Single Outlets.   It is customary to provide a separate circuit for each of the following appliances:

1. Range.
2. Water heater.
3. Clothes dryer.
4. Automatic clothes washer.
5. Garbage disposer.
6. Dishwasher.

7. Each permanently connected appliance rated at 1,000 watts or more (for example, a bathroom heater).

8. Each permanently connected motor rated at $\frac{1}{8}$ hp or more (oil burner, blower on furnace, water pump, garbage disposer, etc.).

According to the Code, an individual circuit is not necessarily required for some of the items listed, but as a practical matter it has become common usage to do so.

The circuits for appliances may be either 115- or 230-volt, depending on the rating of the appliance.   The wire must be of sufficient size to carry the amperage of the device it is to serve.   The amperage rating of the circuit breaker or fuse in the circuit depends on the specific appliance or motor served.   It may not exceed 150% of the amperage rating of the appliance or motor (unless the rating of the appliance or motor is less than 10 amp, in which case the usual 15-amp fuse is acceptable) and naturally may not exceed the ampacity of the wire in the circuit.

Circuits Serving Two or More Outlets.   There are five such circuits: 15-, 20-, 30-, 40-, and 50-amp, based on the ampacities of Nos. 14, 12, 10, 8, and 6 wire.

Note that the ampere rating of a circuit is based on the rating of the fuse or breaker protecting the circuit, not on the size of the wire in the circuit.   While normally the rating of the fuse or breaker is matched to the ampacity of the wire in the circuit, this is not always the case.   A 20-amp circuit is normally wired with No. 12 wire with ampacity of 20 amp, but if you used a 20-amp fuse or breaker and wired the circuit with No. 10 wire (for example for reduced voltage drop, or for mechanical strength in an overhead run), it would still be a 20-amp circuit.

Number of Outlets per Circuit.   For wiring in houses and apartments, the Code places no limit on the number of outlets that may be connected to one lighting circuit.   As a practical matter, if you have provided the recommended number of circuits, you will seldom have need for more than 10 outlets on one circuit.

Fifteen-ampere Branch Circuit.   This is the ordinary circuit used in routine house wiring.   It is wired with No. 14 wire and is protected by 15-amp overcurrent protection.   The receptacles connected to it may not be rated more than 15 amp, which means that only the ordinary household variety of receptacle may be used.   Any type of sockets for lighting may be connected to it.   No *portable* appliance used on the circuit may exceed 12 amp (1,380 watts) in rating.   If the circuit serves lighting outlets or portable appliances, as is usually the case, and also serves fixed or permanently connected appliances, the total of all the fixed appliances may not exceed 7½ amp (863 watts).

Twenty-ampere Branch Circuits.   The special appliance circuits described earlier in this chapter are 20-amp circuits, but they are restricted as to their use, so they are *not* the "garden variety" of 20-amp circuits now under discussion.

Any circuit wired with No. 12 wire and protected by 20-amp overcurrent protection becomes a 20-amp circuit as defined by the Code. It may serve lighting outlets, receptacles, permanently wired appliances, or an assortment of them.   Any kind of sockets for lighting purposes may be used on the circuit.   The receptacles may be either the ordinary 15-amp type or special 20-amp type.   No single portable appliance used on the circuit may exceed 16 amp (1,840 watts) in rating. If permanently connected appliances are also on the circuit, the total of all such fixed appliances may not exceed 10 amp (1,150 watts).

The wise homeowner will demand 20-amp circuits for general lighting purposes throughout the house.

Thirty-, Forty-, and Fifty-amp Branch Circuits.   Circuits of these ratings, each serving a single outlet for range, clothes dryer, or similar appliance, are commonly used and will be discussed in Chap. 21.   However, such circuits serving *more than one outlet* are not used in residential work; they are commonly used in nonresidential wiring and will be discussed in Chap. 29.

Balancing Circuits.   In a house served by a 115/230-volt service (which includes, of course, a neutral wire), the neutral wire of each 115-volt branch circuit is connected to the incoming neutral.   Care must be used in connecting the hot wires of the branch circuits so that they will be divided approximately equally between the two incoming hot wires.   If this is not done, practically all the load may be thrown on only two of the three incoming wires, and a 2-wire service might as well have been used in the first place.   Unbalanced conditions lead to frequent blowing of fuses.

Location of Branch-circuit Overcurrent Protection.   The main switch with main fuses and branch-circuit fuses is usually a small, compact cabinet.   If circuit breakers are used, the same applies.   The Code requires that this equipment be located near the point where the wires enter the building.

In any house, most of the power is consumed in the basement and kitchen appliances, and relatively a small part of it is consumed in the remainder of the house. For minimum cost, plan your layout so that the wires enter the building at a point from which the large, heavy wires to the range, water heater, clothes dryer, and kitchen appliances can be short; let the smaller, less expensive wires to the remainder of the building be the long ones.

In most houses today this equipment is located in the basement. Sometimes it is located in the kitchen, where it is more accessible. Of course, if the house is really adequately wired, fuses will rarely blow, so this point becomes less important.

Branch-circuit Schemes. The most usual scheme of locating all branch-circuit fuses at one point is shown in Fig. 13-1. As far as fusing

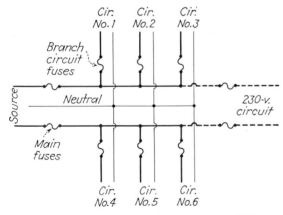

Fig. 13-1 The most common scheme: locate all fuses in one location.

is concerned, the wires beyond the branch-circuit fuses may be as long as desired. Note that 230-volt circuits may be run at any point by simply tapping off the two black wires. For example, the black wire of circuit No. 1 and the black wire of circuit No. 4 together would make one 230-volt circuit, the white wires, of course, being disregarded. One such 230-volt circuit is shown in dotted lines.

In very large houses, occasionally the owner prefers to locate the branch-circuit fuses in various locations throughout the house, the fuses controlling the basement circuits, for example, being placed in the basement, those controlling the first floor being placed in the kitchen, those controlling the second floor being placed on that floor, and so on. The scheme of Fig. 13-2 is used.   Assume that in an installation the

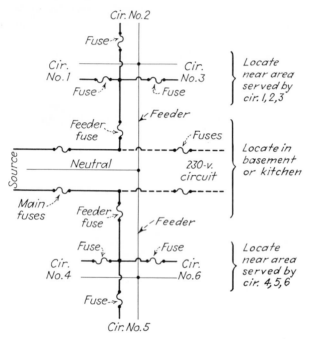

Fig. 13-2   In this scheme, branch-circuit fuses are placed in groups in various locations in the house.   Fuses are required where the feeders to these groups of fuses begin.

service switch and the branch-circuit fuses are in the same cabinet.   If the installation is made in accordance with Fig. 13-1, the fuses are branch-circuit fuses, and the wires beyond the fuses are branch-circuit wires.   If the installation is made in accordance with Fig. 13-2, the same wires (but now running to another fuse box and of a heavier size) become feeders, and the fuses (but now of a higher amperage rating) become feeder fuses.

The wires from the fuse cabinet in Fig. 13-2 to each location where a group of branch-circuit fuses is installed are feeders, as already explained.   The wires of any feeder must be of sufficient size to carry the total load of the branch circuits which that feeder serves.   The ampere rating of the overcurrent protection for the feeder must not be greater than the ampacity of the feeder.   In accordance with Sec. 215-2 of the Code, No. 10 is the smallest wire permitted when a 2-wire feeder serves *more than one* 2-wire circuit, when a 3-wire feeder serves *more than two* 2-wire circuits, or when it feeds *more than one* 3-wire circuit.

If the scheme of Fig. 13-2 is used, one point is important.   The fuse cabinet at the end of the feeder, some distance from the main fuses, will of course contain a neutral bar to which all the white neutral wires are connected.   Normally such a neutral bar is grounded to the cabinet. This is *not* permitted; the neutral *must* be grounded at the service switch, but *may not* be grounded at a fuse or breaker cabinet at the end of a feeder.   Therefore if you are installing such a cabinet at a distance, be sure the neutral is not grounded to the cabinet.

If you are using nonmetallic-sheathed cable with a bare grounding wire in addition to the insulated wires, that bare wire must be grounded to the cabinet, not to the neutral bar which you have just insulated from the cabinet.

Distribution of Outlets on Circuits.   If the usual scheme is used, placing all the branch-circuit overcurrent devices in one location, the question of what particular outlets to place on each circuit must be carefully studied.   It is not wise to place all the first-floor outlets on one circuit, all those on the second floor on another circuit, and so on, for then an entire floor will be in darkness if one fuse blows.   It is best to have on each circuit outlets of several different rooms and preferably different floors, for then when a fuse blows, there will be at least some light on each floor.

Service Entrance.   Every installation includes wires for bringing the electric power into the building, with proper means as required by the Code for disconnecting the service, for grounding, for overcurrent protection, and so on.   While by Code definition the overcurrent protection for the *branch* circuits is not considered part of the service entrance, yet it is a fact that such overcurrent protection is almost always integral mechanically with other parts of the service.   Accordingly in this book, such overcurrent protection will be considered part of the service entrance.

The service entrance then consists of:

1. Service-drop wires (from the power supplier's lines to the building).

2. Service-entrance wires (from the outside of house to equipment on the inside).

3. Meter.

4. Disconnecting means (to disconnect the entire installation completely from the power supplier's lines).

5. Overcurrent protection.

6. Ground.

The details of these several components are dependent on many factors, such as the size of the building, the amount of power required, the number of circuits, and so on.   In this chapter we shall discuss only the general design of the service entrance, and in Chap. 17 we shall discuss the actual installation of the service entrance.

Minimum Size of Service Entrance.   This chapter will concern only houses, not commercial or industrial projects.   The Code in Sec. 230-71 has specific requirements regarding the minimum size of service-entrance equipment.   References here are to the size of the main switch or main circuit breaker.

If the installation is a very small one with one or two 2-wire circuits, the minimum is a 30-amp service.   Under all other circumstances, the minimum is a 60-amp service.   But this minimum will rarely apply because of a most important exception: If the house has more than five circuits, or if the load (calculated in a manner to be outlined) is 10,000 watts or more, then the minimum is a 100-amp 115/230-volt service. As a practical matter, this means that the mimimum for any house is 100-amp, for it is hard to conceive of a house with only five circuits.

When fused equipment is used, switches are available only in ratings of 30-, 60-, 100-, and 200-amp.   Circuit breakers are available in ratings of 30-, 50-, 60-, 70-, 100-, 150-, and 200-amp ratings.   Of course ratings above 200 amp are available for other buildings or exceptionally large houses.

The actual minimum size of wire that may be used for a service of any given amperage will be discussed in the next paragraph.   Do remember that the minimum specified by Code is based on safety, not practicability or convenience.   Look ahead: Instead of considering the 100-amp minimum as acceptable, install 150- or 200-amp equipment.   A few local Codes already require 200-amp as minimum.

Size of Wire for Various Services.   As you learned in Chap. 6, the ampacity of a given size of wire depends not only on its size, but also on the kind of insulation.   Sometimes also the ampacity depends on whether the wire is installed in a wet or dry location; a properly installed service entrance in conduit is considered a dry location even if the conduit is outdoors.   The ampacity of each size and kind of wire is found in Code Table 310-12 (see Appendix).   The minimum size required for a service of any given capacity is as follows:

Amperage of service	Minimum size of wire if Code Type		
	T, TW	RH, RHW, THW, THWN	RHH, THHN, XHHW
30	No. 8	No. 8	No. 8
60	No. 4	No. 6	No. 6
70	No. 4	No. 4	No. 6
100	No. 1	No. 3	No. 3
140	No. 2/0	No. 1/0	No. 1
150	No. 3/0	No. 1/0	No. 1/0
200	250MCM	No. 3/0	No. 3/0
400	750MCM	600MCM	500MCM

If you are using service-entrance cable Type SE, or underground cable Type USE, use the second column above.

Per Code Table 310-12, the ampacity of No. 10 wire is 30 amp; therefore you would expect it to be suitable for a 30-amp service.   In the table above, No. 8 is specified because the Code, in Secs. 230-41 Exc. 2 and 230-71 Exc. 2, requires a minimum of No. 8 for a service.

Remember that the sizes shown are for the service entrance proper, not for the wires from the power supplier's line to your buildings (your power supplier will determine the size of those wires).   On farms, the wires from a yard pole to a building are considered service-entrance wires, and must be sized as shown in the table.   But if you are using weatherproof wire, you may use sizes based on the ampacities shown in the last column of Code Table 310-13 (see Appendix).   However, remember that the smaller the size of the wire, the greater the voltage drop.   You may want to use a size larger than the minimum.

When using fused service-entrance equipment of a size *larger than the minimum required,* it is not contrary to the Code to use service wires smaller in capacity than the rating of the switch, provided that the fuses in the main switch are not rated higher in amperes than the ampacity of the wires.   But if you are using a wire with an ampacity such that there is no standard fuse of that ampere rating, you may use a fuse of the next higher standard rating.   Thus if you use wire with 95-amp ampacity, use a 100-amp fuse, for there is no standard 95-amp fuse.   In spite of all this, it makes sense to use service-entrance wires of such ampacity that the full rated capacity of the switch or circuit breaker can be utilized.

One additional caution may be necessary: The service switch or circuit breaker does not need to be rated at the total number of amperes you would reach by adding up all the individual amperages of all the individual branch circuits.   There never will be a time when *every* circuit in the house will be loaded to its maximum capacity.

To determine the size of the service entrance, follow Code Sec. 220-4.   If it is decided in advance that the entrance will be 100-amp or larger, the Code permits a short-cut method.   However, we shall discuss the original, or long, method first.   The total load that must be carried by the entrance can be broken down into five groups:

1. Lighting.
2. Small appliances.
3. Special laundry circuit.
4. Heavy appliances.
5. Motors.

Lighting.   The total power that the entrance must carry for lighting (and that includes the receptacle outlets for supplying items like radio, television, clocks, vacuum cleaner, and similar portable devices, but not kitchen appliances or laundry) is outlined in Sec. 220-2(a): 3 watts per square foot of area, the same figure you used in determining the number of circuits required.

Small Appliances.   The Code includes in "small appliances" all irons, toasters, coffee makers, and similar portable appliances ordinarily used in the kitchen, as well as washing machines and the like used in the laundry, but excludes all permanently installed appliances like clothes dryers and the like.   To allow for their use, add to the wattage for lighting, as determined by the previous paragraph, 3,000 watts.   This corresponds to an average load of 1,500 watts for each of the two special appliance circuits required by Code Sec. 220-3(b); while each of these circuits is capable of supplying 20×115 or 2,300 watts, it is not likely that both these circuits will be loaded to capacity at the same time.

Also add 1,500 watts for the special laundry circuit which was added by the 1968 Code.

Demand Factor. The larger the house, the less likelihood that it will *all* be lighted at the same time. Considering that fact, the requirement of 3 watts per sq ft plus 3,000 watts for small appliances might be considered excessive from a *safety* standpoint, which is the chief concern of the Code.

Accordingly, the Code in Sec. 220-4(a) establishes a "demand factor" of 35% for that portion of the wattage which exceeds 3,000 watts, computed as above outlined for lighting and small appliances (but excluding fixed appliances and motors). If the computed total is 3,000 watts or less, count all of it. If it is over 3,000 watts, count only 35% of the portion above 3,000 watts. Examples will be given later in the chapter.

It should be remembered that this demand factor applies only in determining the service-entrance equipment and not in determining the number of circuits. The individual circuits must be so planned that any *portion* of the house can be lighted to the point which requires 3 watts per sq ft, but there is no likelihood that the *entire* house will ever be so lighted that the full 3 watts per sq ft will be required throughout the entire area.

Heavy Appliances. The major heavy appliance is the electric range, which will be discussed separately. Other heavy appliances are such permanently installed appliances as water heater, clothes dryer, automatic clothes washer, bathroom heater, and the like. Each of these must be counted at its full rating in watts. No demand factor applies.

Range. The wattage consumed by a range when all burners and the oven are turned on to maximum heat will vary considerably with the size, type, and brand of the range. It may be as high as 15,000 watts or even more. But since it is not likely that all burners and the oven will be turned on at the same time while everything else in the house is also turned on, the Code does not require that it be counted at its full maximum rating. If the range is rated at not over 12,000 watts, use an arbitrary value of 8,000 watts. If it is rated over 12,000 watts, proceed as follows: start with 8,000 watts, plus 400 watts for each thousand watts (or fraction thereof) over 12,000 watts. If you are installing a separate wall-mounted oven (or two of them) and also counter-mounted cooking units, add the wattage of all these together, and use the total as you would for a single self-contained range.   Do note that this applies only for the purpose of calculating the service entrance.

Electric Heating.   In some localities electric heating is becoming quite common.   The Code rules are fairly complicated, but in general, per Sec. 220-4(e), you must allow for the total wattage consumed by all the heating elements at the same time.   If, however, the circuits are so arranged that not all the elements can be turned on at the same time, count only that wattage that can be in use at one time.

Air Conditioning.   Determine the horsepower of the motor on the equipment, then allow the wattage outlined in the next paragraph for a motor of that size.   But if the house is to have both electric heating and air conditioning, the two obviously will not be in use at the same time. Count the larger of the two loads.

Motors.   The Code devotes considerable space to the proper installation of motors, and the subject is complex.   For the purpose of calculating the service-entrance equipment *for homes,* allow the wattages shown below, and the Code requirements will be more than met:

⅙ hp	450 watts
¼ hp	700 watts
⅓ hp	850 watts
½ hp	1,000 watts
¾ hp	1,350 watts
1 hp	1,500 watts
1½ hp and larger	1,200 watts per hp

The wattages shown above are considerably above the actual wattages consumed by the motors, but since alternating-current motors have relatively low power factors, the amperages are considerably more than merely watts/volts.   Also, motors can be overloaded to deliver considerably more than their name-plate horsepower and will then consume more than their rated amperage.

If there are several motors but no likelihood that all will ever operate at the same time, estimate the proper demand factor.   For example, if a home workshop is to have four motors, it is not likely that more than one will be used at a time.   Count only the largest of them.

Calculating a 2,000-sq-ft House.   Assuming a house of 2,000 sq ft area (determined in the same way as outlined for determining number of circuits) without a range and with no motors, the calculations will be as follows:

	Gross computed watts	Demand factor, %	Net computed watts
Lighting, 2,000 sq ft at 3 watts . . .	6,000		
Small appliances (minimum)	3,000		
Special laundry circuit	1,500		
Total gross computed watts	10,500		
First 3,000 watts.	. . . . .	100	3,000
Remaining 7,500 watts	. . . . .	35	2,625
Total net computed watts	. . . . .	. . . . .	5,625

The amperage involved is 5,625/230, or 24.5 amp. Since No. 10 wire has an ampacity of 30 amp it would appear to be large enough, but remember the Code requires a minimum of 100-amp service when there are more than five circuits. Since this house will have a minimum of three lighting circuits, two special appliance circuits, one laundry circuit, a minimum of six altogether, there is no choice: install a 100-amp service.

Assume now that an electric range rated not over 12,000 watts is to be added, and also a permanently installed 1,500-watt bathroom heater. The calculations will then be:

	Gross computed watts	Demand factor, %	Net computed watts
Lighting, 2,000 sq ft at 3 watts . . .	6,000		
Small appliances (minimum)	3,000		
Special laundry circuit	1,500		
Total gross computed watts	10,500		
First 3,000 watts	. . . . .	100	3,000
Remaining 7,500 watts	. . . . .	35	2,625
Bathroom heater	1,500	100	1,500
Range (arbitrary minimum)	. . . . .	. . .	8,000
Total net computed watts	. . . . .	. . .	15,125

At 230 volts, the amperage is 15,125/230, or 65.8 amp. Therefore a 70-amp service would appear to be large enough, but again we have more than five circuits; per the Code a 100-amp service is the minimum required.   The small extra cost will be a good investment.

Let us now go a step further and figure a somewhat larger 3,000-sq-ft house on a more generous basis.   Add a water heater consuming 3,500 watts.   Add two ¼-hp motors, one for the furnace and one for its blower.   Add a clothes dryer rated at 4,000 watts.   Assume that now it is a suburban home with a ½-hp motor on the water pump.   The calculations then will be as follows:

	Gross computed watts	Demand factor, %	Net computed watts
Lighting, 3,000 sq ft at 3 watts . . .	9,000		
Small appliances	3,000		
Special laundry circuit	1,500		
Total gross computed watts . . .	13,500		
First 3,000 watts	. . . . . .	100	3,000
Remaining 10,500 watts	. . . . . .	35	3,675
Range (arbitrary minimum)	. . . . . .	. . .	8,000
Fixed appliances:			
Bathroom heater	1,500	100	1,500
Water heater	3,500	100	3,500
Clothes dryer	4,000	100	4,000
¼-hp motor, oil burner	700	100	700
¼-hp motor, blower on furnace . .	700	100	700
½-hp motor, water pump	1,000	100	1,000
Total net computed watts . . .	. . . . . .	. . .	26,025

At 230 volts, this is equivalent to 26,025/230, or about 113 amp. The 100-amp service is now too small, and you would then install a 150-amp service with circuit breakers or a 200-amp service with fuses.

Optional Method.   Instead of following the procedure already outlined, the Code in Sec. 220-7 permits an optional method.   Include lighting at 3 watts per square foot, two appliance circuits at 1,500 watts each, one laundry circuit at 1,500 watts, but do not apply a demand factor as in previous examples.   Add the range at its full wattage rating, and all other loads at full wattage rating.   Add all items together; count the first 10,000 watts at 100%, and the remainder at 40%.   See the following example:

	Gross computed watts	Net computed watts
Lighting, 3,000 sq ft at 3 watts.	9,000	
Small appliances	3,000	
Special laundry circuit	1,500	
Range, maximum rating	14,000	
Fixed appliances:		
Bathroom heater.	1,500	
Water heater.	3,500	
Clothes dryer	4,000	
¼-hp motor, oil burner	700	
¼-hp motor, blower on furnace	700	
½-hp motor, water pump	1,000	
Total gross computed watts	38,900	
First 10,000 watts at 100% demand factor	. . . . .	10,000
Remaining 28,900 watts at 40% demand factor.	. . . . .	11,560
Total net computed watts	. . . . .	21,560

The total computed is 21,560 watts, which at 230 volts is 94 amp, as compared with 113 amp as computed the other way in the previous paragraph. Obviously per this optional method, a 100-amp service would be acceptable. You would meet the Code requirements by installing a 100-amp service; you would be wise to install a larger service.

Note that if you have four or more permanently installed space-heating units, *each separately controlled*, they may be included in the totals above outlined. But if you have three or fewer, you may not include them, but must add them separately after having determined the total; in the example given, you would add them to the total of 21,560 watts. Likewise, if you have central electric heating, add it after the 21,560 watts. The same applies to all air-conditioning loads. But if you have both heating and air conditioning, naturally you would never operate both at the same time, so add only the larger of the two.

These sample calculations, regardless of whether you use the old method or the optional method, will serve merely as examples. It will be well for you to calculate other houses in the same fashion; for example: (1) the house in which you live as it is now wired, (2) the same house as you would like to have it wired, and (3) the "ideal" house in which you would like to live.   But always remember that the answers you reach by these Code methods tell you only the *minimum* size of wire the Code considers necessary for *safety*.   The actual size of wire you will want to use will be larger than the minimum, to allow for convenience and practicability and above all for the future electrical appliances that you will surely add as time goes on.

Disconnecting Means and Overcurrent Protection.   A device must be provided at or near the point where the service wires enter the building, to disconnect or isolate the entire building from its source of supply. Likewise it is necessary to have devices which disconnect a single circuit at a time from the source of supply.   This is a safety factor, for there are times when it is most necessary to disconnect, or "kill," parts or all of the wiring, as in case of fire or when working on parts of the system.

Likewise it is necessary to provide overcurrent protection which protects the installation as a whole, or parts of it (individual circuits), against short circuits or overloads.   Usually the disconnecting means and the overcurrent protection come in the form of a single metal cabinet that houses all the necessary components.

Fused Equipment.   The "service switch" consists of a main switch by means of which the entire installation can be disconnected at one time, plus main fuses to protect the installation as a whole, plus as many individual branch-circuit fuses as are required to protect those branch circuits.   Occasionally the branch-circuit fuses are in a separate cabinet.

Fig. 13-3   The simplest type of switch.   Occasionally it is used as a service switch in a *very* small installation.   It is more frequently used for other purposes.   (*Square D Co.*)

According to Code the switch must be "externally operable," which condition is fulfilled if the switch can be operated without the operator being exposed to live parts.   Older switches used an external handle as shown in Fig. 13-3.   Most switches today don't have hinged switch blades in the usual sense at all. Instead, the main fuses are mounted on a small block of insulating material which can be pulled out of the switch. When this block is removed, with the fuses, the switch is dead, just as if blades had been operated with a handle. Such switches once installed have no exposed live parts, whether the block with the fuses is in place or removed, for the prongs on the removable block project into the switch through very small openings. A switch of this kind is shown in Fig. 13-4; the pull-out block shown at the bottom holds the main

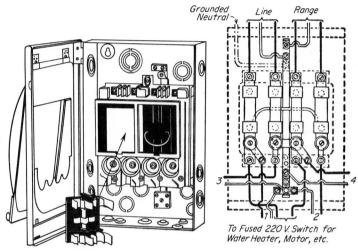

Fig. 13-4 A pull-out type of switch. The main fuses are mounted on an insulating block. When the block is pulled out, no live parts are exposed. Pulling the block out turns off the switch.

fuses and fits into the upper left portion of the switch. A second similar pull-out block at the upper right holds a pair of cartridge fuses to protect a 230-volt range or similar load. At the bottom are four plug fuses to protect four 115-volt branch circuits.

Note that if you insert the pull-out block upside down, the circuit is dead, just as if you had pulled the handle on a switch with a handle.

Also note that some switches of this kind are "parallel wired" so that if you remove the pull-out block with the main fuses, everything is dead *except* the range (or other load) protected by the fuses in the second pull-out block. Use whatever kind is used locally; check with your power supplier or the local inspector.

Solid-neutral Switches.   For a 115-volt installation, use a 2-pole *solid-neutral* switch with one main fuse.   Such installations may not have more than two branch circuits.   In all other cases the installations will be 115/230-volt type, for which use a 3-pole *solid-neutral* switch with two main fuses.   You have already learned that the neutral or grounded wire is never interrupted by a switch[2] or fuse.   "Solid neutral" then means that the neutral wire in the switch is not interrupted by switch or fuse; instead it runs to a "neutral bar" in the switch. This neutral strap or neutral bar is merely a copper strip with a number of terminal lugs to which the neutral of the incoming wires, the neutral wires of all the 115-volt branch circuits, and also the ground wire are connected.   The "3-pole solid neutral" switch then has two blades plus two main fuses to disconnect and protect the two hot wires, the third or neutral wire running to the neutral strap.

In this connection, Sec. 230-70(a) of the Code may prove confusing. This section reads "Means shall be provided for disconnecting *all* conductors in the building from the service-entrance conductors." Section 230-70(i) reads "where the switch or circuit breaker does not interrupt the grounded conductor, other means shall be provided in the service cabinet or switchboard for disconnecting the grounded conductor from the interior wiring."   One statement seems to contradict the other, and both may seem contrary to statements of previous paragraphs that the neutral wire always runs directly to the neutral bar without interruption by a switch.

The answer lies in the fact that *all* connections inside the switch must be by means of *solderless* connectors as required by Sec. 230-72 of the Code.   Being so made, the neutral can be opened using ordinary tools, and that satisfies the Code requirement.   If such wires were *soldered* to the terminal bar, this would not be true.

Ratings of Service Switches.   Service switches are rated at 30, 60, 100, 200, 400, and 600 amp with no in-between ratings.   In residential work the 30-amp size is never used except possibly in a small summer cottage or similar building.   The 60-amp size was formerly the most common but has proved too small in most cases.   The 100-amp is the

---

[2] If we wish to be precise, it must be stated that the grounded neutral *may* be interrupted by a switch blade or circuit breaker (but never a fuse) provided the device is so designed that it is impossible to open the grounded wire without simultaneously opening all the ungrounded wires.   Such devices are not used in homes.

minimum size now specified by Code in most cases, as already explained.   The 200-amp size is becoming fairly common and is already specified as minimum in some local codes.

Selection of Specific Switch.   Switches are available with main fuses plus other fuses for 2 to 20 or more branch circuits.   One fuse is needed for each 115-volt branch circuit; two fuses for each 230-volt branch circuit.   Plug fuses are suitable for either[3] voltage, but cartridge fuses are necessary if the rating is over 30 amp.   Remember that all plug fuses are mechanically interchangeable.   Cartridge fuses rated at 35 to 60 amp all have the same dimensions and fit the same holders.   Those rated 65 to 100 amp are larger but have the same dimensions, and are of the knife-blade type.

The so-called "range combination" shown in Fig. 13-4 was once very popular.   Rated at 60 amp and with 60-amp main fuses, plus a pair of cartridge fuses for a range or other load within the 230-volt 35- to 60-amp size, plus plug fuses for four 115-volt branch circuits, it was considered big enough for the average house.   Time has shown that it was much too small.

Today a 100-amp switch is considered a minimum.   Such switches are available with very many assortments of 115- and 230-volt branch-circuit fuses in addition to the main fuses.   The one shown in Fig. 13-5 has one pull-out block for two 100-amp main fuses.   In addition it has three pull-outs each for two 35- to 60-amp cartridge fuses, and two pull-outs each for two 15- to 30-amp cartridge fuses, all for 230-volt loads.   It also has provision for 16 plug fuses.   In larger houses and on many farms, similar but larger switches with 200-amp main fuses, with branch-circuit fuses as needed, are often used.

Remember that while any one plug fuse will protect a 115-volt circuit, two plug fuses (one on each of the two hot wires, or "legs," in the switch) will protect a 230-volt circuit of not over 30-amp rating.

Many of these switches have a pair of lugs for wires running to a 230-volt water heater.   Wires from these lugs are usually *not* protected by branch-circuit fuses; consequently a separate fused switch of the gen-

---

[3] Do not be confused by Code Sec. 240-6(c), which is often misinterpreted to mean that plug fuses may not be used on 230-volt circuits.   They may be used on 230-volt circuits in farm and residential work, because such circuits are derived from a 115/230-volt service with grounded neutral, so that the *voltage to ground* is only 115 volts.   For definition of voltage to ground, see Chap. 9.

eral type shown in Fig. 13-3 must be used in the circuit to the heater.

For your final selection, choose a switch that has the necessary number of fuses to protect the number of circuits that you have decided are necessary for your house, with a few spares for future circuits.

Fig. 13-5 Typical 100-amp service switch with many fuses for branch circuits. (*Square D Co.*)

After having selected the exact switch you want, it is wise to discuss your selection with your power supplier.   Be prepared with the manufacturer's number of the switch you have in mind.   Some power suppliers are very particular about the switches on their line and have "approved lists" of brands and specific catalog numbers that are acceptable.

Circuit Breakers.   Remember that an individual circuit breaker looks like a switch.   On overload, the circuit breaker opens itself; restore service by flipping the handle, like closing a switch (see Fig. 13-6).   One single-pole breaker protects a 115-volt circuit; one two-pole (double-pole) breaker protects a 230-volt circuit.   The two-pole breaker in some brands has a single handle; in some other brands it looks like two separate single-pole breakers side by side, each with its own handle, but with the two handles mechanically tied together to become one single handle, so far as the Code is concerned.   A two-pole breaker is dimensionally twice as wide as a single-pole.

In some brands, a small pilot light below the handle of the breaker lights up when the breaker is tripped, or turned off.

Circuit breakers and fuses are rated in amperes. The most common ratings are 15-, 20-, 30-, 40-, 60-, 70-, 100-, and 200-amp corresponding substantially to the ampacities of Nos. 14, 12, 10, 8, 6, 4, 2, and 4/0 wire. In-between ratings of 25-, 35-, 45-, 50-, 80-, 90-, 110-, 125-, 150-, and 175-amp ratings are also available. Naturally ratings above 200-amp are available for larger installations.

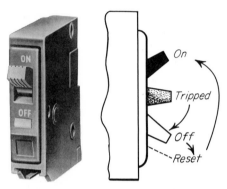

Fig. 13-6 A typical single-pole circuit breaker. If breaker trips because of overload, force the handle *beyond* the "off" position, then move it to "*on*." (*Square D Co.*)

Circuit breaker cabinets are usually sold without the breakers. The cabinet contains a bus-bar type of arrangement into which you plug breakers that you buy separately, in any amperage that you select, and single-pole (115-volt) or two-pole (230-volt), as you choose. The cabinet also contains the neutral bar with lugs for the incoming neutral and the ground wire, and terminal screws for the neutrals of the 115-volt branch circuits. But if the equipment you need includes a main breaker ahead of the branch-circuit breakers, the main breaker is usually already installed in the cabinet as purchased.

Remember that when fused equipment is used, there is always an "externally operable switch" so that the entire installation can be disconnected by one movement of the hand. In the case of circuit breakers, up to and including the 1962 Code, it was not necessary to have a single, main circuit breaker to disconnect the entire installation, provided *not over six* movements of the hand were required to disconnect everything.   Considering that two single-pole breakers for 115-volt circuits placed side by side could be mechanically tied together by a tie-bar or mechanical handle, so that both could be opened by one movement of the hand, it is obvious that up to 12 poles of breakers could be installed without using a main breaker ahead of them.

This was changed beginning with the 1965 Code [Sec. 384-16(a)] and the scheme may be used today only in residential work, and only if all the breakers are rated at *more than 20-amp*.   But in ordinary residential use, ordinary branch-circuit circuit breakers are almost always 15- or 20-amp.   Therefore the old scheme may not be used, and breakers such as shown in Fig. 13-7 may no longer be used without additional overcurrent protection ahead of them.

Fig. 13-7 A breaker cabinet with six branch circuits.   If any of the breakers are 15- or 20-amp this cabinet cannot be used for service entrance, unless a main breaker is installed ahead of it.   (*Square D Co.*)

But a circuit breaker cabinet with only 12 poles is too small for almost any house today; so the old scheme was becoming obsolete in any event.   There are several solutions to the problem.   The simplest method is to use a breaker cabinet which includes a main breaker; one is shown in Fig. 13-8 together with its wiring diagram.   It contains the main breaker, and space for up to 30 poles.   These might be occupied by up to 30 single-pole 115-volt breakers; alternately, two-pole 230-volt breakers may be installed, each occupying the space required by two single-pole breakers.   The diagram in Fig. 13-8 shows only single-pole breakers.   Since the cabinet is designed to accept almost any combination of single-pole and two-pole breakers, it is naturally impossible to show diagrams of all possible combinations.   To convert the diagram for two-pole breakers, remove any two adjacent single-pole breakers; in their place plug in one two-pole.

In the diagram, the block marked "SN" means solid neutral—in other words the neutral bar or strap.

Instead of cabinets containing one main breaker, cabinets containing two main breakers are also in limited use.   The two main breakers per-

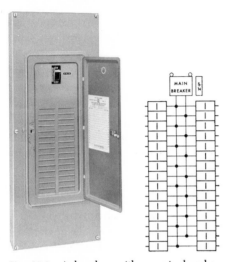

Fig. 13-8   A breaker with a main breaker, and space for many branch-circuit breakers. The wiring diagram shows only single-pole breakers, but 2-pole breakers may also be used.   (*Square D Co.*)

mit the entire installation to be disconnected by two movements of the hand.   The rating of the entire cabinet is the sum of the two individual breakers; if each of them is rated at 70 amp, the cabinet is rated at 140 amp, and so on.

Also in use are the "split-bus" type of breaker cabinets, without a main breaker.   The diagram for one is shown in Fig. 13-9.   The top portion contains *not over six* breakers, and each must be rated at more than 20 amp.   Five of them protect five 230-volt circuits.   The sixth one protects a group of 15- and 20-amp breakers in the bottom of the cabinet; there may be as many as you wish.   They are protected by the one double-pole breaker ahead of them.   Split-bus breaker assemblies are prohibited by local codes, in some localities.

When using split-bus breakers, add together the ampere ratings of all the breakers in the top, not over six. The total of these ratings must not exceed the ampacity of the wires in the service entrance.

The wiring diagrams for breakers shown in this chapter are merely typical, to illustrate the principle. They will vary with the brand of breakers, with the number and combination of circuits, and so on.

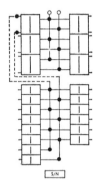

Fig. 13-9 Diagram for a typical "split-bus" breaker cabinet. One of the six 2-pole breakers in the top part becomes the main breaker for all the branch-circuit breakers in the bottom part of the cabinet. (*Square D Co.*)

Each breaker must be selected to match the ampacity of the wire in the circuit, that it is to protect. But in the split-bus type, the one breaker that protects a group of smaller 15- or 20-amp 115-volt circuits, does not have to have a capacity equal to the sum of all those circuits. For ex-

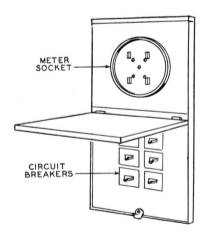

Fig. 13-10 This combination of meter with circuit breakers is favored in some localities.

ample if you have eight 15-amp circuits, four will be on one leg of the 115/230-volt service, four on the other, making a total of 4×15, or 60 amp on each leg. It is not likely that all four will be loaded to capacity at the same moment; so the one breaker protecting the 15-amp circuits may be 60 amp, but does not have to be 60 amp.

In some localities, the trend for residential installations is toward a combination of meter socket, plus circuit breakers for the branch circuits, all in a single weatherproof cabinet.   One such combination is shown in Fig. 13-10.

Concentric   Knockouts. The   cabinets   of   service   switches   and breakers are too small to permit installing knockouts of all the different sizes that might be required, in different installations.   Such cabinets are therefore provided with "concentric" knockouts as shown in Fig. 13-11.   Remove as many parts of this complicated knockout as required,

Fig. 13-11   A concentric knockout.   Remove as many sections as necessary, to   provide   size opening required.

to provide the size you need.   This is a tricky operation; so be especially careful to remove only as many sections as necessary, to provide the size you need.   Remove the center section only to provide the smallest size; remove two sections for the next larger size, and so on.

Installation of Service Entrance.   The equipment having been selected, it must be installed.   The details are discussed in Chap. 17.

14

Good Lighting

In the very early days of electric lighting even one lamp[1] hung in the center of a room was such an improvement over the ordinary kerosene lamp or even the gaslight then in use that apparently little time was spent in considering whether the new illuminant provided really sufficient light for good seeing. This type of thinking persisted too long, and even today too little thought is given to providing good lighting. As a result, entirely too many homes and other buildings that are electrically lighted have not one-half or even one-quarter the illumination that is necessary for good seeing.

This chapter will be devoted to a discussion of the fundamentals of lighting, as well as the selection, installation, and use of lighting fixtures and the lamps that go with them, in order to provide good lighting. Many volumes have been written on the subject, some of them covering

---

[1] Is it a "lamp" or a "bulb"? The complete "light bulb" is properly called a lamp; the glass part of a lamp is the bulb. Fluorescent lamps are often called "tubes."

but one single small aspect of the science, and on some points there is a good deal of disagreement among the authorities.   The author does not flatter himself therefore that he can begin to cover the subject in a single chapter.   He does propose, however, to set forth some of the fundamentals involved, together with what are today considered standards, so as to give some degree of working knowledge to the reader.

Dozens of factors can be enumerated that go to make up a good lighting system.   The more important ones are that the lighting system must:

1. Provide sufficient quantity of light.
2. Provide light free from glare.
3. Provide light free from objectionable shadows.
4. Provide the right kind of light.

Good lighting is important in many ways.   It contributes to personal comfort and reduces eyestrain.   It leads to greater efficiency in all activities.   It tends to promote safety by preventing accidents due to poor visibility.

Extent of Defective Vision.   Broad surveys have shown that an amazing percentage of all people today have defective vision.   One survey shows the percentage of defective vision to be:

Under 20 years	23%
20 to 30 years	39%
30 to 40 years	48%
40 to 50 years	71%
50 to 60 years	82%
Over 60 years	95%

Another survey shows that of all students in the elementary grades of school 9% have defective vision, in high school 24%, and in colleges the figure has risen to 31%.

A different type of survey shows that the percentage of defective vision varies considerably by occupations.   Those who work relatively little under artificial light and those whose work is not of an exacting nature suffer relatively little.   For farmers and common laborers the proportion is under 20%, while for carpenters and painters it has risen to between 20% and 40%.   For machinists and printers the figure is over 40%, whereas for draftsmen and secretaries it is over 80%.

It used to be generally accepted that poor lighting is one of the *causes* of defective vision; medical authorities now question that conclusion.

Whichever theory is correct, it remains a fact that many people do have defective vision; the proportion increases with age. The important point is that those who do have defective vision need better lighting than those with normal vision. Since there are few families or other groups of people among whom there isn't at least one person with defective vision, it follows that lighting should be designed, not for those with perfect vision, but for those with impaired vision.

How Light Is Measured. Inasmuch as the candle was the common method of illumination when this science was first studied, it is not surprising that standards sprang up based on candle light, so that today there are such terms as "candlepower" and "footcandles." Since a big candle naturally gives more light than a little candle, there developed a standardized candle, the definition of which, however, need not concern us here.

Candlepower. This term does not measure the *total* amount of light emitted by the source, for the light may be brighter in one direction than in another. The standardized candle mentioned in the previous paragraph emits 1 candlepower of light in a particular direction; in other directions it may emit more or less than 1 candlepower. Because of this, the term "candlepower" is of relatively little value because the direction must always be taken into consideration when the term is used. For this reason lamps are no longer rated in candlepower. However, many times the term "candlepower" is still used when the term "horizontal candlepower" or "mean spherical candlepower" is meant.

Mean Spherical Candlepower. If a source of light gives off 1 candlepower in *every* direction, it is said to have 1 mean spherical candlepower; if it gives off 10 candlepower in every direction, it has 10 mean spherical candlepower. If it gives off 5 candlepower in one direction, 10 in another, and 12 in a third, but *averages* 10 candlepower, it has 10 mean spherical candlepower. Automobile lamps are still rated in candlepower; the mean spherical candlepower is meant, but this apparently is too long a term for ordinary commercial use, so that the words "mean spherical" are dropped but are implied.

Beam Candlepower. When a reflector is placed behind a source of light, the emitted light rays which ordinarily go off in all directions are crowded into a relatively narrow beam. The brightness of this beam in candlepower is called the "apparent beam candlepower." For example, an ordinary automobile lamp of 21 candlepower, if used in a good reflector, will provide an apparent beam candlepower of 100,000, within a very narrow beam. This simply means that there is enough candlepower within the beam to produce the same intensity of light at the point of observation as would be produced at that same point by a light source of 100,000 *mean spherical* candlepower, without a reflector, when located at the position of the smaller lamp and reflector.

Lumens. The term "candlepower" measures only the light in a given direction, not the total amount of light emitted. The total quantity is measured in lumens.[2] Assume a light source producing 1 candlepower in every direction, located in the center of a sphere which is exactly 2 ft in diameter; this means that the light source is then exactly 1 ft from every point on the inner surface of the sphere. Assume that there is an opening in the sphere and that this opening is exactly 1 sq ft in area (see Fig. 14-1). The lumen is defined as *the amount or quantity*

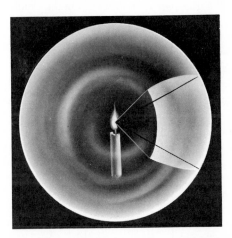

Fig. 14-1 A candle is in the exact center of a sphere which is 2 ft in diameter. An opening with an area of exactly 1 sq ft has been cut out of sphere. (*General Electric Co.*)

of light emitted through an opening of 1 sq ft located at a distance of 1 ft from a light source which emits 1 candlepower in every direction. Note that this opening must be a part of a sphere; if the opening is a hole 12 in. square in a sheet of paper, there is no way of placing the paper so

---

[2] Derived from the Latin word *lumen* meaning light.

that every point is exactly 1 ft from the light source.   This definition is important; study it well.

Simple geometry tells us that the total area of a sphere 2 ft in diameter is 12.57 sq ft ($4\pi \times R^2$).   Since by the definition above 1 lumen of light falls on *each* square foot of area in the sphere, the *total* light falling on the interior of the sphere must be 12.57 lumens.   This was produced by a light source of 1 mean spherical candlepower.   Therefore to find the lumens emitted by a lamp of which the mean spherical candlepower is known, multiply by 12.57; if the lumens are known, divide by 12.57 to determine the mean spherical candlepower.

Law of Inverse Squares.   The candlepower and lumens of a light source are absolute or constant; other circumstances being the same, they do not change.   A candle produces the same amount of light when the observer is 2 ft away from it as when he is 5 ft away from it.   Nevertheless it is much easier to read a newspaper 2 ft distant from the candle than it is when 5 ft distant.   Assume again a light of exactly 1 mean spherical candlepower enclosed in the 2-ft sphere with an opening exactly 1 sq ft in area.   A newspaper placed at the opening will have 1 sq ft of print illuminated by the 1 lumen of light escaping through the opening.   Now move the newspaper so that it is located 1 ft from the sphere or 2 ft from the lamp.   The 1 lumen of light that escapes through the opening will now illuminate an area 2 ft on each side, 4 sq ft altogether.   Move it to 2 ft from the sphere or 3 ft from the lamp.   The area illuminated will now be 3 ft on each side, 9 sq ft altogether (see Fig. 14-2).

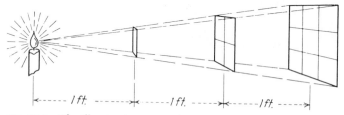

Fig. 14-2   The illumination on an object varies inversely as the square of the distance from the light source.   (*General Electric Co.*)

The area illuminated was first 1 sq ft, then 4, then 9, yet the total amount of light involved remained the same, 1 lumen.   Obviously it will be harder to read at the 3-ft distance than the 1-ft distance, because there is only one-ninth the illumination—the total amount of light has been spread nine times as thin, if we may use the expression. *The illumination of a surface varies inversely as the square of the distance from the light source.* This is known as the "law of inverse squares." If a single light source gives satisfactory illumination for a given job when it is located at a distance of 5 ft, a light source giving four times as many lumens will be required if it is moved to a distance of 10 ft, other conditions remaining the same. To determine the relative amount of illumination, simply divide the square of one distance by the square of the other. For example, comparing the relative illumination of an object 7 ft from a light source as compared with one 4 ft away,

$$\frac{4 \times 4}{7 \times 7} = \frac{16}{49} = 33\% \text{ (approx)}$$

The absolute degree of illumination at any given point, without regard to the power of the source from which the light comes, is measured in footcandles.

Law of Inverse Squares Is Treacherous.   Later in this chapter there is described a simple instrument for measuring footcandles. Assume that such an instrument held 1 ft from a lighting fixture shows 160 footcandles. If it is held 2 ft away, according to the law of inverse squares, it should read 40 footcandles; 4 ft away it should read 10 footcandles; 10 ft away it should read 1.6 footcandles. It does not. Does that mean the law is no law at all, only a theory that does not work? Not at all.

The law is correct when all the light there is comes from one source and that source is dimensionally small (for example, a candle or a very small lamp) and further that none of the light from the source hits a surface and is then reflected back into the lightmeter. The trouble is that such conditions are rarely found. The meter measures not only the light from the fixture that you are thinking about but also light from other fixtures. It measures not only the light that comes directly from the fixture but also light that is reflected from ceilings and walls. Generally speaking the law of inverse squares can be confirmed by a footcandle meter only if the distance from light source to meter is at least five times the maximum dimension of the light source.

All that does not detract from the usefulness of the footcandle meter; it is a very convenient device and serves many useful purposes. More will be said about that later.

Footcandles.   The footcandle is defined as the degree of illumination produced by a light source of one candlepower on a surface exactly one foot distant from the light source.   Remembering the 2-ft sphere of earlier discussions, since every point on the inside is exactly 1 ft from the light source of 1 candlepower, it should be obvious that the surface will be uniformly illuminated to the extent of 1 footcandle.

Since this light source of 1 candlepower emits a total of 12.57 lumens of light and the 2-ft sphere has an area of exactly 12.57 sq ft, it should be equally obvious that it requires 1 lumen of light to produce a uniform illumination of 1 footcandle over an area of 1 sq ft.   This is a most important relation to bear in mind: *One lumen of light per square foot produces illumination of one footcandle.*   Likewise, 10 lumens per sq ft produce 10 footcandles; 100 lumens per sq ft produce 100 footcandles, and so on.

One point that often is misunderstood is the fact that the illumination in footcandles remains the same, no matter what the distance from the light source, as long as the number of lumens of light falling on each square foot does not change.   This at first glance appears to be a complete contradiction of the law of inverse squares, but the following consideration should clarify it.   Assume a reflector so perfect that it condenses *all* the light produced by a lamp producing 100 lumens into a narrow beam so that it illuminates a spot *exactly* 1 *sq ft in area* on a sheet of paper 10 ft from the lamp.   The illumination on the spot will then be 100 footcandles.   If now the sheet of paper is moved to a point 20 ft away, the beam will illuminate a spot 4 sq ft in area and the illumination will be only 25 footcandles.   If, however, a different reflector is then substituted, producing a much narrower beam, so that at the new 20-ft distance the entire 100 lumens will again illuminate a spot only 1 sq ft in area, the illumination on the spot will again be 100 footcandles. As long as *all* the light produced by a source delivering 1 lumen falls on an area of 1 sq ft, that area is illuminated to 1 footcandle no matter what the distance.   It is impossible in practice to concentrate light to this degree, for reflectors are not perfect, absorbing some light and allowing some also to spill in various directions.   As a starting point, however, the relation can be considered correct.   It is a most important rule, and most of this chapter up to this point has been written to help in a clear understanding of this fundamental: 1 *lumen of light on* 1 *sq ft of area produces* 1 *footcandle.*

To illustrate the utility of this rule, assume that an area 12 by 12 ft is to be lighted to 15 footcandles.   Since the total area is 144 sq ft, it will require 144 lumens to provide 1 footcandle. Fifteen footcandles will require 144 × 15, or 2,160 lumens.

The approximate lumens produced by general-purpose incandescent lamps are as follows:

Watts	Lumens	Watts	Lumens
15	126	150	2,730
25	232	200	3,940
40	450	300	6,240
60	855	500	10,500
75	1,170	750	16,700
100	1,750	1,000	23,300

A 150-watt lamp delivering 2,730 lumens in a room 12 by 12 ft should therefore produce about 19 footcandles if all the light produced by the lamp could be directed to the floor and none allowed to fall on the ceiling or walls. But that is entirely a theoretical figure. Instead of 19 footcandles, you will actually attain a much lower figure, between 4 and 6 footcandles, depending on many factors, such as the reflector used with the lamp, the age of the lamp, the reflecting ability of the ceilings and walls, the type of fixture used, and many other factors that will be discussed in more detail in Chap. 30.

Do note, too, that fluorescent lamps produce more lumens per watt than ordinary incandescent lamps. Naturally, then, 150 watts of fluorescent lighting will produce more footcandles of illumination than 150 watts of ordinary lamps. This will be discussed in more detail later in this chapter.

Note that brightness and footcandles of illumination are not the same. A black sheet of paper illuminated to 10 footcandles will not seem so bright as a white one illuminated to 5 footcandles, because the white paper *reflects* a goodly portion of the light falling upon it while the black *absorbs* most of it.

A Few Yardsticks. To provide some starting point of known values in footcandles, bright sunlight on a clear day varies from 6,000 to 10,000 footcandles. In the shade of a tree on the same day there will be somewhere in the neighborhood of 1,000 footcandles. On the same day the light coming into a window on the shady side of a building will be of the order of 100 footcandles, while 10 ft back it will have dropped to something between 7 and 15 footcandles.   At a point 4 ft directly below a 100-watt lamp without a reflector, and backed by a black ceiling and walls which have negligible reflecting power, there will be approximately 6 footcandles.

Measurement of Footcandles.   Candlepower and lumens are not measurable in a simple fashion.   Fortunately footcandles can be measured as easily as reading a voltmeter.   In Fig. 14-3 is shown a direct-

Fig.   14-3 This   instrument reads footcandles of light directly   on   its   scale.   (*General Electric Co.*)

reading footcandle meter, commonly known as a "lightmeter."   Simply set the instrument at the point where the illumination is to be measured, and the footcandles are read directly on the scale.   The device consists of a photoelectric cell, which is a device that generates electricity when light falls upon it.   The indicating meter is simply a microammeter,[3] which measures the current generated, the scale being calibrated to read in footcandles.   The use of this instrument is invaluable, especially in commercial work, as it takes the guesswork out of many lighting problems.

Footcandles Required for Various Jobs.   It must be remembered that there can be no absolute standard.   First of all, the requirements for different individuals vary.   Furthermore, what was considered adequate yesterday is insufficient today; what we by compromise accept today will be considered too little tomorrow.   Accordingly all that can be given are the commonly accepted standards of today.   The more critical the task, the higher the level of illumination required.   The

[3] A microampere is one-millionth of one ampere.

more prolonged the task, the greater the amount of light needed; for example, you can quite easily read a paragraph of a newspaper in the relatively poor light of the dusk of evening, but it is almost impossible to read an entire page.

Today's standards are considerably higher than those of ten years ago. In homes, a minimum of 20 footcandles should be maintained in halls, on stairways, in the general areas of living rooms and dining rooms, and in similar locations. But that isn't enough for locations in those parts of rooms where reading or similar activities are pursued. For casual reading of newspapers, magazines, and books 50 footcandles should be provided; if there is much reading of handwriting, the minimum should be 75 footcandles. For prolonged studying, 75 footcandles would be a wise minimum. Sewing requires from 50 to 200 footcandles; the finer the work and the darker the material, the more light is needed.

In the kitchen and laundry consider 30 footcandles the minimum, with about 50 footcandles in areas where cooking, dishwashing, and ironing are done. The bathroom should have 30 footcandles for general illumination and at least 75 footcandles at the mirror for make-up and shaving.

In nonresidential locations the requirements vary greatly from one type of work to another. In offices the recommended minimum varies from 75 to 200 footcandles, depending on the particular type of work; in stores it varies from 50 to 500 footcandles, depending on the particular area under consideration. This will be discussed in more detail in later chapters concerning various types of buildings and kinds of work.

Will these levels be considered correct ten years from now? Probably not. Much progress has been made in recent years in bringing the general level of illumination upward. Many thousands of offices, for example, now provide 75 footcandles, but still more thousands struggle along with lower levels. Those who have changed have discovered that the added cost is many times offset by increases in efficiency of the work force.

Glare. Because so much space has been devoted to the amount or degree of illumination, do not for a moment think that this is the one all-important factor in good lighting. It is only one of four factors. Another important factor is that the light must be free from glare.

Glare, generally speaking, is caused by a relatively bright area within an area illuminated to a lower level. An exposed lamp in the lobby of a motion-picture theater is not particularly noticeable when one enters during daylight, because the eyes are accustomed to the outdoor brightness; when one leaves, after the eyes are accustomed to the relatively dark interior, the lamp will appear very bright, will glare and hurt the eyes.   Similarly a lighted automobile headlight is barely noticeable in daylight but may be blinding at night.   General Electric Co. in their pamphlet "Fundamentals of Illumination" (from which many of the illustrations of this chapter have been borrowed) define glare as "any brightness within the field of vision of such a character as to cause annoyance, discomfort, interference with vision, or eye fatigue."

Glare is usually caused by exposed lamps so placed that they can be seen while we look at the object we primarily want to see; the lamp may not be directly visible, but even if it is so placed that it can be seen by merely moving the eyes without moving the head, it still produces glare.   Bright automobile headlights constitute a good example of this type of glare.   Figure 14-4 shows an example of extreme glare while

Fig. 14-4   A manufacturing plant lighted to about 30 footcandles.   Efficiency of workers is low.   (*General Electric Co.*)

Fig. 14-5 shows what can be accomplished by modern lighting.   Note the absence of glare and shadows.

Glare of an equally objectionable type may be caused by reflection, for example from glass tops on desks.   Glass is a good reflector, and the reflected image of a lamp may appear almost as bright as the lamp itself. For this reason glass tops have disappeared from the desks of executives. Likewise fewer and fewer highly polished and plated parts are being used on typewriters, office machinery, and other devices because

Fig. 14-5   The plant of Fig. 14-4 now lighted to about 175 footcandles. The efficiency increased greatly with the improvement in lighting. (*General Electric Co.*)

manufacturers have learned that glare causes the evils recounted above with resultant lower efficiency. For this reason, too, printers have learned to avoid papers that are extremely glossy, reflecting too much undiffused light (see the examples shown in Fig. 14-6).

How to Avoid Glare in Lighting.   Everyone has tried reading in direct sunlight and found it difficult. On the other hand it is not difficult to read in the shade of a tree on a bright day, even if there the footcandles are much lower than in the direct sunlight. Therefore the answer cannot lie only in the footcandles of illumination prevailing at any given moment.

The answer does lie in the fact that direct sunlight comes essentially from a single point, the sun, and causes glare. In the shade of a tree the light comes from no point in particular but rather from every direction —north, south, east, west, and above. Not coming from one point, it does not cause harsh shadows or glare. It is *diffused light*.

In lighting a home or office, the more the lighting can be made to duplicate the conditions of the shade of a tree, the better the lighting will be.   Someday in the future there will be a way of having at least the ceilings of rooms give off light of low intensity but sufficient total volume to eliminate the need for lighting fixtures.   A reasonable approach to that solution is the "luminous ceiling" consisting of translucent plastic squares or rectangles, with fluorescent lamps properly spaced above them so that there is uniform, even light from the entire ceiling.

Fig. 14-6   Glass desk tops and glossy paper are both sources of glare.   (*General Electric Co.*)

Surface Brightness.   When looking at an exposed 300-watt lamp of the clear-glass type, you see a concentrated filament less than an inch in diameter.   The lamp itself is a little over 3 in. in diameter, but because the filament is so bright, the glass bulb itself is almost invisible.   That is why practically all lamps today are frosted—the filament is not seen, but rather the entire bulb.   Since the bulb has a diameter of about 3 in., there is exposed to the eye an apparent area equivalent to the area of a circle of the same diameter, or about 8 sq in.   The same total amount of light, which in the case of the clear-glass bulb was concentrated in an area of about 1 sq in., is now distributed over a much larger area of about 8 sq in.   Obviously, then, while the total amount of light is the same, the brightness of the larger area is greatly reduced—the "surface brightness" is lower. It is still uncomfortably bright if looked at directly.

Put the lamp inside a globe of translucent, nontransparent glass about 8 in. in diameter; this has an apparent area of about 50 sq in. altogether instead of the 8 sq in. observed before. The total amount of light is still the same, but it is far more comfortable to the eye because the surface brightness of the light source has been reduced still more. Consider the most common fluorescent lamp: It is the 40-watt type and is 48 in. long. The surface brightness is very low because the total surface area of the tube is very large for the amount of light emitted and there are no bright spots.

For comfortable lighting, use light sources that have low surface brightness.

Direct and Indirect Lighting. When light in a room is produced by fixtures that do not let any of the light fall on the ceilings and walls but instead throw all the light directly on the area to be lighted, the method is called direct lighting. When the light is produced by fixtures that throw all the light on the ceiling, which reflects it to the areas to be lighted, the method is called indirect lighting.

An extreme but completely impractical example of *direct* lighting would be to mount automobile headlights on the ceiling, shining down toward the floor. Such lighting would create extreme glare, sharp shadows, and uneven levels of light and would be very poor lighting— direct lighting of the worst kind. But it should not be supposed that direct lighting can't be made quite effective in stores and similar locations, particularly with flush fixtures in the ceiling, provided that objects below and the floor are of a light color so as to reflect the light upward, and provided that the ceiling is of a light color.

An extreme and again totally impractical example of *indirect* lighting would again involve automobile headlights, now mounted high enough on the wall so that they could not shine directly into your eyes, and with their beams thrown against the ceiling, to be reflected downward. It would provide very bright spots on the ceiling and very uneven lighting in the area below—it would be indirect lighting of the worst kind. Indirect lighting of all kinds is very inefficient, but in establishments other than homes can be made quite effective by properly designed fixtures that distribute the light properly on the ceiling, which must be of a light color.

Good lighting in homes is neither completely direct nor completely indirect.   Most light sources let some of their light fall on the ceiling, some on the areas below.   Indeed that is what we are looking for in good lighting.

Figure 14-7 shows what happens in the case of a single-lamp fixture

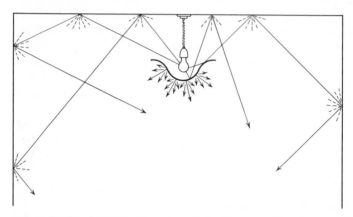

Fig. 14-7   This diagram shows how light becomes diffused as it is reflected from ceilings and walls.

with a translucent shade.   A goodly share of the light goes through the shade; it is diffused as it goes through, so that there are no dark spots below the fixture, or any extremely bright spots on the shade, to produce glare.   Much of the light is reflected to the ceiling, thence is reflected through the room, thus approaching the desired shade-of-a-tree type of light.   In the illustration, note the paths of individual rays of light. The solid lines show what the paths would be if the ceiling and walls were absolutely smooth like a mirror.   While a plaster wall may appear to be very smooth, it will if examined under a magnifying glass be found to be full of hills and dales.   Therefore the rays as they strike any one point, strike those hills and dales and are reflected and dispersed in every direction; what would otherwise be individual rays are split into a multitude of rays (as shown in the dotted lines) giving diffused lighting. While fixtures of the exact type shown in the illustration are little used today, the picture does illustrate a principle.

In general, the entire area of a room (including ceilings and walls) must be lighted to a reasonable degree; smaller areas must be lighted to much higher intensities for reading and other activities that require more light for comfort and efficiency. In some rooms ceiling fixtures provide light for general seeing and floor and table lamps provide the light for reading and similar work; in other rooms only such lamps are used. Such lamps, however, do not throw all their light downward; a considerable part goes upward to the ceiling. Whatever the light source, a reasonably even distribution of light is important for comfort and a room of pleasing appearance: no very dark areas, nor areas of extreme brightness as compared with other nearby areas.

Fixtures for Homes. When a room is lighted with fixtures using exposed lamps, most people think of it as direct lighting. It is a combination of direct and indirect, for a good deal of the light produced falls on the ceiling and is then reflected. But for most purposes it is not good lighting because the exposed lamps have high surface brightness, leading to great contrast with their immediate surroundings and much glare. Because the light comes from one point, such fixtures do not produce diffused shade-of-the-tree light that is so comfortable to the eyes.

A simple one-lamp fixture with an exposed lamp on the other hand is justifiable in attics, halls, and other little-used area, where light is used for casual illumination and not for reading or other work where good lighting is essential.

In general, fixtures so designed that you cannot see the lamps directly are to be preferred. Naturally the room in which the fixture is to be used will determine the general type of fixture. For bedrooms and kitchens, fixtures with one to three lamps and glass shades are often used, installed on the ceiling, or suspended a short distance below the ceiling. Some of the light falls on the ceiling, some is reflected from the glassware to the ceiling, and some of the light goes directly downward. The combination eliminates dark areas on the ceiling, eliminates extremely bright spots on the fixture, and produces generally diffused light which is easy on the eyes. Whether any given fixture will produce enough light in any given area will depend on the wattage of the lamps used, and the design of the fixture.

All that has been said refers to fixtures using ordinary incandescent lamps. In kitchens and other areas where their appearance is acceptable, fluorescent fixtures will provide up to three times as much light per watt; be sure to select a type which permits part of the light to strike the ceiling, so that there will be no dark, dismal areas on the ceiling.

What has been said about lighting fixtures has to do with their function as producers of good lighting. Fixtures also have a decorative function, to carry out a decorative scheme in a home. Indeed, most residential fixtures seen in showrooms seem to be designed with the aesthetic or decorative function in mind. Choose the kind of fixtures that you like, but if they are largely decorative, be sure to supplement them with floor and table lamps so that an adequate level of illumination may be available, where reading or similar activities are carried on.

Reflectance. If the ceiling and walls of the room were mirrors, obviously a very great percentage of the light striking the ceiling and walls would be reflected downward. However, there would be direct reflections of the lamp, and the glare would be as bad as from an exposed lamp. On the other hand, if the ceiling were black, most of the light would be absorbed and little reflected, and the lighting would be inefficient indeed. Accordingly, for best results, the ceiling and walls should be a color that reflects as much light as possible, and at the same time the finish should be dull or flat rather than glossy, so as to avoid the mirror effect, with bright spots producing glare.

Experiment has shown that various colors of paint reflect light in various degrees. The percentage of light that they reflect is known as the "reflectance" of the surface. The table below shows the approximate reflectance of various colors of painted surfaces.

White	80 to 85%
Pale pink or yellow	75 to 80%
Ivory	70 to 80%
Cream	65 to 75%
Buff	55 to 65%
Gray	35 to 50%
Light blue	35 to 50%
Light green	30 to 40%
Dark green	15 to 25%
Red	15 to 25%
Dark blue	5 to 15%
Brown	8 to 12%
Black	2 to 5%

Woods in natural finish seldom reflect over 50%, and the darker woods may fall as low as 15%.

Shadows. When we read in direct sunlight, the extreme glare makes reading uncomfortable. Moreover, the reader's own shadow on the reading material is extremely sharp; the contrast between bright light and sharp shadow is most annoying and tiring.   In the shade of a tree the footcandle level is much lower, but reading is more comfortable: The light is diffused; shadows are soft and not objectionable.   Two sources of light, for example a floor lamp and an overhead fixture, lead to greater comfort in reading.   Use fixtures or lamps that provide diffused light, light that seems to come from many points, as in the shade of a tree.

Color of Light.   Sunlight has come to be accepted by most people as a standard, although the color and the quantity vary greatly, depending on the hour of the day and on the season.   It may seem strange to speak of the "color" of sunlight, which appears colorless, yet sunlight is composed of a mixture of all colors.   The rainbow is simply a breaking down of sunlight into its separate colors.

What makes one object red and another object blue when this mixture of all colors which we call sunlight hits these objects?   The explanation is simple.   When "white" light strikes certain objects, the component colors are all reflected equally, and we call such objects white. When white light hits other objects, the light instead of being reflected is absorbed and we see no light; we then say that such objects are black.

Still different objects may absorb some of the colors of the spectrum, reflect the others.   For example, they may absorb all except the red and reflect that.   We then see only the red part, so we say such objects are red.   So with every other color: whatever the color of the object, that is the color that the object can reflect; all the other colors are absorbed and destroyed.

Everybody has observed that the color of an object appears different when observed in natural light as compared with artificial light.   What causes this?   Light produced by incandescent lamps does not have the same proportion of colors as exists in natural light; it has more orange and red and less of the blue and green.   It is not surprising, then, that efforts have been made to produce special lamps that more nearly duplicate the mixture and proportions of colors existing in natural light. The effort has led to lamps that approximate the light from a northern sky rather than direct sunlight.   Such lamps are called "daylight type" and in one brand are called "coloramic sky blue."   Use such lamps when color comparisons are important; don't use them for general lighting because they are not as efficient as ordinary lamps—they produce less light per watt.   In the fluorescent type the "de luxe cool white" is best suited for the purpose.

Life of Lamps.   The approximate average life of lamps, when burned at their rated voltage, varies a great deal with the size and type of lamp. For ordinary lamps of various types, the approximate average life will be found in the table on page 242.   It is impossible to make lamps so that all will have exactly their rated life.   Assume a large group of identical lamps with 1,000-hr rated life, all turned on at the same time and never turned off.   At 800 hr about 20% will have burned out.   At 1,000 hr 50% will have burned out; at 1,200 hr about 80%.   At 1,400 hr some will be still burning.   Their *average* life will be 1,000 hr. These ratios hold substantially true for fluorescent as well as incandescent lamps.

It is a simple matter to make lamps that last longer, but in doing so their efficiency is reduced; the lamp will produce less light, fewer lumens per watt.   A 60-watt lamp costing 25 cents uses 60 kwhr of energy during its 1,000-hr life.   At 3 cents per kilowatthour, the cost of the energy used is therefore $1.80, as compared with the 25-cent cost of the lamp itself.   If, then, to obtain longer life in a 25-cent lamp, we reduce the efficiency, an extra 50 cents may be spent for electric power to produce the same amount of light, and expense rather than economy results.

Where no great amount of light is needed and the lamp serves merely as a signal, for example in pilot lights, the lamp may be designed to last 2,000 hr or more; on the other hand, where a great deal of light is needed and where it is important to limit the heat, for example in lamps for motion-picture projectors, the lamp may be designed for a relatively short life but with corresponding increase in efficiency, thus permitting smaller-wattage lamps producing less heat to be used.   For example, a 1,000-hr 1,000-watt lamp produces about 20 lumens per watt, but a 50-hr lamp of the same wattage produces about 28 lumens per watt, while a photoflood lamp of the same approximate wattage, but only 10-hr life, produces roughly 31 lumens per watt.

Effect of Voltage on Lamp Life.   If a lamp is operated at a voltage below that for which it was designed, its life is prolonged considerably, but the watts, the lumens, and the lumens per watt drop off rapidly.   If it is operated at a voltage above normal, its life is greatly reduced, although the watts, the lumens, and the lumens per watt increase.   For lowest over-all cost of illumination use lamps on a circuit of the voltage for which they were designed.

Careful study of the following table will confirm these statements.

If a lamp is burned at the voltage for which it is designed, it will have characteristics shown in the line labeled 100%. If it is burned at a voltage higher or lower than its design voltage, the various factors will change to a new value shown as a percentage of normal.

Circuit voltage as percentage of rated voltage of lamp	Total average life, hours	Total output, lumens	Actual watts	Lumens per watt
85%	800%	58%	78%	72%
90%	400%	68%	85%	80%
95%	200%	83%	93%	90%
100%	100%	100%	100%	100%
105%	58%	118%	108%	109%
110%	37%	140%	116%	120%
115%	18%	162%	124%	131%

As an example, an ordinary 100-watt lamp designed for use on a 120-volt circuit will when burned at 120 volts have a life of about 750 hr, produce 1,750 lumens, and consume 100 watts, resulting in 17.5 lumens per watt. If burned at 108 volts (90% of 120 volts), it will have a life of about 400%, or 3,000 hr; produce 68% of 1,750, or 1,190 lumens; it will consume 85% of normal, or 85 watts, resulting in 80% of normal efficiency, or about 14 lumens per watt.

A lamp burned at its rated voltage represents the lowest total cost of light, considering the cost of the lamp itself, the cost of the power it consumes during its life, and the labor of replacing it. A 100-watt lamp during its average 750-hr life consumes 75 kwhr of power, costing perhaps ten times as much as the lamp itself. Using a lamp at a voltage lower than that for which it was designed does indeed prolong its life, but the amount of light produced is greatly reduced, as already discussed. You will have to use a higher wattage lamp than if you use a standard lamp at its rated voltage, to get the same amount of light.

"Long-life" Lamps. You will see advertised *for use in the home,* lamps made by unknown manufacturers, guaranteed for 5,000 hr, or 5 years, or some other period far beyond the life of ordinary lamps. Naturally their price is very high, compared with ordinary lamps. Don't buy them; it would be a poor investment. There is no secret or "gimmick" that makes such lamps last longer.   They are neither more nor less than lamps designed to burn at 130 or 140 volts, but marked to pretend they should be used on ordinary 115- or 120-volt circuits.   Naturally they will last a long time, but deliver far less light *per watt* than ordinary lamps; sometimes they consume more watts than the size stamped on them, giving the illusion of producing just as much light as ordinary lamps.   You might as well use ordinary lamps of a smaller wattage.   Remember that the cost of the power consumed during the life of a lamp, is very much more than the cost of the lamp itself.

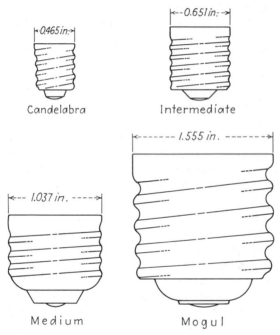

Fig. 14-8   The screw-shell bases used on lamps are standardized to the dimensions shown.   The illustrations are actual size.

Nevertheless there are locations where it is quite proper to use lamps made by legitimate manufacturers and labeled "extended-service." Such lamps have an average life of two or three times that of ordinary lamps, but produce less light *per watt*.   Their price is only a trifle higher than that of ordinary lamps.   They are properly used in industrial and similar locations where it is extremely difficult to reach them: high up or similar locations where the cost of labor in replacing them is very high.

Occasionally there will be applications where it is entirely in order to use lamps at a voltage considerably *higher* than that for which they were designed. There may be conditions when a great deal of light is needed, but only for short periods, as for example in lighting athletic fields. By using 115-volt lamps on a 130-volt circuit (or 110-volt lamps on a 125-volt circuit) the amount of light produced *per watt* is increased by about 25%. The fact that the lamp will last only about 165 hr as compared with its normal 750-hr life is immaterial when weighed against the extra cost of larger or more numerous floodlighting reflectors, and heavier power lines and transformers, to accommodate the larger lamps that would otherwise be needed.

Bases. There are various standardized sizes and kinds of bases in use, matched to the watts, the physical size, and the purpose of the lamp. In the screw-shell type the largest is the mogul, used mostly on lamps from 300 watts upward; its dimensions are shown in Fig. 14-8, together with the dimensions of other standard bases. The medium is used on ordinary household lamps. Smaller lamps use the intermediate and the candelabra. Flashlight and similar lamps use the miniature, which is still smaller.

Fig. 14-9 The bipost type of base is used on larger lamps. (*General Electric Co.*)

Fig. 14-10 The prefocus type of base is used on lamps for projectors. (*General Electric Co.*)

Another base used on high-wattage lamps is the bipost of Fig. 14-9. The prefocus base of Fig. 14-10 is used mostly on lamps for projectors and maintains the lamp in one particular position which gives maximum light in one direction. Both of these types are available in two sizes: medium and mogul. A three-light lamp is merely a lamp with

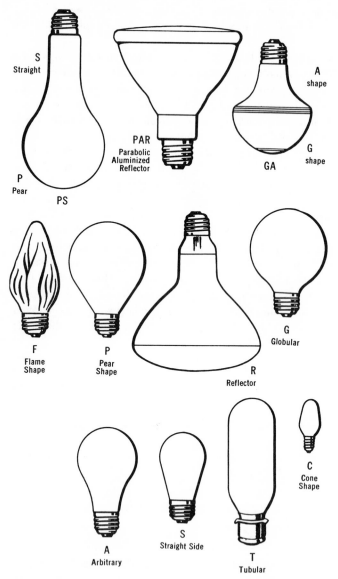

Fig. 14-11 The shape of lamps is well standardized. (*General Electric Co.*)

two separate filaments, say 100 and 200 watts, with a special base so arranged that either filament alone or both at the same time can be used, producing 100, 200, or 300 watts as desired.

Lamp Designations. The mechanical size and shape of a lamp are designated by standardized abbreviations such as A-19, PS-35, F-15, etc. The letter designates the shape in accordance with the outlines shown in Fig. 14-11. The numeral designates the diameter in eighths of an inch. Thus an A-19 lamp has the simple A shape and is $^{19}\!/_8$, or $2^3\!/_8$, in. in diameter.

Lumiline Lamps. This type of lamp, shown in Fig. 14-12, has a special contact at each end, and is available in 30-, 40-, and 60-watt sizes.

Fig. 14-12 Lumiline lamps have a contact at each end. (*General Electric Co.*)

At one time it was in quite common use, but it is an inefficient lamp, producing less than 10 lumens per watt. For that reason it is little used today in new installations, but is available for replacements.

Reflector Lamps. Special lamps with a silver or aluminum reflector deposited on the glass bulb are available for floodlighting or spotlighting. While they have some use in homes, they are more commonly used in commercial or industrial lighting, and therefore will be discussed in more detail in Chap. 30.

Efficiency of Various Sizes of Lamps. Larger lamps produce more light *per watt*, so that generally speaking, if there is a choice, it is better to use one large lamp rather than several smaller ones. Study the table on page 242, which is based on general-purpose lamps of present manufacture.

Study of this table will show you that one 150-watt lamp produces approximately as much light as twelve of the 25-watt size consuming 300 watts in total; one 300-watt lamp produces more light than seven of the 60-watt size.

Efficiency of Various Colors of Lamps. There is no difference in effi-

Size of lamp, watts	Average life, hours	Total lumens	Lumens per watt
25	2,500	232	7.3
40	1,500	450	11.2
60	1,000	855	14.3
75	750	1,170	15.6
100	750	1,750	17.5°
150	750	2,730	18.2
200	750	3,940	19.7
300	750	6,240	20.8
500	1,000	10,500	21.0

° The first lamps made by Edison in 1888 produced less than 2 lumens per watt, in the 100-watt size.

ciency between a lamp with a clear-glass bulb and one of the inside-frosted type.   The latter type produces light that is better diffused than in the case of the clear-glass type.   A new trend is away from the inside-frosted to a still newer type with a translucent, milky-color internal silica coating that diffuses the light still more, resulting in less glare with no appreciable difference in efficiency.

However, incandescent lamps that produce colored lights are very inefficient.   The coloring material simply absorbs most of the light produced by the lamp; only that portion which matches the color of the bulb is transmitted.   For this reason colored lamps should not be used except for their decorative value.

Fluorescent Lighting.   This form of lighting is relatively new, having been introduced commercially in 1938.   Present-day lamps produce vastly more lumens per watt, and last several times longer.   The fluorescent installation is familiar to all, but few understand its method of operation, which is quite complex as compared with that of an ordinary incandescent (filament-type) lamp.

In the ordinary incandescent lamp, a filament made of tungsten wire is heated by an electric current flowing through it until it reaches a high temperature, when it emits light.   It operates by its terminals being connected to two wires of an electric circuit of the proper voltage.

The fluorescent lamp consists essentially of a glass cylinder called the "tube," with a filament at each end.   The filaments are not connected to each other in the lamp.   Each filament is brought out to two pins on the end, as shown in Fig. 14-13.   The inside of the glass tube is coated with a whitish or grayish powder, called a "phosphor." The air has been pumped out of the tube, and a carefully determined amount of a gas, usually argon, sometimes argon and neon, is introduced. A very small amount of mercury is also put into the tube. That is the basic machinery of a fluorescent lamp, but if it is connected directly to the ordinary 115-volt circuit, it will not operate.

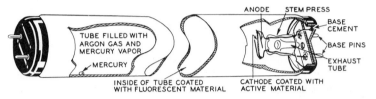

Fig. 14-13 Construction of a fluorescent lamp. (*General Electric Co.*)

If a fluorescent lamp is connected to a source of high-voltage current, the lamp will light up but will be quickly destroyed. If it is connected to a source of high-voltage current and just at the instant of starting the voltage is reduced to 115 volts, it will still be destroyed. Apparently then some special accessories are required to make the fluorescent lamp operate.

Figure 14-14 shows the basic scheme. The secret of operation con-

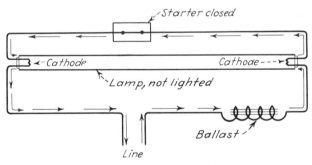

Fig. 14-14 Flow of current in a fluorescent lamp circuit, at the instant when the lamp is turned on.

sists of two devices: a ballast or choke coil, and an automatic switch, called a starter. Any coil of wire wound on an iron core has two peculiarities: (1) When connected to an alternating-current circuit, it tends to resist any change of current flowing through it, and (2) when a current flowing through it is cut off, it delivers momentarily a voltage much higher than the voltage applied to it.   The ballast for a fluorescent lamp is just such a coil.   The automatic switch or starter (part of the lighting fixture) is so designed that it is ordinarily closed (while the lamp is turned off), but when the lamp is turned on, the starter opens a second or so after the current starts to flow and then stays open until the lamp is turned off again.

Visualize then what happens.   Start with Fig. 14-14, which shows the circuit just as the lamp is turned on.   Current flows as indicated by the arrows, through the ballast, through one filament, or "cathode" as it is called in the case of the fluorescent lamp, through the automatic switch or "starter," through the other filament or cathode, and back to the line.   During this period the lamp glows at each end but does not light.   Then the automatic switch (the starter) opens, and the ballast does its trick—it delivers a high voltage as mentioned in the previous paragraph, a voltage considerably above 115 volts and high enough to start the lamp.   The current can no longer flow through the starter because it is open; it then flows through the tube, jumping the gap and forming an arc inside the glass tube, following the arrows of Fig. 14-15

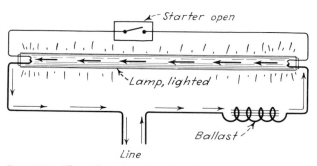

Fig. 14-15   Flow of current through a fluorescent lamp after it has started. The automatic "starter" has opened and current cannot flow through it.

(in both Figs. 14-14 and 14-15 the current flows first in the direction indicated by the arrows, then in the opposite direction, because the current involved is alternating current).   The ballast then performs its other function: It limits the current flowing through the lamp to a predetermined safe value.

How does the current jump from one end of the tube to the other? It is a complicated story, and yet it is, in some ways, simple. The filaments or cathodes of coiled tungsten wire are coated with a chemical which when heated emits electrons, particles so small that billions of them laid side by side would still be invisible, being actually basic negative charges of electricity. They shoot out into space as popcorn does in a corn popper; they travel more easily through argon gas than through a vacuum, which is the reason that gas is introduced. A stream of these particles constitutes an electric current, which heats the argon, which heats the mercury to become mercury vapor, which then becomes the path for a heavier electric current.

If a fluorescent lamp such as just described had a wall of clear glass, an insignificant amount of light would be visible, and yet the fluorescent lamp as installed produces a great deal of visible light. The answer lies in the fact that the electric arc through the mercury vapor produces only a slight amount of *visible* light but a great deal of *invisible* ultraviolet light.

The inside of the tube is covered with a layer of chemicals that become fluorescent or light-producing when exposed to ultraviolet light. In other words, invisible ultraviolet light striking fluorescent chemicals makes the chemicals glow brightly, producing visible light. The particular chemical used determines the color of the light.

The exact scientific principles that govern the emission of electrons from a heated coated filament, the creation of the arc, the production of ultraviolet light by the arc, and the creation of visible light when invisible ultraviolet light strikes certain chemicals had best be left to the chemists and engineers.

Advantages of Fluorescent Lighting. The greatest single advantage of fluorescent lamps lies in their efficiency. Per watt of power used, they produce two to three[4] times as much light as ordinary incandescent lamps. Their life is much longer than that of incandescent lamps. Being more efficient, they produce much less heat, which is important when larger amounts of power are used for lighting, especially if the lighted area is air conditioned.

Another major advantage is that the fluorescent lamp, being of relatively large size (in terms of area in square inches compared with the total light output), has relatively low surface brightness, which in turn

[4] In the case of certain colors, the fluorescent lamp produces over 100 times as much light per watt as an incandescent lamp.

leads to less glare, less shadow, all contributing to better seeing, and less eyestrain.   The surface brightness being low, there is less need for enclosing glassware; thus the cost of installation and upkeep is reduced.

Disadvantages of Fluorescent Lighting.   Ordinary incandescent or filament-type lamps operate in any temperature.   Fluorescent lamps are somewhat sensitive to temperature.   The ordinary type used in homes will operate in temperatures down to about 50°F.   For industrial and commercial areas other types are available that will operate at lower temperatures, some as low as −20°F.

In the case of filament lamps, as the voltage is reduced, the light output drops off much faster than the voltage.   However, even at greatly reduced voltages the filament still glows, producing some light.   In the case of fluorescent lamps, as the voltage is reduced, the light output also drops off, but not as fast as in the case of ordinary filament lamps.   But if the voltage is greatly reduced, the lamps go out completely, long before a filament lamp ceases to glow.   Therefore fluorescent lamps cannot be used where there are frequent and violent fluctuations in voltage.   Special voltage-regulating ballasts however are available for such locations, so that fluorescents need not be ruled out completely.

Life of Fluorescent Lamps.   The life of an ordinary incandescent or filament lamp varies from 750 to 2,500 hr depending on the size, type, and purpose of the lamp.   It does not make any difference how often the lamp is turned on and off.   A *fluorescent* lamp that is turned on and never turned off will probably last 20,000 hr.   If it is turned on and off every 5 minutes it may not last 500 hr.   The reason lies in the fact that there is a specific amount of electron-emitting material on the filaments or cathodes; a specific amount of it is used up every time the lamp is turned on; when it is all gone, the lamp is inoperative.   It is not possible to predict the exact number of starts the lamp will survive, and ordinary operation of the lamp also consumes some of the material, but the fact remains that the oftener a fluorescent is turned on, the shorter its life will be.

The published figures for approximate life are based on the assumption that the lamp will burn 3 hr every time it is turned on.   The average life varies a good deal with the size and type of lamp; for the most ordinary 40-watt lamp the average is about 15,000 hr.   Under ordinary operation, fluorescents last five to ten times as long as ordinary incandescent lamps.

Rating of Fluorescent Lamps.   An ordinary incandescent lamp marked "40 watts" will consume 40 watts when connected to a circuit of the proper voltage. A *fluorescent* lamp rated at 40 watts will also consume 40 watts within the lamp, but an additional wattage is also consumed by the ballast; this additional wattage is from 10 to 20% of the wattage consumed by the lamp itself, and must be added to the wattage of the lamp to arrive at the total wattage of the combination.

Power Factor of Fluorescent Lamps. Ordinary *incandescent* lamps have a power factor of 100%. A single *fluorescent* lamp connected to a circuit has a power factor of somewhere between 50 and 60%. Assume 100 lamps rated at 40 watts, connected singly to a 115-volt circuit. The ballast for each lamp can be expected to consume approximately 8 watts. The total for each combination is 48 watts, and for 100 such lamps, the total is 4,800 watts. The amperage consumed by these 100 lamps and their ballasts is, however, not 4,800/115, or 42 amp, as might be expected (and as is the case using ordinary incandescent lamps), but rather (assuming a power factor of 60%) 4,800/(115 × 0.60), or 4,800/69, or approximately 70 amp. Therefore the wiring serving this load must be capable of carrying 70 amp rather than a theoretical 42 amp.

Fortunately this is not so serious as it sounds. The common method is to have either two or four lamps per fixture and, in addition to the usual ballast, to use power-factor-correction devices built into the same case with the ballast, which bring the power factor up to about 90% or better.

Fortunately, most two- or four-lamp fixtures on the market are the high-power-factor type, and especially so in the case of fixtures for commercial or industrial use. But if you buy "bargain" fixtures for the home, be sure they are of the high-power-factor type.

Sizes of Fluorescent Lamps. The lamps most commonly used in homes are those listed below. Note that the wattage consumed by the ballast is *not* included in the column headed "watts."

Designation	Length, inches	Diameter, inches	Watts	Total lumens	Lumens per watt
T-8	18	1	15	870	58
T-12	24	1½	20	1,220	61
T-8	36	1	30	2,100	70
T-12	48	1½	40	3,120	78

In addition to those listed above, the Circline lamps of the type shown in Fig. 14-16 are frequently used in fixtures and lamps where a long, straight lamp would not fit. They are available in three sizes: 22-watt, 8¼-in. outside diameter; 32-watt, 12-in. diameter; and 40-watt, 16-in. diameter.

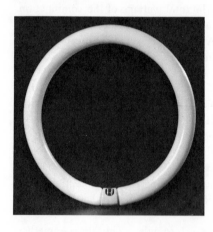

Fig. 14-16 The "Circline" fluorescent lamp is used in floor lamps and fixtures. (*General Electric Co.*)

Other Types of Fluorescent Lamps.   The lamps just discussed are the kind commonly used in homes, and, of course, they are used in other locations too.   For nonresidential use there are several other types of fluorescent lamps, and these will be discussed in Chap. 30.   Let it be noted too that the circuit of Figs. 14-14 and 14-15 is that of a single lamp.   More complicated circuits are used when the fixture has two or more lamps.

Color of Light from Fluorescent Lamps.   Everyone has noticed that the colors of flowers, clothing, and so on look different under ordinary filament lamps than in natural sunlight.   Ordinary filament lamps produce light that is rich in red.

The color of the light produced by fluorescent lamps is determined entirely by the chemical and physical composition of the "phosphor" or powder which is deposited on the inside surface of the tube.   The most ordinary color is "white," but there are many varieties of it, among them de luxe warm white, warm white, white, cool white, and de luxe cool white.   The differences lie in the proportion of red and blue in the light produced by the respective lamps.   The "warm" varieties emphasize the red and yellow (like light from ordinary incandescent lamps), while the "cool" varieties emphasize the blue (more like natural sunlight). The kind of "white" to use depends on the effects desired.

If you look at a "black-and-white" printed page under each of the "white" colors in turn, you will see little difference. But if you look at a colored page, colored fabrics, or food under each kind in turn, you will see much difference.

Under de luxe cool white, people's complexions will appear much as in natural light, but the de luxe warm white will flatter them a bit, adding a ruddy or tanned appearance (much as is done by ordinary incandescent lamps). The cool white gives quite a good appearance also, but with a slight tendency toward paleness.

As to home furnishings, if your preference leans toward those that have warm colors (red, orange, brown, tan), use de luxe warm white; if you prefer the cool colors (blue, green, yellow), use the de luxe cool white.

It must be mentioned here that the "de luxe" varieties are not nearly so efficient as the others; they produce about 30% less light per watt of power used. If then you want maximum efficiency and are willing to sacrifice good color rendition a bit, use the warm white in place of the de luxe warm white, and the cool white in place of the de luxe cool white. This difference in efficiency is probably the major reason for the fact that about 75% of all fluorescents sold are of the cool-white variety.

Most sizes of fluorescent lamps are available also in such colors as blue, green, pink, gold, and red. Their efficiency in colors is extraordinarily high; for example, the fluorescent lamp produces about 100 times as much green light per watt as is produced by incandescent or filament lamps. These colored lamps find a particular application where spectacular color effects are needed, for example, in theater lobbies, lounges, stage lighting, advertising, and similar purposes.

Luminescent Lighting. A new kind of lighting called "electroluminescent" is under development. No bulbs or tubes are involved. It has been found that if two conducting surfaces (one of them transparent) are laid parallel and very close together and the space between filled with the correct chemical powder, the powder will glow and produce light when an alternating-current voltage is applied to the two surfaces. See Fig. 14-17, which shows the basic construction. Flat light-producing areas of considerable size can be made in this way. The color and intensity of the light can be controlled by the voltage and

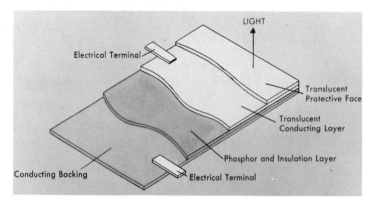

Fig. 14-17   Cross section of a luminescent panel.   (*General Electric Co.*)

the frequency of the applied power.   The output in lumens per watt is almost insignificant, so that it is not likely that this kind of lighting will ever be used for general lighting.   Nevertheless, it is practical where only a minute quantity of light is needed, as, for example, on clock faces or radio dials.

15

Residential and Farm Motors

A working man can deliver no more than $\frac{1}{10}$ to $\frac{1}{8}$ hp continuously over a period of several hours.   If the man is paid $3 an hour, it costs at least $24 for a horsepower for an hour.   At average rates an electric motor will deliver a horsepower for an hour for 5 cents.   The motor costs little to begin with.   It will operate equally well on a hot day or a cold day.   It never gets tired, and costs nothing except while running.   It uses electricity only in proportion to the power it is called upon to deliver. With reasonable care, it will last for many years.

A later chapter will cover the wiring of industrial motors; this chapter will cover only ordinary wiring of the type needed in homes and on farms.   However, a discussion of the characteristics and limitations of the types in common use should be in order here.

How Motors Are Rated.   A motor is rated in horsepower.   This means that (except for special-purpose motors) it will deliver the horsepower stamped on its name-plate hour after hour, all day and all week without a stop, if necessary.

Starting Capacity. Motors deliver far more power while starting than after they are up to full speed. The proportion varies with the type of motor; some types have starting torques four or five times greater than at full speed. Naturally the watts consumed during the starting period are much higher than while the motor is running at full speed. That means the motor will heat up quickly if it does not reach full speed because of too heavy starting load. Therefore the right kind of motor must be used for each machine, depending on how easy or how hard it is to start that machine. This will be explained later in this chapter.

Overload Capacity. Almost any good motor will for short periods develop from 1½ to 2 times its normal horsepower without harm. Thus a 1-hp motor is usually able to deliver 1½ hp for perhaps a quarter of an hour, 2 hp for a minute, and usually even 3 hp for a few seconds without harm. Of course, no motor should be deliberately overloaded continuously, for overloading leads to overheating, and that in turn leads to greatly reduced life of the windings. Rewinding a burned-out motor is very expensive; in smaller sizes it usually is less expensive to replace the motor. Nevertheless this ability of a motor to deliver more than its rated horsepower is most convenient. For example, in sawing lumber, ½ hp may be just right, but when a tough knot is fed to the saw blade, the motor will instantly deliver, if needed, 1½ hp, dropping back to its normal ½ hp after the knot has been sawed. A water-pressure system in a suburban home or on a farm may be properly equipped with a ½-hp motor. When the motor is first turned on, with no pressure in the water tank, the pump may require a good deal less than ½ hp. As the pressure in the tank builds up, the horsepower required also gradually increases. Just before the automatic pressure switch cuts off the motor as a 40-lb pressure is reached in the tank, the motor will be delivering a good deal more horsepower than in the beginning of the running cycle, possibly as much as ¾ hp. The motor with its overload capacity automatically takes care of the problem.

Gasoline Engines vs. Electric Motors. A gasoline engine is rated at the maximum power it can deliver *continuously*. Thus, an engine rated at 5 hp can deliver 5 hp continuously, but unlike an electric motor, it cannot, even for a short time, deliver more than 5 hp. This explains why it is often possible to replace a 5-hp engine with an electric motor sometimes as small as 3 hp. If the engine being replaced always runs smoothly with little effort, if it seldom labors and slows down, it can be replaced with an electric motor of much smaller horsepower. If, on the other hand, the engine is always laboring at its maximum output, then the motor that replaces it should be of the same horsepower as the engine, because no motor should be expected *continuously* to deliver more than its rated horsepower.

Power Consumed by a Motor.   The watts drawn from the power line by a motor are in proportion to the horsepower it is delivering.   The approximate figures for a 1-hp motor are as follows:

While starting	4,000 watts
While idling	200 watts
While delivering ¼ hp	400 watts
While delivering ½ hp	600 watts
While delivering ¾ hp	800 watts
While delivering 1 hp	1,000 watts
While delivering 1½ hp	1,500 watts
While delivering 2 hp	2,000 watts
While delivering 2½ hp	2,600 watts
While delivering 3 hp	3,300 watts

Motors are designed to operate at greatest efficiency when delivering their rated horsepower; while they are delivering more or less power, the efficiency usually falls off.   In other words, it costs a little more per hour to run a 1-hp motor at half load than it does to run a ½-hp motor at full load.   The 1-hp motor, while delivering only ½ hp, as per table above, consumes about 600 watts; a ½-hp motor delivering ½ hp consumes about 525 watts.   In the long run therefore, especially if the motor must run many hours, it will be less costly to have a small motor for a small load than to run a small load with a big motor that you may happen to have handy.

Speed of AC Motors.   The most common 60-cycle AC motor runs at a theoretical 1,800 rpm but at an actual speed of 1,725 to 1,750 rpm while delivering its rated horsepower (the horsepower stamped on its nameplate).   When the motor is overloaded, the speed drops further.   If the overload is increased too far, the motor will stall; it will burn out if not quickly removed from the line.   Low voltage also reduces the speed.

Motors of other speeds are built, running at theoretical speeds of 900, 1,200, and 3,600 rpm, actual speeds a little lower.   These cost more; in case of motor failure, they are hard to replace quickly because they are seldom stocked, which makes it necessary to change pulleys and belts and leads to similar inconveniences.   Use an 1,800-rpm motor wherever it is at all possible.   If a very slow speed is needed, use a motor which runs at 1,800 rpm but has built-in gears which cut the speed of the drive shaft to a much lower figure.

The speed of ordinary alternating-current motors *cannot be controlled* by rheostats, switches, or similar devices.   Special variable-speed motors are obtainable, but they are very expensive special-purpose motors and will not be discussed here.

Service Factors on Motors.   At one time motors were rated on the basis of a temperature *rise*, over and above the ambient temperature (the temperature at the motor location).   Ordinary motors were based on a rise of not over 40°C or 72°F,[1] which meant that the motor could be used *continuously* at its rated horsepower, without its temperature increasing more than 40°C or 72°F above the ambient temperature, without harm.   The actual temperature of the motor installed in a hot location, for example on a farm in a pump house where the temperature on a hot day might be 115°F, would then increase by 72°F for a final temperature of 187°F, not far below the boiling point of water.   This would feel exceedingly hot to the hand, but would not harm the motor.

Over the years, the heat-resisting qualities of the wire used to wind the motor, and the insulating materials in the motor, have been vastly improved.   This in turn has made it possible to reduce the physical size of motors a great deal—today's 10-hp 3-phase motor is no larger than a 3-hp motor made in 1950.   But these smaller motors run much hotter than the old larger size, but will not be harmed by temperatures that would have burned out the 1950 motor.   They often exceed 212°F (the boiling point of water) but are not damaged.

Motors are no longer rated based on a temperature rise.   Instead, each motor has a "service factor" stamped on its name-plate, ranging from 1.00 to 1.35.   If for a specific motor the service factor is 1.00, it means that if installed in a location where the ambient temperature is not over 40°C or 104°F, it can deliver its rated horsepower continuously, without harm.   But if its service factor is for example 1.15, it can

---

[1] Do not confuse *change* in readings of thermometers with their *actual* readings.   When a centigrade thermometer reads 40°, a Fahrenheit thermometer reads 104°.   While a centigrade thermometer changes by 40°, the Fahrenheit changes by 72°.   Thus, if the centigrade changes from 40 to 80°, the Fahrenheit changes from 104 to 176° (by 72°).

be used at up to 1.15 times its rated horsepower, under the same conditions.   Multiply the rated horsepower by the service factor: a 5-hp motor with a service factor of 1.15 can be used continuously as a 5.75-hp motor.

Most open-type motors larger than 1 hp have a service factor of 1.15. Fractional-horsepower motors have a service factor of 1.25, some as high as 1.35.   If the ambient is higher than 104°F, the motor should not be used continuously at its full rated horsepower.   Regardless of temperature, install motors where plenty of air is available for cooling.

All this does not mean that the motor can't be used to deliver more than its rated horsepower, for short periods.   The overload capabilities discussed elsewhere in this chapter remain true, despite the change from the temperature-rise to the service-factor method of rating.

Types of Single-phase Motors.   There are three common types of motors designed for single-phase work, and these will be described separately in the following paragraphs.

Split-phase Motors.   This type of motor operates only on single-phase alternating current.   It is the most simple type of motor made, which makes it relatively trouble-free; there are no brushes, no commutator.   It is available only in sizes of ⅓ hp and smaller.   It draws a very heavy amperage while starting.   Once up to full speed, this type of motor develops just as much power as any other type of motor, but it is not able to start heavy loads.   Therefore it should never be used on any machine which is hard to start, such as a deep-well water pump or an air compressor that has to start against compression.   It can be used on any machine which is easy to start or on one where the load is thrown on after the machine is up to full speed.   It is entirely suitable for washing machines, grinders, saws and lathes, and general utility use.   It can be used on paint sprayers which do not start against compression but not on those that start against pressure.

Capacitor Motors.   This type of motor also operates only on single-phase alternating current.   In construction it is quite similar to the split-phase type, with the addition of a "capacitor" or a "condenser" which enables it to start much harder loads.   There are several grades of capacitor-type motors available, ranging from the home-workshop type, which starts loads from 1½ to 2 times as heavy as the split phase, to the heavy-duty type, which will start almost any type of load whatever.   Capacitor motors usually are also more efficient than split phase, using fewer watts per horsepower. The amperage consumed *while starting* is about half that of the split-phase type. Capacitor motors are made in any size but are commonly used only in sizes up to 7½ hp.

Repulsion-start, Induction-run Motors. Most people call this type of motor simply an R-I motor. It operates only on single-phase AC current. It has a very large starting ability and should be used for the heavier jobs; it will "break loose" almost any kind of hard-starting machine. The starting current is the lowest of all types of single-phase motors. It is available in sizes up to 10 hp.

Dual-voltage Motors. Most AC motors rated at ½ hp or more are so constructed that they can be operated on either of two different voltages: 115 or 230 volts. Single-phase motors of this type have four leads: Connected one way, the motor will operate at 115 volts; connected differently, the motor will operate at 230 volts (see Fig. 15-1).

Fig. 15-1   Larger motors are made so that they can be operated at either 115 or 230 volts, depending on the connections of the four leads shown in the diagram above. It is always wise to operate such motors at 230 volts if possible.

If the motor is of the dual-voltage type, that is, if it will operate on either 115 or 230 volts, two different voltages and amperages are shown on the name-plate. For example, it may be marked: "Volts 115/230, Amps 24/12." This simply indicates that, while delivering its rated power, it will consume 24 amp if operated at 115 volts or 12 amp if operated at 230 volts.

Three-phase Motors. This type of motor, as the name implies, operates only on 3-phase alternating current. Three-phase motors in sizes ½ hp and larger cost less than any other type, so by all means use them if you have 3-phase current available. Do not assume that because you have a 3-wire service you have 3-phase current; more likely you have 3-wire 115/230-volt single-phase current. If in doubt, see your power supplier.

Direct-current Motors. Direct current is found in parts of downtown sections of such cities as New York and Chicago and also in a few small towns. All 32-volt farm lighting plants are direct current, as are a few of the 110-volt models. There are numerous types of direct-current motors, such as the series, the shunt, the compound, and others. However, too few direct-current motors are in use to warrant a detailed discussion of each type here.

One important difference in operating characteristics between an alternating-current and a direct-current motor should be discussed. An ordinary alternating-current motor is designed to run at a specific speed, usually between 1,725 and 1,750 rpm, even with considerable fluctuations of voltage and load. Regardless of voltage, the motor will not run faster than its design speed. Its speed cannot be controlled either by a rheostat or by any other device.

A direct-current motor on the other hand, even if designed for 1,800 rpm, will run at that speed only with a particular combination of voltage and load. If the voltage changes, the speed changes; if the load changes, the speed changes. Excess voltage will make it run at more than 1,800 rpm. However, its speed can easily be controlled by a rheostat, although once adjusted, it will again change with any change in voltage or load.

Universal Motors. This type of motor operates on either direct current or single-phase alternating current of 60 cycles or any other frequency. However, it does not run at a constant speed but varies over an extremely wide range. Idling, a universal motor may run as fast as 15,000 rpm, while under a heavy load the same motor may slow down to 500 rpm. This, of course, makes the motor totally unsuitable for general-purpose work. It is used only when built into a piece of machinery where the load is constant and definitely predetermined. For example, you will find this motor on your vacuum cleaner and your sewing machine, on some types of fans, on electric drills, etc.

Reversing Motors. The direction of rotation of a repulsion-induction motor can be changed only by shifting the position of the brushes. On other types of single-phase alternating-current motors it is changed by reversing two of the wires coming from the inside of the motor. If a motor must be reversed often, a special switch can be installed for the purpose.

Problems With Large Motors. The size of farms is constantly increasing, leading to the use of larger sizes of farm machinery, requiring bigger motors: 10 hp, 25 hp, and even larger. In turn this reduces the number of man-hours of labor for a given output, contributing to higher efficiency of labor on the farm, the food factory, that is continuously being demanded to make the farm profitable.

But most farms have only a *single-phase* 3-wire 115/230-volt service. That requires only two high-voltage lines to the farm, and only one transformer. Single-phase motors are not usually available in a size larger than 7½ hp, although a few larger ones are made. But before buying even a 5-hp single-phase motor, check with your power supplier, to see whether the line and the transformer serving your farm are big enough to operate such a motor.

Single-phase motors 5 hp and larger require an unusually high number of amperes *while starting,* and the line and transformer often are too small to start such a motor. If you operate the motor only a comparatively few hours per year, your power supplier will object to installing a heavier line and transformer, just as the farmer would not buy a 10-ton truck to haul 10 tons a few times per year, while using it for much smaller loads most of the time.

In a few localities, at least some of the farms are served by a 3-phase line, requiring three (sometimes four) wires to the farm, and three transformers. If you are fortunate enough to have 3-phase service, your problems are solved. Simply use 3-phase motors, which cost considerably less than single-phase; being simpler in construction, there is rarely a service problem. (Note: If you have 3-phase service there will be available 3-phase power usually at 230 volts, and also the usual 115/230-volt single-phase for lighting, appliances, and other small loads.)

But if you have only the usual single-phase 115/230-volt service, and still need larger motors, what to do? One solution is to use smaller machinery requiring motors not over 3 hp, but that is a step backward, for it increases the cost of labor, the number of expensive man-hours that must be expended in operating the farm. There is another solution. During recent years there have been developed "phase converters" that permit *3-phase* motors to be operated on *single-phase* lines. The phase converter changes the single-phase power into a sort of modified 3-phase power that will operate ordinary 3-phase motors, and at the same time greatly reduces the number of amperes required *while starting.* In other words when operating a 3-phase motor with the help of a phase converter, the same single-phase line and transformer that would barely start a 5-hp single-phase motor, will start a 7½-hp or possibly even 10-

hp 3-phase motor; a line and transformer that would handle a 10-hp single-phase motor (if such a motor could be found) would probably handle a 15-hp or 20-hp 3-phase motor.

Phase converters are not cheap, but their cost is partially offset by the lower cost of 3-phase motors, and in any event they make possible the operation of larger motors than would be possible without the converter.   There are two types of converters: static with no moving parts except relays, and the rotating type.   The static type must be matched in size and type with the one particular motor to be used with it; generally there must be one converter for each motor.

The rotating type of converter looks like a motor, but can't be used as a motor.   Two single-phase wires run into the converter; three 3-phase wires run out of it.   Usually several motors can be used at the same time; the total horsepower of all the motors in operation at the same time can be at least double the horsepower rating of the converter. Thus if you buy a converter rated at 15 hp, you can use any number of 3-phase motors totaling not over 30 to 40 hp, but the largest may not be more than 15 hp, the rating of the converter.   The converter must be started first, then the motors started, the largest first, then the smaller ones.

But some words of caution are in order.   A 3-phase motor of any given horsepower rating will not *start* as heavy a load when operated from a converter, as it will when operated from a true 3-phase line. For that reason it will often be necessary to use a motor one size larger than is necessary for the *running* load.   This will not significantly increase the power required to run the motor, once it is started.   The converter must have a horsepower rating at least as large as that of the largest motor.

The voltage delivered by the converter varies with the load on it.   If no motor is connected to the converter, the 3-phase voltage supplied by it is very considerably higher than the input voltage of 230 volts.   Do not run the converter for significant periods of time, without operating motors at the same time, or it will be damaged by its own high voltage. Do not operate only a small motor from a converter rated at a much higher horsepower, for the high voltage will damage the motor or reduce its life.   To be prudent, the total horsepower of all the motors operating at one time should be at least half of the horsepower rating of the converter.

Last but not least, check with your power supplier; some do not favor or permit converters. If they do permit converters, the line and the transformer serving your farm must be big enough to handle all the motors you propose to use.

Care of Motors.   Motors require very little care. The most important is proper oiling of the bearings. Use a very light machine oil, such as SAE No. 10, and use it sparingly—most motors are oiled too much. Never oil any part of the motor except the bearings; under no circumstances put oil on the brushes, if your motor has brushes.

If your motor has a commutator and brushes, occasionally while the motor is running, hold a very fine piece of sandpaper (never emery) against the commutator to remove the carbon that has worn off the brushes. *Be sure you are standing on something absolutely dry to avoid shock.*

Bearings.   In most commonly used motors, there is a choice of sleeve bearings or ball bearings. If the motor is to be operated with the shaft in the ordinary horizontal position, there is no need for ball bearings. If the motor is to be operated with the shaft in an up-and-down position, ball bearings should be used because the usual sleeve-bearing construction lets the oil run out. Ball bearings are also better able than sleeve bearings to absorb the weight of the rotor. Ball bearings of some types are filled with grease and permanently sealed, thus doing away with the nuisance of greasing. If this type is purchased, be sure the bearings are double-sealed, that is, with a seal on each side of the balls. Some bearings are sealed on only one side, so that the grease can still get out on the other side.

Pulleys and Belts.   Even though the figuring of proper pulley ratios is not a wiring problem, a short discussion of this subject should not be amiss. In such calculations four factors are involved:

Motor pulley diameter.
Machine pulley diameter.
Motor speed.
Machine speed.

If any of the four is unknown, it is a simple matter to figure it from the three known factors, using the formulas below:

$$\textbf{Machine pulley diam} = \frac{\textbf{motor pulley diam} \times \textbf{motor speed}}{\textbf{machine speed}}$$

$$\text{Machine speed} = \frac{\text{motor speed} \times \text{motor pulley diam}}{\text{machine pulley diam}}$$

$$\text{Motor pulley diam} = \frac{\text{machine pulley diam} \times \text{machine speed}}{\text{motor speed}}$$

$$\text{Motor speed} = \frac{\text{machine speed} \times \text{machine pulley diam}}{\text{motor pulley diam}}$$

In making the calculations indicated above, remember that there is always some belt slippage for which allowance must be made. For motor speed use the actual speed that the motor will develop under full load.

The ratio between the diameters of the driving and the driven pulleys should be kept within reasonable limits. If one pulley is a great deal larger than the other, especially if they are close together, the belt will make contact with but a small portion of the total circumference of the smaller pulley and slippage will be increased. The ratio should not exceed 12 to 1 in small motors; for a 1-hp motor, 10 to 1 is usually considered the practical limit, decreasing to 8 to 1 in the case of a 5-hp motor and 5 to 1 in the case of a 25-hp motor.

Especially for small fractional-horsepower motors the use of V belts is very common. They have the advantage of being relatively inexpensive, have little slip even on small pulleys, and carry substantial loads on even the smallest size. As the load increases above $\frac{1}{2}$ hp, it is often the

Diam motor pulley	Diameter of pulley on machine, inches													
	1¼	1½	1¾	2	2¼	2½	3	4	5	6½	8	10	12	15
1¼	1,725	1,435	1,230	1,075	950	850	715	540	430	330	265	215	175	140
1½	2,075	1,725	1,475	1,290	1,140	1,030	850	645	515	395	320	265	215	170
1¾	2,400	2,000	1,725	1,500	1,340	1,200	1,000	750	600	460	375	315	250	200
2	2,775	2,290	1,970	1,725	1,530	1,375	1,145	850	685	530	430	345	285	230
2¼	3,100	2,580	2,200	1,930	1,725	1,550	1,290	965	775	595	485	385	325	255
2½	3,450	2,870	2,460	2,150	1,900	1,725	1,435	1,075	850	660	540	430	355	285
3	4,140	3,450	2,950	2,580	2,290	2,070	1,725	1,290	1,070	800	615	515	430	345
4	5,500	4,575	3,950	3,450	3,060	2,775	2,295	1,725	1,375	1,060	860	700	575	460
5	6,850	5,750	4,920	4,300	3,825	3,450	2,865	2,150	1,725	1,325	1,075	860	715	575
6½	8,950	7,475	6,400	5,600	4,975	4,480	3,730	2,790	2,240	1,725	1,400	1,120	930	745
8	.....	9,200	7,870	6,900	6,125	5,520	4,600	3,450	2,750	2,120	1,725	1,375	1,140	915
10	.....	.....	9,850	8,620	7,670	6,900	5,750	4,300	3,450	2,650	2,150	1,725	1,430	1,140
12	.....	.....	.....	.....	9,200	8,280	6,900	5,160	4,130	3,180	2,580	2,075	1,725	1,375
15	.....	.....	.....	.....	.....	.....	8,635	6,470	5,170	3,970	3,230	2,580	2,150	1,725
18	.....	.....	.....	.....	.....	.....	.....	7,750	6,200	4,770	3,880	3,100	2,580	2,070

custom to use two or more such belts side by side on multiple-groove pulleys. Avoid the use of very small diameter pulleys. Their use leads to excessive belt slippage, short belt life, and loss of power.

A common mistake is to run belts too tight. This only increases the load on the motor, causes excessive bearing wear and short belt life.

The table on page 261 should be used in determining pulley sizes for any given machine. It is based on a motor speed of 1,750 rpm, and some allowance has been made for belt slippage. Figures show speed of machine with each combination of pulley diameters.

Wiring for Motors.   The Code sections that govern the installations of motors are extremely complicated, for they cover all motors from the tiniest to those developing hundreds of horsepower. The wiring of motors *for homes and farms*, however, can be covered by a few simple rules. Three points especially must be observed:

1. Overcurrent protection.
2. Disconnecting switch for motor.
3. Wire sizes.

Overcurrent Protection in Motor Branch Circuit.   Bear in mind the fundamental fact that any motor consumes more amperes while *starting* than while *running*. For example, an ordinary washing-machine motor may consume 25 amp while *starting*, but only 5 amp while *running* and delivering its normal horsepower. It is capable of delivering for short periods considerably more than its rated horsepower, but if it is overloaded to do so, it will consume correspondingly more current. If considerably overloaded, it may draw as much as 10 amp. It will not be harmed if it delivers normal horsepower (consuming 5 amp) all day long, nor will it be harmed if it delivers considerably in excess of its normal horsepower (and consuming 10 amp) for short periods, but it will burn out if it is required to deliver considerable overloads (and consuming 10 amp) for a considerable period of time.

The wire to the motor must then be big enough to carry its *starting* current momentarily, its running current continuously, and its normal running current plus a considerable overload for short periods. It must be protected accordingly. Consider then a washing-machine motor which normally consumes about 5 amp. While starting, it consumes about 25 amp. The ordinary branch circuit is wired with No. 14 wire and is protected with 15-amp fuses. This fuse will frequently blow on 25 amp while the motor is starting. Therefore, for those circuits which serve motors, it is wise to use circuit breakers, or time-delay fuses, of an amperage determined by the size of the wire in the branch circuit. Fuses of this type were shown in Figs. 5-2 and 5-11.   As a matter of fact, the Code in Sec. 430-42(d) requires that when fuses are used to protect a circuit to which a motor is connected, they must be of the time-delay type.

Components of Motor Circuit.   In addition to the overcurrent protection in the branch circuit serving the motor, three additional components are required:

1. *Controller:* to start and stop the motor.

2. *Running-overcurrent protection:* to protect the *motor* against overload.   The fuse or circuit breaker in the branch circuit that serves the motor, protects the circuit, not the motor; its rating depends on the size of wire in the circuit, not the amperage of the motor.

3. *Disconnecting means:* to totally disconnect the motor and its controller from the circuit.   This becomes necessary when working on the motor, or the machinery that it drives, to guard against accidental starting, which would be most dangerous.

Two or all three of these requirements are often combined into a single device.   The Code requirements for these things are quite complicated when considered for the total range of motors from the smallest to the very largest, and Chap. 31 will discuss them.   For ordinary motors as found in homes and on farms, the essential points will be discussed here.

Controller.   This may be an automatic device, part of a refrigerator, water pump, or any other equipment that starts and stops automatically.   In that case you can safely assume that the proper controller comes with the equipment.   The following discussion will be about manually operated controllers.

If the motor is portable and rated ⅓ hp or less, the cord and plug serve as the controller.   If however it is over ⅓ hp, proceed as in the case of stationary motors, as discussed in following paragraphs.

If the motor is 2-hp or less, you have a choice.   You may use a general-use AC-only toggle switch as used to control lights, if the switch has an amperage rating at least 125% of the ampere rating of the motor. Alternately, you may use an enclosed switch of the general type shown in Fig. 15-2.   Such switches are sometimes rated in amperes, sometimes in horsepower, sometimes both.   If rated only in horsepower, it must have a rating not less than the horsepower of the motor. If it is rated in amperes, it must have a rating at least double the ampere rating of the motor.   Such switches need not be fused, but as a practical matter unfused switches are hard to find.   Use the fused variety, with one fuse for a 115-volt motor, two fuses for a 230-volt motor; the fuses, if of the proper size, will then serve as the motor-running overcurrent protection which will be discussed later.

Fig. 15-2 A switch of this type may be used with small motors.   If it has two fuses, it is for a 230-volt motor; if it has only one fuse, it is for a 115-volt   motor. (*Square D Co.*)

If the motor is larger than 2 hp the switch you use must be rated in horsepower, not less than the horsepower of the motor.

Fig. 15-3   Controls of this type not only start and stop a motor, but protect it against overload.   (*Square D Co.*)

Regardless of the size of the motor, instead of using switches as described, special motor starters of the type shown in Fig. 15-3 are generally used. These are rated in horsepower; use one rated at not less than the horsepower of the motor. They have built-in motor-running protection. For smaller (fractional-horsepower) motors use the kind shown in the left-hand part of the illustration; it isn't much larger than an ordinary toggle switch, and controls the motor just as an ordinary switch would. For larger motors, use the kind in the right-hand part of the illustration. This has push buttons in the cover for starting and stopping the motor. It is also available without the push buttons, which are then installed separately at a distance in some convenient location. Chapter 31 will discuss such starters, and their method of operation, in more detail.

An ordinary circuit breaker may also be used as a controller, but its amperage rating must be carefully selected to meet the requirements of motor-running overcurrent protection, as will be discussed later.

One important point: If the motor is located more than 50 ft from the controller, or is located where the controller can't be seen from the motor (regardless of distance), the controller must be so designed that it can be locked in the open or "off" position.

Motor-running Overcurrent Protection. Your branch-circuit fuses or circuit breakers, if of an amperage rating sufficient to carry the *starting* current of the motor, rarely will protect the motor against damage from continued overloads. Nevertheless, the Code does permit 115-volt *portable* motors, manually controlled, and rated at 1 hp or less, to be plugged into any ordinary 15- or 20-amp circuit.

Often, and especially on smaller horsepower motors, the motor-running protection is built into the motor, then called a "thermal protector integral with the motor" in the Code. Such protectors are called by various names such as Thermoguard, Thermotron, Klixon, and so on. They are thermostatic contrivances that open the circuit when the motor temperature exceeds a safe value. They sense the heat from the amperes flowing through the motor, as well as the actual temperature of the motor. In just about every case they must be reset by hand when they trip. Motors equipped with such protection must be marked "Thermally Protected" on the name-plate.

If the motor is permanently installed and not provided with built-in protection, separate motor-running protection must be provided. For motors of the kind ordinarily used in homes and on farms, it may be rated at not more than 125% of the amperage of the motor. If you use a switch of the general type shown in Fig. 15-2, be sure to use time-delay fuses or they will blow during the starting period. The switch then serves as the controller, the motor-running overcurrent protection, and also the disconnecting means which will be discussed a bit later.

You may also use a circuit breaker (not the one protecting the branch circuit, but a separate one) rated at not over 125% of the motor amperes, in which case it will also serve all three purposes.

More likely you will use a starter of the general type shown in Fig. 15-3, to serve as your controller. Such starters have inside them special "heater coils" which must be selected to match the amperage of the motor. If properly selected, they will carry the running current of the motor indefinitely, will carry a small overload for some time, but will trip the starter and shut off the motor quickly in case of heavy overload, or if the motor fails to start. Such a starter will then serve the combined purpose of controller and motor-running overcurrent protection, but *not* as the disconnecting means. For a more detailed discussion of how such motor starters operate, see Chap. 31.

Disconnecting Means. If your motor is portable, then regardless of size, the plug and receptacle will serve as the disconnecting means. The following paragraphs will concern permanently installed motors.

If the motor is ⅛ hp or less, no separate disconnecting means is required; the branch-circuit fuses or circuit breaker will serve the purpose. On larger motors, if you have used a switch of the general type shown in Fig. 15-2 (or a larger one with cartridge fuses for a larger motor) as the controller, it will also serve as the disconnecting means.

If your motor is protected by built-in thermal protection, or if you have used a motor starter of the general type of Fig. 15-3, you must provide separate disconnecting means. This can be a switch of the type shown in Fig. 15-2, rated in horsepower not less than the horsepower of the motor. Such switches need not be fused, but since unfused switches are hard to find, use the fused variety. Use any kind of fuse that will not blow while starting; the fuses are not required, and are used only because you were not able to locate an unfused switch.

If your motor is 2 hp or less, you may also use a switch of the type of Fig. 15-2 rated in amperes (not horsepower) at least double that of the motor amperes; you may also use a general-use AC-only toggle switch of the kind you use to control lights, provided it is rated at least 125% of the motor amperes. In any case, use a single-pole switch for a 115-volt motor, double-pole for a 230-volt motor.

Other Considerations. This chapter concerns only single-phase motors. For three-phase motors and a more complete discussion of motors in general, see Chap. 31.

Wire Sizes. First of all, extension cords made of ordinary No. 18 or 16 lamp cord should never be used, even on small fractional-horsepower motors. A short cord on the motor is in order, but if a longer extension is added, the voltage drop in the cord during the starting period while the amperage is high is apt to be so great that the motor never gets off its starting windings. A damaged motor may easily result. The wire must be heavy enough to carry the starting amperage, and the horsepower of the motor and the distance involved must also be taken into

Motor			Wire sizes								
Horse-power	Volts	Am-peres	14	12	10	8	6	4	2	1/0	2/0
1/4	115	5.8	55	90	140	225	360	575	900	1,500	1,800
1/3	115	7.2	45	75	115	180	300	450	725	1,200	1,500
1/2	115	9.8	35	55	85	140	220	350	550	850	1,100
3/4	115	13.8	25	40	60	100	150	250	400	600	800
1	115	16.0	—	35	50	85	130	200	325	525	650
1 1/2	115	20.0	—	25	40	65	100	170	275	425	550
2	115	24.0	—	—	35	55	85	140	225	350	450
3	115	34.0	—	—	—	40	60	90	160	250	325
1/4	230	2.9	220	360	560	900	1,450	2,300	3,600		
1/3	230	3.6	180	300	460	720	1,200	1,600	2,900		
1/2	230	4.9	140	220	340	560	875	1,400	2,200		
3/4	230	6.9	100	160	240	400	600	1,000	1,600	2,400	
1	230	8.0	85	140	200	340	525	800	1,300	2,100	
1 1/2	230	10.0	70	110	160	280	400	675	1,100	1,700	2,200
2	230	12.0	60	90	140	230	350	550	900	1,400	1,800
3	230	17.0	—	65	100	160	250	400	650	1,000	1,300
5	230	28.0	—	—	60	100	160	250	400	650	800
7 1/2	230	40.0	—	—	—	70	110	175	275	450	550
10	230	50.0	—	—	—	—	90	140	225	350	450

Figures below the wire sizes indicate the *one-way* distance in feet (not the number of feet of wire in the circuit) that each size wire will carry the amperage for the size motor indicated in the left-hand column, with 1½% voltage drop. A dash indicates that the wire size in question is smaller than the minimum required by the Code for the horsepower involved, regardless of circumstances. Figures are based on single-phase AC motors.

consideration. The Code requires that the wire must have ampacity equivalent to at least 125% of the name-plate amperage of the motor.

For convenience the table on page 267 has been worked out. Under each size of wire is given the maximum distance for which that size of wire should be used if maximum efficiency of the motor is expected. This table is calculated for *single-phase* AC motors. The table is not applicable to 3-phase motors.

The table is based on 1½% voltage drop, and assumes that the motor is not overloaded. Most motors are overloaded to some degree during some portion of their normal running cycle, so that the drop will probably be more nearly 2% during part of the cycle. Moreover, all motors consume many more amperes while starting than they do while running at full speed. During the starting cycle the drop may be as much as 6 or 8%. Too much drop during the starting period prolongs the starting period, leading to excessive heating. If the motor must start exceptionally hard-starting loads, too much drop may prevent the motor from reaching full normal speed, which could lead to motor burnouts.

If in doubt as to the right wire size, use the next larger size.

Actual Wiring: Residential and Farm

Part 2 of this book explains the actual wiring of houses and farm buildings of every description. The author feels that it will be much easier for you to study first these relatively simple installations than it would be if one chapter pertaining to one particular phase of the work included everything from a simple cottage up to an elaborate project.

All the fundamentals covered by Part 2 will in practice also be used in the bigger projects. They are the foundation for the more complicated methods to be covered by Part 3, which covers the wiring of larger industrial and similar projects.

16

Planning an Installation

The plans for an electrical installation usually consist of outline drawings of the rooms involved, with indications where the various outlets for fixtures, receptacles, and other devices are to be located. Obviously a *picture* of a switch, a receptacle, or other device cannot be shown on the plans at each point where one is to be used. Standardized symbols are used instead.

Symbols. In order to cover the wiring of all sorts of buildings from the smallest to the very largest, a great many symbols are required. The more ordinary ones, likely to be encountered in the kinds of wiring discussed in this book, are shown in Figs. 16-1 and 16-2. Note that *round* symbols designate outlets served by full voltage (115 volts or higher) and *square* symbols designate low-voltage outlets. Study these symbols until each one means as much to you as a picture would.

Sometimes a symbol is supplemented by additional letters near the symbol to more fully define the outlet, as follows:

WP	Weather proof	DT	Dust tight
VT	Vapor tight	EP	Explosion proof
WT	Water tight	G	Grounded
RT	Rain tight	R	Recessed

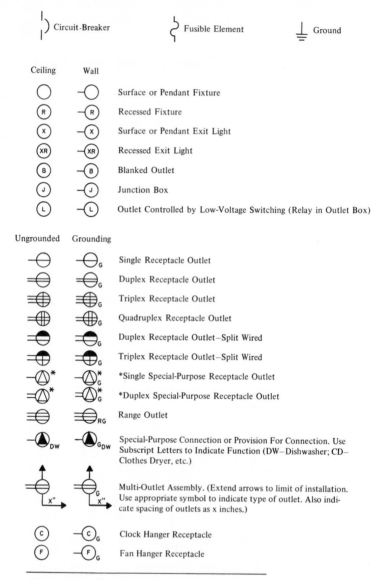

* Use numeral or letter either within the symbol or as a subscript alongside the symbol keyed to explanation in the drawing list of symbols to indicate type of receptacle or usage.

Fig. 16-1 Symbols as used in architectural drawings, to designate electrical outlets.

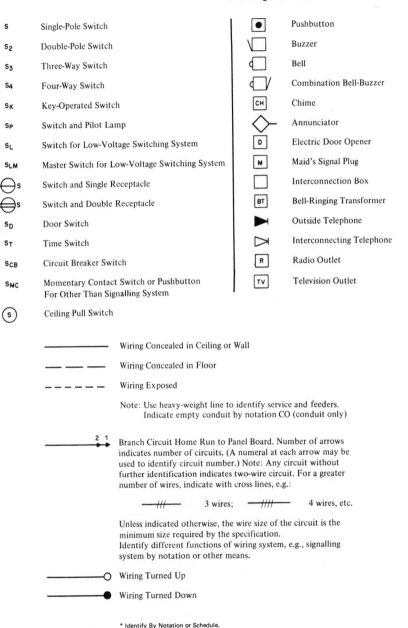

S	Single-Pole Switch	●	Pushbutton
S₂	Double-Pole Switch		Buzzer
S₃	Three-Way Switch		Bell
S₄	Four-Way Switch		Combination Bell-Buzzer
Sₖ	Key-Operated Switch	CH	Chime
Sₚ	Switch and Pilot Lamp		Annunciator
Sₗ	Switch for Low-Voltage Switching System	D	Electric Door Opener
Sₗₘ	Master Switch for Low-Voltage Switching System	M	Maid's Signal Plug
S	Switch and Single Receptacle		Interconnection Box
S	Switch and Double Receptacle	BT	Bell-Ringing Transformer
Sᴅ	Door Switch		Outside Telephone
Sₜ	Time Switch		Interconnecting Telephone
Sᴄʙ	Circuit Breaker Switch	R	Radio Outlet
Sᴍᴄ	Momentary Contact Switch or Pushbutton For Other Than Signalling System	TV	Television Outlet
Ⓢ	Ceiling Pull Switch		

─────────── Wiring Concealed in Ceiling or Wall

── ── ── Wiring Concealed in Floor

─ ─ ─ ─ ─ Wiring Exposed

Note: Use heavy-weight line to identify service and feeders.
Indicate empty conduit by notation CO (conduit only)

Branch Circuit Home Run to Panel Board. Number of arrows indicates number of circuits. (A numeral at each arrow may be used to identify circuit number.) Note: Any circuit without further identification indicates two-wire circuit. For a greater number of wires, indicate with cross lines, e.g.:

──/// ── 3 wires; ──/// ── 4 wires, etc.

Unless indicated otherwise, the wire size of the circuit is the minimum size required by the specification.
Identify different functions of wiring system, e.g., signalling system by notation or other means.

──────O Wiring Turned Up

──────● Wiring Turned Down

* Identify By Notation or Schedule.

Fig. 16-2 More symbols. All are from American National Standards Institute "Standard Y32.9."

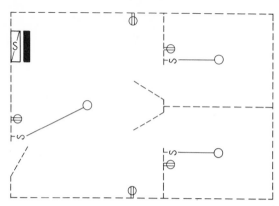

Fig. 16-3 Layout of a simple 3-room project.

Typical Plans. Consider first a very simple plan, covering a small three-room cottage with two circuits, involving one ceiling outlet controlled by a wall switch for each of the three rooms, with three receptacle outlets for the larger room and one for each of the smaller rooms. The service entrance is 2-wire 115-volt only. The plan for this installation is shown in Fig. 16-3. Note that this does not provide adequate wiring, nor does it meet all Code requirements; it is shown merely as an exercise in solving problems.

To make it easier to interpret this plan, Fig. 16-4 shows the same lay-

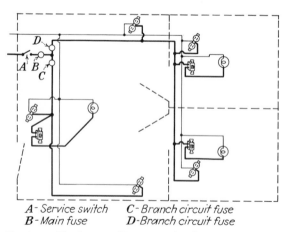

A - Service switch   C - Branch circuit fuse
B - Main fuse   D - Branch circuit fuse

Fig. 16-4 The layout of Fig. 16-3, but here shown in pictorial fashion.

out in pictorial fashion, with all the wires shown in detail.   The neutral wire is shown as a light line; the "hot" wires as heavy lines.   Note how the neutral wire runs without interruption from the point where it enters the building to each device where current is to be used.   The black wires run from their fuses direct to each receptacle outlet and to each switch; an additional length runs from each switch to the light it controls, and that completes the wiring.

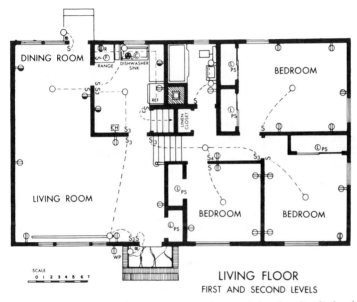

Fig. 16-5 Wiring diagram for first and second levels of split-level house. (*Industry Committee on Interior Wiring Design.*)

A represents the main switch. *B* represents the main fuse. *C* and *D* represent the two fuses, one for each branch circuit. The first branch circuit comprises all the wiring served by the current that flows through fuse *C*; the second circuit comprises all wiring served by the current that flows through fuse *D*.

The wiring plans for a larger house are shown in Figs. 16-5 and 16-6. These diagrams[1] may at first sight seem rather formidable, but with study they become simple. Such plans are supplemented by detailed written specifications which give such information as size and type of

[1] Reproduced by permission from the "Handbook of Interior Wiring Design," by the Industry Committee on Interior Wiring Design.

service entrance, number of circuits, type of materials to be used, and similar data.

Make some plans of a similar nature of other installations, for example, your own home as it is wired and as you would like to see it wired. Do this until symbols are as clear to you as the printed words of a book.   Remember that *round* symbols always denote outlets served by the full voltage of the wiring system; *square* symbols always denote outlets operating at low voltage, for example, bells, buzzers, etc.   Solid lines denote wires in ceilings or walls; dotted lines denote wires under the floor.

Fig. 16-6   Wiring diagram for basement of house shown in Fig. 16-5.   (*Industry Committee on Interior Wiring Design.*)

Making Plans.   Often the electrician may be called upon to make the plans for a job instead of finding them ready-made.   In that case include all those details found in plans of the type shown in Figs. 16-5 and 16-6.   Likewise include in the specifications such things as the size of service-entrance wires and switch, the number of circuits, the location of circuit breakers or service switch and fuse panels, material to be used, and similar details.

In making such plans there is usually little choice except to follow the general ideas of the owner as to the number of outlets and similar details.   On the other hand, the average homeowner knows very little about things electrical, with the result that his specifications may result in an installation that is far from adequate.   Explain to your customer what advantages there are for him in an adequate installation, and it will mean a larger sale for you and a better satisfied customer.

17

Installation of Service Entrance and Ground

Chapter 13 covered the selection of the proper service-entrance wires, the rating of the service switch, and similar essentials.   This chapter will cover the actual installation of the materials selected.   Many variations are possible in the selection of mechanical arrangement and the different parts; the service-entrance wires may come in through conduit, in the form of service-entrance cable, or underground; the meter may be outdoors or indoors; the branch-circuit overcurrent equipment may be circuit breakers or fuses; main overcurrent protection may or may not be required.

We shall now have to assume that you have already decided the question of 2- or 3-wire, the amperage capacity of the service and the size of the wire you are going to use, the number of branch circuits, whether circuit breakers or fused equipment are to be used, and similar details. We shall discuss first factors that pertain to all services.   Then we shall discuss the entrance using conduit, then using service-entrance cable, and lastly using underground cable.   Finally we shall discuss the ground.

Solderless Connectors. The Code in Sec. 230-72 prohibits the use of soldered joints in connecting wires to service-entrance switches or circuit breakers. The reason for this is not hard to understand. It is not particularly hard to learn how to make good soldered joints when using small wires; it is, however, quite an art to solder a joint when a heavy wire and a tubular solder lug are involved. In the smaller sizes of wire, solder joints are frequently made, giving the workman much opportunity for practice; in the heavier sizes they are infrequently made, affording little opportunity for practice. In connecting wires to service equipment, use only joints made with solderless or "pressure" connectors or similar clamp-style terminals using no solder. The same requirement holds with respect to connections involving the ground wire.

Uninsulated Wire in Service Entrance. In the wiring of houses and farms, the neutral wire is grounded. As was pointed out in an earlier chapter, this means that whenever you touch a water pipe or radiator, you touch the neutral wire. That being the case, why insulate the neutral? The Code in Sec. 230-40 does permit uninsulated or bare wire for the grounded neutral wire of the service entrance, but only if the voltage *to ground*[1] is not over 300 volts, which condition is met with a 2-wire 115-volt, or 3-wire 115/230-volt single-phase, or a 4-wire 120/208- or 277/480-volt 3-phase service. Insulated wires are entirely acceptable, but you may use bare wire. In case of a service using conduit, you may use a bare, uninsulated neutral wire. In case of a service using service-entrance cable, the neutral wire will be automatically bare.

Size of Neutral Wire. In a 3-wire 115/230-volt installation, the 230-volt loads are connected only to the two hot wires and impose no load whatever on the neutral wire. Such loads will operate just as well whether there is a neutral wire or not. Therefore in a 3-wire installation, when both 115- and 230-volt loads are installed, the neutral of the service entrance carries a smaller amperage than the two hot wires. As a matter of fact, even if only 115-volt circuits are installed, properly balanced between the two legs of the service, the amperes carried by the neutral will be less than the number of amperes in the hot wires (this will be discussed in detail in Chap. 20 in connection with Figs. 20-31 to 20-35).

[1] See Chap. 9 for definition of "voltage to ground."

That being the case, the neutral wire in the service may be smaller than the hot wires. The rules for figuring the neutral are a bit complicated [Secs. 220-4(d) and 230-41 Exc. 5 of the Code] but as a practical matter the neutral wire may be one size smaller than the hot wires, provided those are No. 6 or larger. In other words, you may use:

Two No. 6 with one No. 8
Two No. 4 with one No. 6
Two No. 2 with one No. 4
Two No. 1/0 with one No. 2
Two No. 2/0 with one No. 1
Two No. 3/0 with one No. 1/0
Two No. 4/0 with one No. 2/0

Meter. The power supplier decides whether the meter is to be located indoors or outdoors. In most cases it will be the outdoor type shown in Fig. 17-1 with its socket. Both are installed exposed to the weather. The power supplier furnishes the meter. The socket is sometimes furnished by the power supplier, sometimes by the owner; in any event it is installed by the contractor.

If the meter is to be the indoor type, provide a board, preferably plywood, big enough so that the meter plus the circuit breakers or fused equipment can be mounted on it.

Service-drop Wires. The Code defines service-drop wires as "the

Fig. 17-1 A detachable outdoor meter and the socket on which it is installed. Meters of this type are installed exposed to the weather. (*General Electric Co.*)

overhead service conductors between the last pole or other aerial support, to and including the splicer, if any, to the service entrance conductors at the building or other structure." If the wires are not overhead, they are not classed as service-drop wires but become service-entrance wires, covered by the next paragraph. In general, service-drop wires are supplied by and installed by the power supplier, although the owner or contractor furnishes the insulators by which the wires are supported at the building. The power supplier also determines the size of service-*drop* wires, which often are smaller than the service-*entrance* wires. If this should lead to what you consider too much voltage drop, remember that drop is ahead of the meter, so that the owner of the building does not pay for wasted power.

Service-entrance Wires. The wires from the point where the wires supplied by the power supplier end, up to the service switch, are service-entrance wires. They may be Type T or Type R wires brought in through conduit, or wires made up into special cables designed for the purpose. Often service-entrance wires are run underground. The kind of wire to use for that purpose will be discussed later in this chapter. Service-entrance wires, even if in conduit, should not be run inside the hollow spaces of walls in frame buildings. The Code does not prohibit it, but in Sec. 230-44(b) specifies that it "should" not be done, unless the wires are protected by overcurrent protection at the starting point.

Openings into Buildings. If the building is of frame construction, it is a simple matter to make an opening through the wall. If the building is of brick or concrete construction, hard labor will be involved. For an occasional job, use a large star drill of the type that will be described in Chap. 22. If there are to be many openings, use an electric drill and carbide-tipped drills (or hole saws) that will reasonably easily go through concrete.

Entrance Using Conduit. A typical installation is shown in Fig. 17-2. The size of the conduit and fittings will depend on the size of the entrance wires and has already been discussed in Chap. 11. Only galvanized conduit may be used.

Service Insulators. Insulators for supporting the power supplier's wires where they reach the building (point A in Fig. 17-2) must be provided. These may be simple screw-point insulators, shown in Fig. 17-3; according to Code they must be kept a minimum of 6 in. apart. More often, however, the type shown in Fig. 17-4 is used; choose one with individual insulators 8 in.[2] apart.   These are known as service brackets or secondary racks.

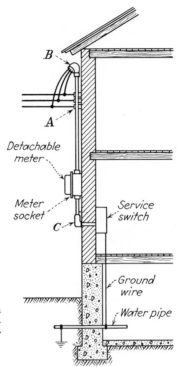

Fig. 17-2 Cross section of a typical service entrance. Locate the service head *B higher* than the insulators.

These insulators must be installed almost as high above the ground as the shape and structure of the building will permit.   Note that point *A* in Fig. 17-2 (the location of the insulators) must per Code Sec. 230-51 be lower than point *B* (the upper end of the service conduit).   A difference of a foot or so is sufficient.   If however because of the structure of the building this can't be done, the same section permits locating the insulators and the service head at approximately the same level, but not more than 24 in. apart.   In that case drip loops as shown in Fig. 17-5 must be installed, and the splice between the service-drop and the service-entrance wires must be at the lowest part of the loop.   All this

[2] In some localities a separation of 6 in. is permitted.

is required to minimize the danger of water following the wires into the conduit.

Fig. 17-3 Screwpoint insulators of this type are used to support outdoor wires.

Fig. 17-4 A bracket with two or three insulators may be used instead of separate insulators.

In Sec. 230-26 the Code defines the clearances that service-drop wires must maintain above ground, as follows:

10 ft—above finished grade, sidewalks, or from any platform or projection from which they might be reached.

12 ft—above residential driveways and commercial areas such as parking lots and drive-in establishments not subject to truck traffic.

15 ft—above commercial areas, parking lots, agricultural or other areas subject to truck traffic.

18 ft—above public streets, alleys, roads and driveways on other than residential properties.

In Sec. 230-24(c) the Code further provides that the wires must be kept 36 in. from windows (unless installed *above* windows), doors, porches, fire escapes, and similar locations from which the wires could be touched. If they pass over a roof, they must have a clearance of at least 8 ft above the highest point of the roof; however, if the roof has a slope of at least 4 in. per foot, the required clearance is reduced to 3 ft.

In wiring the rambler or ranch-house type of residence, it is impossible to maintain these minimum clearances.   Use a mast which will be described later in this chapter.

In installing wires on insulators, install the topmost wire first, then the lower ones, taking care that you have the same tension on each; then they will all run equally spaced over the entire span.   Be sure each wire is securely anchored on its own insulator.   See Fig. 17-6 for the general method.

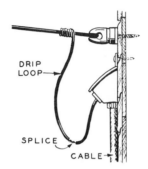

Fig. 17-5   If it is impossible to locate the service head higher than the insulators, install a drip loop in each wire.

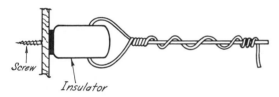

Fig. 17-6   How to install wire on insulator.

Service Head.   At the top end of the service conduit (*B* in Fig. 17-2) the Code requires a fitting that will prevent rain from entering the conduit.   A fitting of this type is shown in Fig. 17-7; it goes by various names such as service head, entrance cap, and weather head.   It consists of three parts: the body, which is attached to the service conduit; an insulating block to separate the wires where they emerge; and the cover, which keeps out the rain and holds the parts together.

Entrance Ell.   At the point where the conduit enters the building (*C* in Fig. 17-2) it is customary to use an entrance ell of the type shown in Fig. 17-8.   With the cover removed, it is a simple matter to pull wires around the right-angle corner.   This device also must be raintight.

Service Conduit.   We shall have to assume that you have already installed the service insulators, that you have cut an opening into the building for the entrance conduit to enter, and have marked the location of the meter, about 5 to 6 ft above the ground.   Cut a piece of conduit

Fig. 17-7   A service head is installed at the top of the conduit through which wires enter the building.   (*Killark Electric Mfg. Co.*)

Fig. 17-8   An entrance ell is used at the bottom of the conduit, where it enters the building.   (*Killark Electric Mfg. Co.*)

to reach from the meter location to a point a foot or two *above* the insulators.   Cut a piece to reach from the service equipment inside the house to a point just outside the wall, where you will install the entrance ell (point *C* of Fig. 17-2).   Cut a third piece long enough to reach from this ell to the meter socket.

Ream the cut ends carefully and thread them, as outlined in Chap. 11.   Assemble the whole "stack" on the ground: service head at top, meter socket in the middle, ell at the bottom, then the short end to run into the house.   Install the wires as outlined in the next paragraph. Then install the whole assembly on the side of the house: Push the short stub of conduit through the opening in the wall, and anchor the conduit to the side of the building, using pipe straps such as shown in Fig. 11-8, or similar.

Inside the building, the conduit must be anchored to the service switch, using the locknut and bushing procedure shown in Fig. 10-6, but instead of using an ordinary bushing, use a "grounding bushing," which will be described later in this chapter.

Pulling Wires.   The conduit will contain three wires.   The neutral if insulated must be white.   The other wires should be black and red if the size of the wire you are using is available in red (heavier sizes come only in black).   Cut three lengths to reach from the meter socket to the top, plus about 3 ft extra to stick out of the top.   Cut three additional lengths to reach from the socket to the service equipment inside the house; don't forget to allow length to reach the neutral strap inside this equipment.

Pull the wires into the preassembled stack while it is still on the ground.   First remove the covers from the entrance ell and the entrance cap.   For occasional jobs you can push the wires from the top to the meter socket, and the shorter pieces from the socket to the end of the run inside the house.   At the ell there is a sharp bend, but you can manipulate the wires around the corner with a little effort.   For easier pulling, first pull a fish wire into the conduit, anchor the three wires to the conduit, and pull all into place.

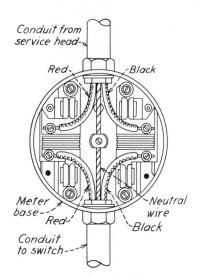

Fig. 17-9 How wires are connected to the meter base, or socket.   The neutral wire is always connected to the center terminal of the socket.

Connect the ends of the wires to the proper terminals in the meter socket as shown in Fig. 17-9.   At the top end replace the cover on the entrance cap, letting each wire come out through a separate hole in its insulating block.   At the bottom replace the cover on the entrance ell. Then install the preassembled stack on the side of the building.

Thin-wall Conduit.   This material may be used instead of rigid conduit.   The procedure is as outlined except that connectors are used at all ends of conduit, instead of threads.   Entrance caps and ells of a slightly different type are used to clamp to the conduit.   Caps and ells designed for rigid, threaded conduit may be used employing adapters that were described in connection with Fig. 11-13.

Service-entrance Cable.   Instead of separate wires inside conduit, special service-entrance cable is very commonly used.   The most usual type is shown in Fig. 17-10 and is known as Underwriters' Type SE,

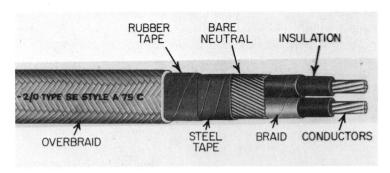

Fig. 17-10   Service-entrance cable, Type SE, Style A.   The neutral consists of a number of bare wires wrapped concentrically around the insulated wires.   Over all the wires comes a protective steel tape.   (*General Cable Co.*)

Style A.   It consists basically of two insulated wires, black and red. The bare, uninsulated wire consists of many strands of tinned copper wire spiraled around the insulated wires.   Next comes a wrapping of galvanized steel tape for mechanical protection (the "A" of "Style A" stands for "armored") followed by a rubber tape for protection against moisture, with a final fabric jacket, usually painted gray.

All the fine wires in the neutral are collectively equal to the same size as the insulated wires; alternately, one size smaller as permitted in services.   Both types are popular.

If the steel tape is left out of the assembly, as shown in Fig. 17-11, the cable becomes Type SE, Style U (the "U" stands for "unarmored"). Whether to use Style A or Style U depends on local custom; check with your power supplier or the local inspector.

The wires used in service-entrance cable are not ordinary Type R, but are usually Type RHW.   Therefore in the larger sizes, their ampacity is a little higher than for ordinary wire.   See Code Table 310-12 and use the 75° column to determine the ampacity of each size.

Fig. 17-11   Service-entrance cable, Type SE, Style U.   This cable is the same as that shown in Fig. 17-10 except that the steel tape is omitted. (*General Cable Co.*)

Regardless of the type you use, all the fine wires of the neutral must be bunched together, twisted, to become a single neutral wire, all as shown in Fig. 17-12.   It is then handled just as if it had been an ordinary stranded wire throughout the cable.

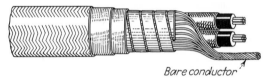

Bare conductor

Fig. 17-12   The concentric conductor is bunched together where it is connected to a terminal.

Fittings for Entrance Cable.   The service head for cable is slightly different from the type used for conduit, in that it is anchored to the building, instead of being supported by the conduit; see Fig. 17-13.   The

Fig. 17-13   A typical service head for service-entrance cable.   (*Killark Electric Mfg. Co.*)

cable itself is secured to the building using one of several types of straps that were shown in Fig. 11-8.

Cable is anchored to the meter socket by means of waterproof connectors, shown in Fig. 17-14. These connectors incorporate soft rubber glands; as the locking nut or the locking screws are tightened, the rubber is compressed, making a watertight seal around the cable. In

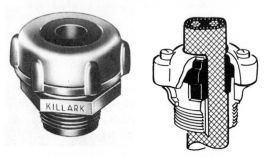

Fig. 17-14 A watertight connector for service-entrance cable. The rubber gland in the connector squeezes against the cable, making the installation watertight. (*Killark Electric Mfg. Co.*)

use, the connector is screwed into the threaded opening of the meter socket, the thread being treated with waterproofing compound. Next the cable is slipped through the rubber gland and the locking screws or nut taken up, making a complete watertight connection. Inside of buildings watertight connectors are not required; armored-cable connectors of appropriate size are used instead.

Fig. 17-15 Install a sill plate where the cable enters the building. Soft rubber compound seals the openings, to keep out water. (*Killark Electric Mfg. Co.*)

At the point where the cable enters the building, a sill plate is used to prevent the rain from following the cable into the inside of the building. One type is shown in Fig. 17-15; usually a soft rubber compound is supplied with it to seal any opening that might exist.

Service-entrance cable in most brands comes with a grayish paint finish, which permits painting to match the building on which it is installed.

Service Wires on Side of Building.   In residential work it is usually possible to plan the installation so that the service conduit or cable drops straight down from the insulators to the point where it enters the building.   But if that can't be done, make a special effort to make the installation neat.   Conduit or cable running at an angle is most unsightly.   Try to run the conduit or cable directly down from the insulators, and then at a point level with the point of entry, let it run horizontally to the opening.

Underground Services.   When the service wires run underground, the wires must be protected against moisture and against mechanical damage.   At one time the common method was to use lead-sheathed cable.   This consists of two or three Type T or Type R wires encased in a continuous sheath of lead.   The Code requires that it be enclosed in conduit for mechanical protection.   The lead is soft and easily damaged, with the result that when it is pulled into conduit, the sheath is often damaged, even punctured, defeating the purpose of the lead. Moreover, it is a costly type of installation because the cable itself is expensive, it must run inside conduit, and installation labor is high. For these reasons, this method is rarely used today.

Several styles of cable are available that can be buried directly in the ground, no conduit being required.   The carrying capacity of buried wires is the same as for wires in conduit.   See Code Table 310-12 in the Appendix.

Type USE Cable.   This material has a basic layer of insulation that is especially moisture-resistant, and over that another layer of insulation that is mechanically very tough and sturdy, besides being water-resistant.   This cable is shown in Fig. 17-16.   The letters "USE"

Fig. 17-16 Type USE cable.   Bury it directly in the ground.   (*General Cable Co.*)

stand for "underground service entrance."   The conductors are Type RHW insulation, so use the 75° column of Table 310-12 to determine their ampacity.

Type UF Cable.   This cable is shown in Fig. 17-17.   It resembles

Type USE except that the insulated conductors are embedded in a plastic compound that is very moisture-resistant, and also tough mechanically. It may be buried directly in the ground but *must be protected by fuses or circuit breakers* at the starting point; the letters "UF" stand

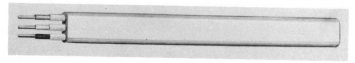

Fig. 17-17   Type UF is used like Type USE, but must be protected by fuses or circuit breaker at the starting point.

for "underground fused." Consequently it may *not* ordinarily be used for services, except on farms *provided* there is overcurrent protection at the starting point: the meter pole. However, it is useful for runs from one building to another as from house to garage, or between farm buildings. The insulation of the individual conductors is Type TW, and the ampacity is the same as for ordinary Type T.

This material is almost identical with nonmetallic-sheathed cable Type NMC, discussed in Chap. 6. If the cable is marked "Type NMC" it may not be buried in the ground. If it is marked "Type UF" it may be buried directly in the ground, and may also be used for general wiring, even in wet locations [Code Sec. 339-3(d)]. A continuous length may therefore be installed underground and then into a building, and continued from there.

Installing Underground Cables. Bury these cables directly in the ground. No splices are permitted—use continuous lengths. The Code in Sec. 339-3(c) requires a minimum depth of 18 in., which may be reduced to 12 in. "provided supplemental protective covering such as a 2-in. concrete pad, metal raceway, pipe or other suitable protection is used." Especially on farms, common sense suggests a depth greater than 18 in. if there is any likelihood of future disturbance by digging, or if located where there might be deep ruts in roads or driveways. It is wise to place a board or similar obstruction over the cable, to serve as warning should future digging become necessary in the area of the cable.

If you use single-conductor cables, bunch them instead of having them spread far apart; at the same time, do not let one cross another. Crossing wires will permit pressure in the insulation between two wires, which in the course of time might lead to weakening of the insulation. Do not pull the cables tight from one end to the other; "snake" them a bit to permit expansion and contraction under the influence of temperature and frost.   It is best to have a layer of sand above and under the cables, to guard against the danger of sharp rocks or similar objects bearing against the insulation, after filling the trench.

Underground conductors usually begin at the power supplier's pole. The Code in Sec. 230-32(b) requires that the cable be given good mechanical protection at all points above the ground.   Run a length of conduit starting a minimum of 8 ft above the ground.   At the top provide a service head.   At the bottom install a bushing at the end of the conduit, which should end about halfway between the bottom of the trench and the ground level, and should point straight down rather than in a horizontal sweep.   Then be sure to provide an S loop in the cable. This will tend to prevent damage to the cable (especially in northern climates) due to movement of the earth in changes from winter to summer.   All this is shown in Fig. 17-18.

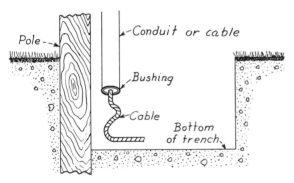

Fig. 17-18 Provide an "S" loop where an underground run ends, to protect against frost damage.

At the house, if the meter is the outdoor type, follow the same procedure, using a piece of conduit from the meter socket down into the trench.   But if the cable is to run through the foundation to an indoor meter, follow the procedure shown in Fig. 17-19.   Provide a length of conduit as shown, and after the cable has been installed, fill the openings around the conduit and inside the conduit with asphalt compound to prevent rain water or melted snow from following the cable and running into the building.

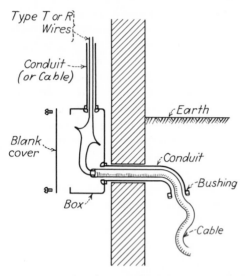

Fig. 17-19   If underground conductors cannot be run directly to the entrance switch, use the construction shown in changing over from underground cable to the usual interior wiring.

Bare Underground Wires.   If the underground entrance is made using conduit, the neutral wire may be bare, provided it is copper, not aluminum.   If you are using multiconductor cable, it may be bare.   If you are using individual single-conductor cables, the neutral must be insulated just like the hot wires.

Service Switch or Circuit Breakers.   The service wires end at the service-switch or circuit breaker cabinet.   From this cabinet you must run the ground wire, also all the wiring for the individual branch circuits.

The switch or breaker cabinet must be located as close as possible to the point where the wires enter the building.   It may be mounted directly on the wall or preferably on a piece of plywood that has been installed for the purpose.   If the meter is to be the indoor type, be sure the plywood is large enough to hold the meter also.   Observe the local custom in your locality.

The neutral wire of the service runs directly to the neutral strap in the cabinet. The two hot wires run to the main lugs. They are usually marked "Mains." The proper connections are usually shown on a wiring diagram that comes with the equipment.

Spend a few extra minutes to run the wires *neatly* within the cabinet. Instead of running them this way and that at random, bend them to run parallel to the sides of the cabinet, or to each other, where this is practical. You can then take pride in your workmanship.

Neutral in Service. If the installation includes equipment operating at 115 volts, and other equipment operating at 230 volts, naturally the neutral must run from the transformer to the service switch.

Occasionally an installation will consist of a single-phase load such as a motor operating at 230 volts, but nothing operating at 115 volts. Alternately the load might consist of a 3-phase motor and nothing else. In either case, the neutral is not necessary for the operation of the load. It would then seem logical that the neutral need not be brought from the transformer (or transformers) serving the load, to the service switch.

However, if an installation of the type described is served from a supply system that includes a ground at the transformer or transformers, as is usually the case, then the Code in Sec. 250-23(b) requires that the neutral be brought to the service switch, where it ends, even if the neutral is not necessary for the proper operation of the equipment. At the service switch it must be grounded as in other installations. But the size of the neutral in the service need be only as large as the wire required for grounding, regardless of the size of the hot wires.

A complete technical explanation for this requirement is beyond the scope of this book; accept the fact that it is a very important factor in safety in case of accidental grounds in the wiring system or the load.

Grounding. As already discussed in Chap. 9, the Code differentiates between two types of grounds: system grounds and equipment grounds. The system ground consists of grounding the white neutral incoming wire as well as the neutral wires of the branch circuits. The equipment ground consists of grounding the metal parts of the service entrance, such as the metal box of the service switch, as well as the service-entrance conduit or the armor[3] of service-entrance cable. If the branch-circuit wiring is metallic (any type of conduit or armored

---

[3] According to Code definition "armored" service-entrance cable is a type not shown in this book and rarely used. The cable shown in Fig. 17-10 has a flat steel protective tape (which is not armor in the sense of the Code definition) which becomes automatically grounded when the neutral wire is grounded.

cable), the metal raceway is automatically grounded because it is anchored to the service switch. The equipment ground also includes motor frames, switchboard frames, and similar equipment, which, however, need not be considered in residential wiring *if metal raceway or armored cable is used for branch circuits,* because then such equipment is automatically grounded. If you use nonmetallic-sheathed cable in residential wiring, use the kind with the extra, bare grounding wire discussed in Chap. 11.

System and equipment grounds are combined and handled by a single grounding wire.

Method of Grounding. The usual ground connection is to a water pipe of a city water system. Use the cold-water piping, not the hot, because the former runs more directly to the ground. Make the ground connection as near as practical to the meter or, if at all possible, to the street side of the meter; on the other hand, keep the ground wire as short and direct as possible. Other things being equal, it is probably best to run the ground wire to the nearest cold-water pipe.

If there is no underground *city* water system, any underground piping system at least 10 ft long may be used. This will be discussed later in this chapter. A metal well casing (but not a drop pipe in a dug well) is considered part of such a piping system.

In the absence of underground water pipe, the metal framework of a building may be used or even gas piping. The local inspector should be consulted. In the absence of all such means, use a "made electrode" (artificial ground) as described later.

Ground Wire. The ground wire does not need to be insulated, although there is no objection to using insulated wire. It may never be lighter than No. 8, which serves the purpose when the largest service conductor is not heavier than No. 2. If the largest service conductor is No. 1 or 1/0, use No. 6 wire. If the largest service conductor is No. 2/0 or 3/0, use No. 4 wire. (For heavier service conductors, see Chap. 28.) There is no objection to using a ground wire larger than the minimum required by the Code. Note, however, that when a "made electrode" (as described later) is used, the grounding wire never need be larger than No. 6 [Code Sec. 250-94(a)].

If No. 4 ground wire is used, it requires no further protection such as conduit; it may be run open or concealed; it may run directly to ground without following the exact contour of the building. It may be stapled to the building, but this is not required. Common sense, of course, dictates that it be guarded against mechanical injury if in a location where it might be disturbed.

If No. 6 wire is used, it requires no further protection such as conduit, provided that it closely follows the surface of the building and is rigidly stapled to it, assuming further that it is free from exposure to mechanical injury. If these conditions are not met, as is quite usual, then it must be given mechanical protection in the same way as No. 8, which the next paragraph covers.

If No. 8 wire is used (or No. 6 wire neither following the surface of the building nor stapled to it), it must be run inside of conduit, or it may be the armored type of wire shown in Fig. 17-20, consisting of bare wire

Fig. 17-20   Armored ground wire.   The conductor is not insulated.

plus armor. If the protection is conduit, it must be attached to the service switch by locknut and bushing; if it is armor, with a connector of the type used on armored cable. At the water pipe end, either conduit or armor must be fastened to the same clamp by which the ground wire is connected to the pipe.

If the ground wire is enclosed in either conduit or armor, that conduit or armor *must* be securely clamped at *both* ends. At the service switch or breaker cabinet, use an ordinary cable clamp or connector for cable, or a locknut and bushing for conduit. At the ground end, select a ground clamp that securely clamps the armor, or conduit. This is an important point to watch. If the cable armor, or conduit, is clamped at one end only, or not at all, the resultant ground is very much less effective than it would be, had a bare or insulated wire without armor or conduit, been used. The explanation of this fact is beyond the scope of this book.

Ground Clamps. There are many varieties of ground clamps on the market, all serving the same purpose. Figure 17-21 shows a common clamp fitting $\frac{1}{2}$-, $\frac{3}{4}$-, or 1-in. water pipe. For the smaller sizes reverse the lower jaw. Larger sizes are available for larger size pipe. Use it with bare ground wire without further protection.

The type in Fig. 17-22 is very similar except that there is an extra clamp to which the armor of armored ground wire is clamped. The type in Fig. 17-23 is also similar except that it has a still larger fitting in which conduit is clamped if used as protection for the ground wire.

The type of ground clamp must be carefully selected. The ordinary metal-strap type is prohibited by Sec. 250-116. If the pipe to which the ground connection is made is iron pipe, the ground clamp should be

Fig. 17-21 Ground clamp for bare wire without armor. (*All-Steel Equipment, Inc.*)

Fig. 17-22 Ground clamp for armored ground wire. (*All-Steel Equipment, Inc.*)

Fig. 17-23 Ground clamp for use with ground wire run through conduit. (*All-Steel Equipment, Inc.*)

made of iron. If the ground connection is made to copper or brass pipe or to a copper or copper-coated rod, the clamp should be made of copper or brass. Unless this point is observed, electrolytic action is likely to set in, resulting in a high-resistance ground, which is not much better than no ground at all.

Solder Connections Prohibited. The Code in Sec. 250-115 prohibits joints that depend on solder, as far as ground wires are concerned. Use only joints dependent upon solderless or "pressure" connectors; do not use solder.

Water Meters. It is not unusual for a water meter to be removed from a building, at least temporarily while testing. If the ground connection is made to a water pipe between the meter and some other part of the building, removing the meter then leaves the system ungrounded, which leaves a temporary hazardous condition.

In some cases the joints between the water pipes and the meter are very poor joints, electrically speaking—practically insulated. If the ground of the electrical system is between the water meter and some other part of the building, it results in a "ground" apparently meeting Code requirements but actually is no ground at all. This is very dangerous. Therefore the Code in Sec. 250-112(a) requires that when the ground connection is not made on the street side of the water meter, a jumper be installed across the meter, as shown in Fig. 17-24.   Two ground clamps are used; the size of the jumper wire is the same as used for the ground proper.

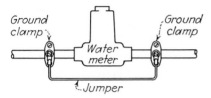

Fig. 17-24   A jumper must be installed around the water meter.

A similar jumper is required around any other object in the incoming water pipe if there is likelihood that the electrical continuity of the piping is or could be disturbed.   For example, water softeners are sometimes installed on a rental basis, and the installation made with non-metal hose to carry water into and out of the softener.   A jumper then is required around such an installation unless the ground wire is connected to the pipe at a point between the softener and the street.

Most homeowners have not the slightest conception of the importance of a good ground.   Ground connections have even been removed by misguided or uninformed people.   Read again Chap. 9 concerning the importance of a good, permanent ground.

Connections within Service Switch.   Service switches and breakers contain a brass or copper grounding strap, on which are installed two or more heavy connectors, and a number of small terminal screws.   The heavy connectors are for the neutral of the service wire, the ground wire, and the neutral of the wires to a range.   The smaller screws are for the neutrals of 115-volt branch circuits.   Usually this ground strap is solidly grounded to the metal cabinet; if it is not, see to it that you ground it yourself.

In the finished installation the service conduit, as also all lengths of conduit or armored cable in the branch circuits, are anchored and grounded to the cabinet.   Therefore it would seem that running a ground wire from the ground strap in the cabinet should effectively ground the entire installation.   In practice however it has been found that the resistance created by various connectors or locknuts and bushings, is too high.   The Code therefore requires additional measures.

Grounding Bushings.   Where the service wires or the ground wire enter the cabinet, use a grounding bushing of the type shown in Fig. 17-25, instead of an ordinary bushing.   The particular bushing shown has a setscrew on the side which bites into the conduit; other brands have a setscrew that bites into the metal of the cabinet.   Either type prevents the bushing from turning, and the setscrew bites into the metal, leading to good continuity of ground that is so important in safety.   The bushing has a lug on the side, as is required if a jumper is necessary, as will be explained in the paragraph concerning jumpering.   It also has an insulated throat, which is required only on some larger installations, as will be explained in Chap. 29; it may be used in any installation, but for ordinary residential and farm jobs, all-metal bushings are entirely acceptable.

Fig. 17-25 A grounding bushing.

Fig. 17-26 A grounding locknut.

Fig. 17-27 A grounding wedge.

If you have used thin-wall conduit, or armored cable, use a grounding-type locknut shown in Fig. 17-26 in place of the locknut on the connector.   You may also use a grounding wedge shown in Fig. 17-27; install it on the inside of the cabinet, between the wall and the locknut of the connector.

Jumpering.   Your switch or breaker cabinet probably has concentric knockouts, which were described in Chap. 13.   If you have removed *all* parts of this knockout so that no part remains in the box, grounding bushings as just described are all that is required.   Often some part of the knockout remains in the cabinet, and in that case you must install a jumper from the grounding bushing to ground, as shown in Fig. 17-28, which shows only the neutral wires and the ground wire.   The jumper wires must be the same size as the ground wire. Sometimes the switch or breaker cabinet contains a separate grounding screw, in which case the jumpers from *A* and *B* in Fig. 17-28 are run to that screw, point *C*. If the cabinet does not contain such a screw, run the jumpers to the ground strap as shown in dotted lines.

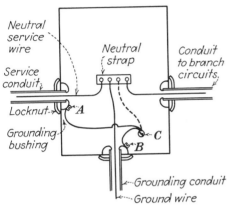

Fig. 17-28 Typical grounding scheme at entrance switch of residential job. Only the grounded neutral wire is shown.

If you have used service-entrance cable with a concentric bare neutral *and* metal tape over that (Type SE, Style A), the grounding bushing is not required where the cable enters the cabinet; see Code Sec. 250-73. It is not necessary to use grounding bushings where conduits or cables serving branch circuits enter the cabinet.

Made Electrodes. When a city water system is not available, a substitute must be used. Early Codes called this an "artificial ground"; more recent codes call it a "made electrode." Grounding to the pipe in a *dug* well is not recommended. Often this leads to exceptionally long ground wires; furthermore, wells sometimes go dry, resulting in no ground at all. On the other hand if the well is a driven well located close to the logical point for the ground connection, its casing makes an excellent ground.

The usual made electrode is a driven pipe or rod driven at least 8 ft into the earth. Pipe may be used if it is galvanized and at least

¾-in. trade size. Solid rods may be used and if of iron, must be galvanized and at least ⅝ in. in diameter. The most common ground is copper, or an approved substitute, and need be only ½ in. in diameter. The most common substitute which the Underwriters have listed is the Copperweld type—a rod of steel with a layer of copper welded to the outside. For more information about proper installation of ground rods, see Chap. 24.

The Code in Sec. 250-83(d) requires that the ground rod or other made electrode be entirely independent of, and kept at least 6 ft from, any other ground rod of the type used for radio, telephone, or lightning rods.

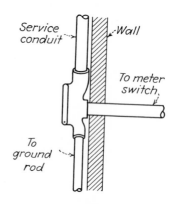

Fig. 17-29 A grounding elbow, generally used with a "made electrode" when the entrance wires run through conduit. (*Killark Electric Mfg. Co.*)

Fig. 17-30 Installation showing the grounding elbow of Fig. 17-29.

When a made electrode is used, the ground rod is often out of doors; it is then necessary to bring the ground wire out of the building. In many localities, the inspector will require a special type of entrance ell, as shown in Fig. 17-8 but with an extra opening at the bottom, for the ground wire. This is shown in Fig. 17-29 and its use should be clear from Fig. 17-30.

When service-entrance cable is used, a meter ring of the type shown in Fig. 17-31 is sometimes used. This is slipped between the service-entrance-cable connector and the meter socket. The ground wire is connected to it and from there run to ground. Since the neutral of the cable is connected to the socket, this connection then grounds the neutral.

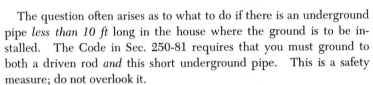

Fig. 17-31 A meter ring. (*Killark Electric Mfg. Co.*)

The question often arises as to what to do if there is an underground pipe *less than 10 ft* long in the house where the ground is to be installed. The Code in Sec. 250-81 requires that you must ground to both a driven rod *and* this short underground pipe. This is a safety measure; do not overlook it.

Television Antennas. It is a very wise move to ground any metal support for a television antenna; the Code requires it, in Sec. 810-15. In cities it is not unusual to see such supports grounded to a separate ground rod. That is wrong and dangerous. Run the TV ground wire to the water pipe, grounding it there with a separate clamp next to the clamp used to ground the electrical system. On farms you may use a separate ground rod, but must run a wire from that rod to the one used for grounding the wiring system. Use wire of the same size as used for grounding the wiring system.

Separate Meter for Water Heater. In many localities, the power supplier provides a separate meter and an automatic time switch for the water heater. The automatic switch connects the heater to the line only during "off-peak" periods. In other words, the heater is disconnected during the periods when there is maximum demand for power, as, for example, in the periods before meals when many electric stoves are in operation, during early evening hours, and so on. In return, there is usually a lower rate per kilowatthour for the heater, than would otherwise apply.

There are many ways of handling the second meter. Usually a pair of wires is tapped to the service-entrance wires where they enter the first meter socket. They run on to a second meter, then to a fused switch or circuit breaker, which serves as overcurrent protection as well as the disconnecting means for the heater. The size of the wire from the first meter to the second is determined by the method outlined in Chap. 5 for taps under 25 ft long. The size so determined is more than enough for the amperage required by the heater. The size of the wires from the second meter to the heater depends on the size of the heater and has already been discussed.

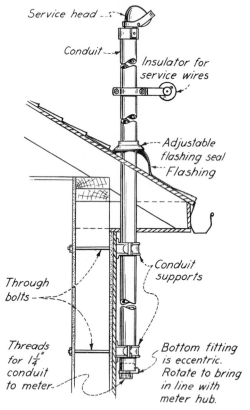

Fig. 17-32   Typical mast installation.   (*M. & W. Electric Mfg. Co.*)

Service Masts.   Many houses being built today are of the so-called rambler or ranch-house type. They are so low that it is impossible to run the service drop directly to the building and still maintain the minimum clearance required by Code above ground, sidewalks, and driveways.   The usual way of solving this problem is to use a service mast of the general construction shown in Fig. 17-32.   The service conduit itself becomes the supporting member, but you must use 2 or 2½ in. to provide sufficient rigidity.   Some of the fittings are so designed that they will fit either size of conduit without threading.   Flashing is usually provided with each kit to make a watertight opening through the overhang of the roof.

If the mast is within 4 ft of the edge of the roof, the wires may have a clearance of as little as 18 in. from the roof at the nearest point [Sec. 230-24(a) Exc. 2].

18

Installation of Specific Outlets

In previous chapters installations of electrical devices were considered in rather general fashion; in this chapter the exact method of installing a variety of outlets using assorted materials will be discussed in detail. Only methods used in *new work* (buildings wired while under construction) will be explained. *Old work* (the wiring of buildings *after* their completion) will be described in a separate chapter.

When buildings are wired with conduit, the conduit is installed, all the outlet and switch boxes are mounted, and all similar details handled while the building is in the early stages. This is termed "roughing-in." The wires are not pulled into the conduit until after the lathing, plastering, papering, and similar work is finished, and obviously the switches, receptacles, fixtures, and other devices cannot be installed until the building is practically completed. However, to avoid repetition later, the pulling of the wires will be included in this chapter. If cable is used, the wires are automatically installed in the roughing-in process.

Each type of outlet will be treated separately. In the pictures each outlet will be shown in different ways, as follows:

1. As it would appear on a blueprint.
2. As it would appear in diagrammatic or schematic fashion.
3. In pictorial fashion using conduit. However, in the case of receptacles, in addition to the wires shown, you must run a grounding wire from the green terminal of the receptacle to the box. That grounding wire will not be shown in the illustrations (to avoid "cluttered" illustrations), but if you have studied Chap. 11 well, you will have no problem.
4. In pictorial fashion using nonmetallic-sheathed cable. Again the grounding wire will not be shown, but cable *with* the grounding wire will be required in many cases, and always if receptacles are involved. Again, if you have studied Chap. 11 well, you will have no problem. The diagrams are also correct for armored cable. The grounding strip inside the armor must be properly handled as outlined in an earlier chapter, and the green terminal of each receptacle must be grounded to the box.

The same basic details shown in Fig. 18-1 will be used in other drawings, and since these will be smaller, the individual parts cannot be named as in the first picture, but you will easily recognize them.

In each case the two wires over which the current comes are labeled SOURCE. *The white or neutral wire always runs, without interruption by a switch or fuse or other device, up to each point where current is to be consumed at 115 volts.* Joints are not interruptions. Switches consume no current; hence the white wire does not run to a switch (except when using cable, which will be discussed later). Black wires may be joined to black or red or other colors as the need requires, but not to white (except sometimes when using cable, which will be discussed later).

Ceiling Outlet, Pull-chain Control. This is the type of outlet generally used in closets, basements, attics, and similar locations, with the wires ending at the outlet. It is the simplest possible outlet to wire. Figure 18-2 pictures it. At *A* is shown its designation as found on blueprint layouts; the wire running up to it is not shown, for, in these blueprints, wires are shown only to connect switches with the outlets that they control. At *B* is a wiring diagram for the same outlet; the light lines desig-

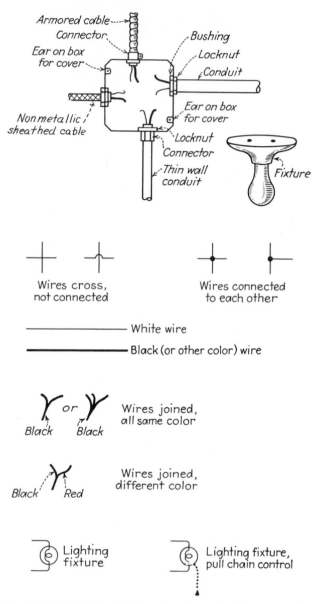

Fig. 18-1 Study these details well, so that other diagrams in this chapter will be clear.

nate white wire, the heavy lines black wire or some color other than white.

Assume that you have already mounted the outlet box in the ceiling by one of the methods covered in a previous chapter; if conduit was used, you have provided a box with a plaster-ring cover of the type that was shown in Fig. 10-12.

Installing this outlet with conduit as in Fig. 18-2, C, assume that you

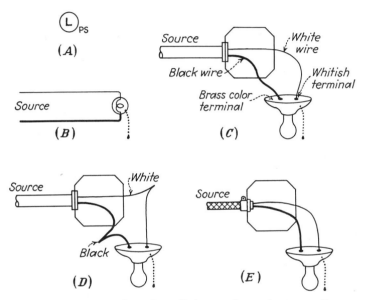

Fig. 18-2 A simple hookup of a pull-chain outlet, with wires ending at that outlet.

have properly reamed the cut ends of the conduit, that you have properly anchored it to the outlet box with a locknut and bushing, and that you have pulled two wires, one black and one white, into the conduit with about 6 in. sticking out of the box. To complete the outlet all you have to do is to connect the wires to the fixture and mount the fixture in one of the ways discussed in the next chapter.

If the fixture is one which has two wire leads (instead of two terminals to which the wires from the outlet box can be connected), the wiring is still the same, except that a couple of splices must be made in the box, as shown in Fig. 18-2, D.

If you plan to use thin-wall conduit (EMT) instead of rigid conduit, follow the same procedure as with rigid conduit, except use the threadless fittings such as were shown in Figs. 11-10 and 11-11 instead of the locknuts and bushings used with rigid conduit.

How to install this outlet using nonmetallic-sheathed cable is shown in *E* of Fig. 18-2. If you are using cable without the grounding wire, merely run it to the outlet, anchoring it to the box with a connector, and connect the fixture. If using cable with the grounding wire, follow the same procedure except connect that grounding wire to the box as discussed in Chap. 11. The cable must be supported within 12 in. of the box, and at intervals not over 4½ ft.

If you are using armored cable, again follow *E* of Fig. 18-2. You have inserted a fiber bushing between the armor and the wires, you have bent the grounding strip under the clamps of the connector that you have installed on the cable. You have anchored the cable solidly to the box. There is nothing further to do except to connect and install the fixture. Again the cable must be supported within 12 in. of the box, and at intervals not over 4½ ft.

If you plan to use flexible metal conduit ("greenfield"), follow the procedure for armored cable, except that the wires are pulled into it after the conduit is installed. You must also install a separate grounding wire inside the conduit, as explained in Chap. 11. This separate grounding wire serves the same purpose as the grounding strip inside the armor of armored cable.

Same Outlet, Wires Continue to Next Outlet.   This combination is as common as the first and practically as simple. From the first outlet the wires continue to the next; it makes no difference what may be used at that next outlet. The only problem is how to connect at the first outlet the wires running on to the second.

This combination is pictured in Fig. 18-3, and again *A* is its designation as found on blueprint layouts. Comparing this with Fig. 18-2, *A*, you will find no difference. This may be confusing, but in blueprint layouts the wires between different outlets are not indicated except those wires between switches and the outlets they control. It is up to the one making the installation to use his own good judgment as to the exact fashion of hooking together different outlets. At *B* is shown the wiring diagram for the combination.

At *C* is shown the outlet using conduit. Compare it carefully with Fig. 18-2, *D*. The new outlet, Fig. 18-3, *C*, is the same as the former except for the addition of two new wires running on to the second outlet, at the right, and these two wires have been shown in dotted lines to distinguish them from the former wires. Simply join all the black wires together, also all the white wires; solder and tape, or use "wire nuts," and the job is finished.

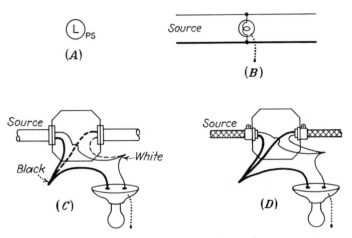

Fig. 18-3 The hookup shown in Fig. 18-2, but with wires running on to another outlet.

Receptacle Outlets. In Fig. 18-4 at *A* is shown the designation for this outlet as found in blueprint layouts; at *B* is a wiring diagram. To wire this outlet using conduit as at *C*, merely connect the wires to the receptacle, *and* run a grounding wire from the green terminal of the receptacle, to the box.

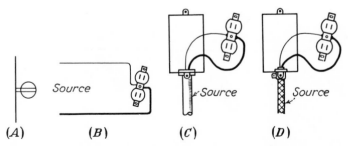

Fig. 18-4 Installation of a baseboard receptacle outlet.

At *D* is shown the same outlet wired using nonmetallic-sheathed cable. The grounding wire in the cable is not shown, but it must run to the green terminal of the receptacle, and must also be grounded to the box. Study Chap. 11.

If you are using armored cable, proceed as with nonmetallic-sheathed cable. A grounding wire must run from the green terminal, to the box.

Receptacle Outlet, Wires Continue to Next Outlet. In Fig. 18-5, *A*, is shown the wiring diagram of the circuit in question. At *B* and *C* is

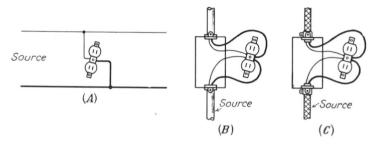

Fig. 18-5 Receptacle outlet, with wires running on to another outlet.

shown the actual installation using conduit and cable; little explanation should be necessary. Simply run the incoming white wire to one of the terminals of the receptacle, and from that terminal continue with another white wire to the next outlet. Do the same with the black wire.

Because it is frequently necessary to do this, receptacles are provided with double terminal screws so that two different wires can be connected to the same terminal strap, as shown in *A* of Fig. 18-6. This is

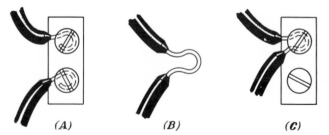

Fig. 18-6 Duplex receptacles have double terminal screws on each side, for convenience in running wires on to another outlet.

much simpler than connecting two different wires under the same terminal screw, which the Code prohibits in Sec. 110-13.   If conduit is used, you may find it far more convenient to pull a continuous wire from SOURCE to box 1 to box 2 to box 3 than to pull one length from SOURCE to 1, a second length from 1 to 2, and another from 2 to 3.   If you use one continuous length, allow a loop of wire to project about 3 in. at each box, as shown in Fig. 18-6, *B*.   Remove the insulation, as shown in *B* of the picture, and clamp the conductor under one of the terminal screws, as shown in *C*.   This method is the preferable one.   Remove just enough insulation from the wire so that the bare, uninsulated conductor will not be exposed after installation for any distance from the terminal screw.

The grounding wire from the green terminal of the receptacle must be installed as already explained in Chap. 11.

Outlet Controlled by Wall Switch.   Naturally this is a very common outlet and fortunately very simple to wire.   Figure 18-7, *A*, shows the blueprint symbol, and *B* shows the wiring diagram.   At *C* is shown the outlet wired with conduit.   Run the white wire directly to the fixture.   Run two blacks from the outlet box to the switch box, and connect them to the switch; connect the upper ends, one to the black incoming wire from SOURCE, the other to the fixture; and the job is finished.   At *D* is shown the optional method whereby one continuous black wire is brought all the way through to the switch box instead of having a joint at point *X* in *C*.

When this outlet is wired with cable, either armored or nonmetallic, as at *E*, a difficulty is encountered.   According to everything learned up to this point, not only the wire from the outlet box to the switch but also the wire running from the switch back to the fixture should be black, since neither one is a grounded wire.   On the other hand, the cable containing these two wires has one black and one white wire.   Should manufacturers, distributors, and contractors then be forced to stock a special cable containing two black wires just for this purpose?   That would be impractical.

The Code in Sec. 200-7 permits an exception to the general rule that the white wire may be used only as a grounded neutral wire and never as an ungrounded wire.   In the case of a switch loop (as the wiring between an outlet and the switch which controls it is called) this section permits the use of a 2-wire cable containing one black and one white wire, even if its use does not fulfill the general rule.   Under this excep-

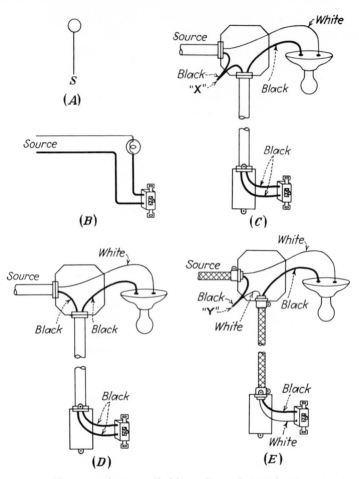

Fig. 18-7   Fixture outlet controlled by wall switch (see also Fig. 18-10).

tion the cable may be used provided that the black wire is made to run directly to the fixture.   This leaves only one place to connect the white wire of the switch loop, and that is to the black wire in the outlet box. This is the only case where it is permissible to connect a black wire to a white wire, and it pertains only if cable is used.

Study this rule well: When cable is used for a switch loop, the wiring up to the outlet box upon which the fixture is mounted is standard, including connection of the white wire from SOURCE to the fixture.   Connect the black wire of the switch loop to the fixture.   The fixture will then have two wires connected to it: one black and one white. Then connect the white wire of the cable of the switch loop to the black wire in the outlet box contrary to the general rules but permitted in this one case by Code Sec. 200-7.

This same procedure is acceptable when using 3-wire cable for 3-way or 4-way switches.

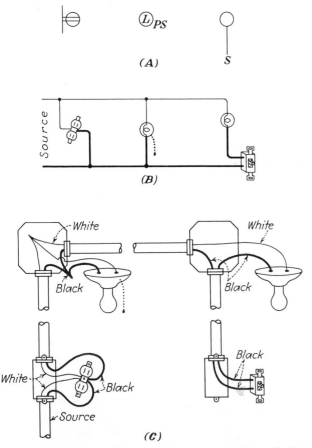

Fig. 18-8 Combining the outlets of Figs. 18-2, 18-4, and 18-7 into one three-outlet combination.

Combining Three Outlets. Three different outlets having been wired, you can combine them into one combination of three outlets, as shown in Fig. 18-8.   As usual, *A* shows the blueprint symbols, *B* the wiring diagram, and *C* the three outlets as wired with conduit.

All the points have been gone over in detail, and if they have been carefully studied, this combination should present no problem.   If any point is not clear, go back to the idea of messengers chasing each other along the different wires, in on the white wire from source, along that wire up to every point where current is to be consumed, and from each such point back over the black wires through switches, if used, until they emerge at the black source wire.

Taps.   At times an outlet box is used merely to house a tap where one wire branches off from another.   Sometimes, especially in conduit work, it is difficult to pull wires into a long length of conduit, so a "junction box" or "pull box" is installed which will serve as an intermediate pulling point.   Whenever a box is used only for connections, it is always covered with a blank cover.   The Code requires all junction or pull boxes to be placed in locations permanently accessible without removing any part of the building (Sec. 370-19).

Figure 18-9, *A*, shows how taps are indicated in blueprint layouts, and at *B* is shown the wiring diagram.   At *C* is shown the method using conduit; merely join all the white wires and then all the blacks.   At *D* is shown an optional method, when, instead of three ends of wire of each

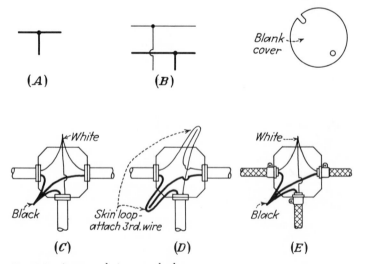

Fig. 18-9   A tap made in an outlet box.

color, one loop of each color is used, formed by pulling a continuous wire through the box, the ends of the loop being skinned and joined to the remaining ends of the wires to the next outlet. *E* shows the same outlet using cable.

Outlet with Switch, Feed through Switch. Switches are connected in different ways to the outlet which they control. The source wires do not always run through the ceiling to the outlet box on which the fixture is installed and from there on to the switch, as in Fig. 18-7, already discussed. Sometimes the source wires come in from below, run through the switch box and then on to the ceiling outlet, where they end, as in Fig. 18-10. At *A* is shown the usual blueprint symbol, and *B* shows the

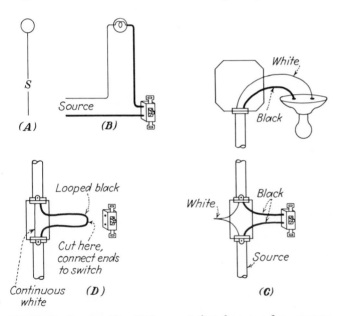

Fig. 18-10 Same as Fig. 18-7, except that the wires from source enter the switch box, instead of through the outlet box on which the fixture is installed.

wiring diagram. At *C* is shown the outlet using the conduit system, and little explanation is needed. At *D* is shown an optional method of handling the wires at the switch box by pulling the white wire straight through. The black wire is also pulled straight through but with a loop which is later cut, the two ends being connected to the switch.

The diagram for this outlet when wired with cable is not shown because there is no new problem. When the cable feeds through the switch box, there is no difficulty with the colors of the wires in the cable as in a previous example. The wires can be run through to completion of the outlet without the need of joining black to white, exactly as when using conduit (C of Fig. 18-10).

Outlet with Switch, with Another Outlet. This is simply a combination of two outlets that have already been discussed separately. The wiring of an outlet with wall-switch control was covered in connection with Fig. 18-7. The wiring of an outlet with pull-chain control was covered in connection with Fig. 18-2. The two have been combined in Fig. 18-11, the left-hand portion of which is identical with Fig. 18-7; to it have been added in dotted lines the wires of Fig. 18-2, making the new com-

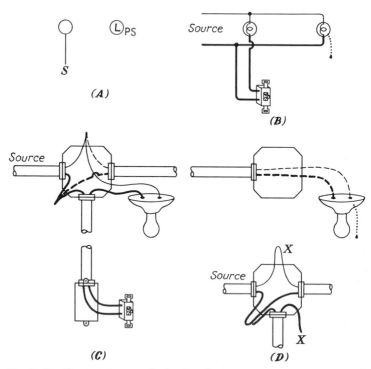

Fig. 18-11 The switch controls the first fixture; wires run on to a second fixture which is controlled by a pull chain.

bination as shown. At *D* is shown an optional method of handling the wires through the outlet box, the wires from the fixture being connected at the points marked *X*. No new problems are involved in the cable methods; for that reason they are not shown.

If the wires from SOURCE (instead of coming in through the ceiling outlet) come in from below through the switch box, the problem is different in that three wires instead of two must run from the switch box to

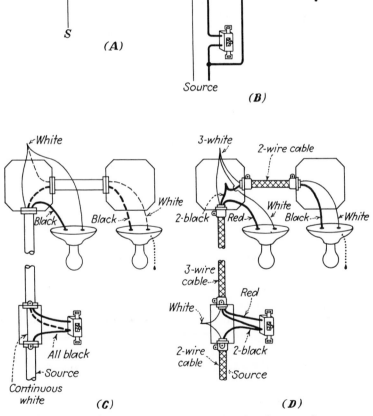

Fig. 18-12 This is the same as Fig. 18-11, except that the wires from SOURCE enter through the switch box.

the first outlet box, as shown in Fig. 18-12, *B*.   A good way to analyze this combination is to consider first the right-hand outlet with the pull chain.   Both a white and a black wire *must* run to this from SOURCE, uninterrupted by any switch, so that the light can always be controlled by the pull chain.   Run the third wire from the switch to the left-hand outlet to control that.

Analysis of Fig. 18-12, *C*, shows that it consists of a combination of Figs. 18-10, *C* (wires in solid lines), and 18-2 (wires in dotted lines).

If instead of conduit you use cable, as in Fig. 18-12, *D*, the problem is equally simple.   If the wires are fed through the switch box, there is no problem in connection with the colors of the wires.   Merely connect white to white in the switch box, and then continue the white to each of the two fixtures.   The remaining colors are as shown in the picture.

Switch Controlling Two Outlets.   When a switch is to control two outlets at the same time, the connections are most simple.   Merely wire the switch to control one fixture, then continue the white wire from the first fixture to the second, and do the same with the black.

As in the other pictures, in Fig. 18-13, *A* shows the blueprint symbols, *B* the wiring diagram, and *C* the combination using conduit. Compare this with Fig. 18-7, *C;* there is no difference except that these wires have been continued as shown in the dotted lines.   To avoid all the joints shown in the outlet box for the first fixture, several of the wires may be pulled through as continuous wires, making a neater job.

If you use cable, there is again the problem of having to use in the switch loop a cable that has one black and one white wire, instead of the two blacks that should be used.   Handle it as in Fig. 18-7, *D*.   The white wire from SOURCE goes to each of the two fixtures.   The black wire in the switch loop also goes to the first fixture, then on to the second.   That leaves only two unconnected wires, the black wire from SOURCE and the white wire in the switch loop; connect them together as before, as permitted in the Code's exception.

Three-way Switches.   When a pair of 3-way switches controls an outlet, there are many possible combinations or sequences in which the SOURCE, the two switches, and the outlet may be arranged; a great deal depends on where the SOURCE wires come in.   The most common are

SOURCE—Switch—Switch—Outlet.
SOURCE—Outlet—Switch—Switch.
SOURCE—Switch—Outlet—Switch.

A fourth is the sequence where the SOURCE comes into the outlet box, from which point two runs are made, one to each switch.

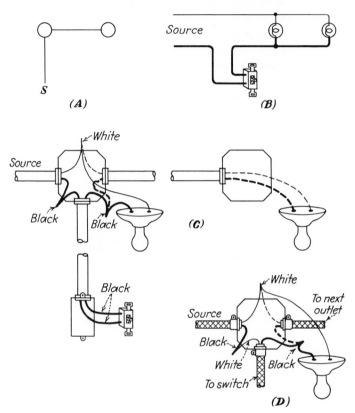

Fig. 18-13 A switch controlling two separate fixtures.

These sequences are shown in Fig. 18-14, parts A1, A2, A3, and A4; the wiring diagrams are shown in B1, B2, B3, and B4. Comparing B1, B2, B3, and B4, you will see little difference except the exact location of the light which the switches control.

Reviewing the subject of 3-way switches, remember that one of the three terminals on such a switch is a "common" terminal, corresponding to the middle terminal of an ordinary porcelain-base single-pole double-throw switch shown in Fig. 4-20; this terminal is usually identified by being of a different color from the other two. The exact location of this common terminal with relation to the other two varies with different brands; for the purposes of this chapter, where 3-way switches are shown in pictorial fashion, the terminal which is alone on one *side* of the switch will always be the common terminal.

Reviewing the subject a bit further, run the incoming black wire from SOURCE direct to the common or marked terminal of either switch. From the corresponding common or marked terminal of the other switch, run a black wire direct to the proper terminal on the fixture. From the remaining two terminals on one switch, run wires to the corresponding terminals of the other switch, which are the only two termi-

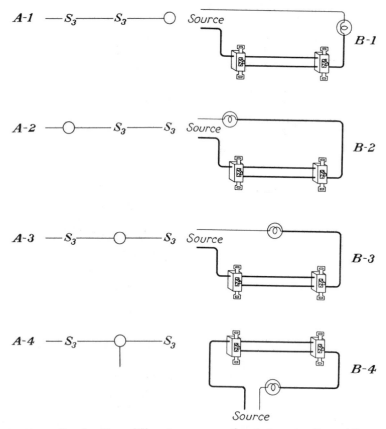

Fig. 18-14 Part 1. Four different sequences of parts in a circuit consisting of one fixture and a pair of 3-way switches.

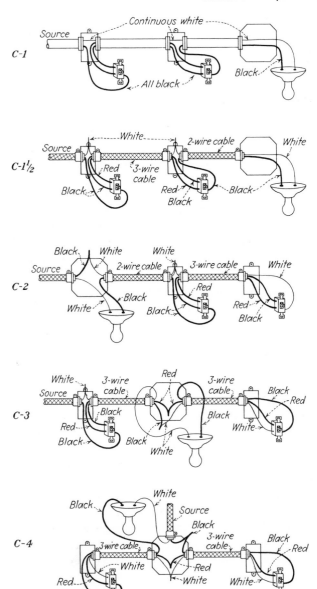

Fig. 18-14 Part 2. Wiring methods of the combinations shown in Part 1 of this figure.

nals on that switch to which wires have not already been connected. To complete the circuit, connect the incoming white wire from SOURCE to the fixture.

With these facts in mind, the wiring of any combination of 3-way switches with conduit becomes quite simple. Assuming that the boxes and conduit have been properly installed, as shown in Fig. 18-14, C1, which covers the sequence of A1, pull the white wire from SOURCE through the switch boxes up to the outlet where the fixture is to be used; run the black wire from SOURCE to the common terminal of the nearest 3-way switch. From the two remaining terminals of this switch, run two black wires to the corresponding two terminals of the second switch. From the common terminal of this switch, run a black wire to the fixture. All wires are black, but sometimes one red wire is used for identification purposes; any color may be used except white or green.

The wiring of the other combinations or sequences with conduit is equally simple if you remember the points of the two previous paragraphs, and no diagrams are shown. It will be good practice to draw the circuits in fashion similar to Fig. 18-14, C1.

This same combination wired with cable is shown in Fig. 18-14, C1½. Note how the white wire from SOURCE is continued from box to box until it reaches the fixture. The black wire from SOURCE, as in the case of conduit, goes to the common terminal of the nearest switch. The 3-wire cable between the switches contains wires of three different colors, of which the white has already been used, leaving the black and the red. Therefore run these two wires from the two remaining terminals of the first switch to the corresponding terminals of the second; the red and the black may be reversed at either end, being completely interchangeable. This leaves only one connection to make, and that is the black wire from the common terminal of the second switch to the fixture.

When some of the other sequences, such as A2, A3, or A4 in Fig. 18-14, are wired with cable, the usual difficulty in connection with the colors of the wires in standard 2-wire and 3-wire cables is met. The red wire of 3-wire cable is interchangeable with the black. Many times, however, you must take advantage of the Code's exception permitting, in switch loops, white wire to be connected to black.

C2 shows the sequence A2 and B2 of Fig. 18-14 wired with cable. The incoming cable from SOURCE contains one black, one white wire; run the white direct to the fixture. The Code requires that the other wire on the fixture may not be white; consequently the black wire of the 2-wire cable that runs to the first 3-way switch is connected to the fixture; connect the opposite end of it to the "common" terminal of the first 3-way switch.   That leaves in the outlet box on which the fixture is mounted only two unconnected wires: the black wire from SOURCE and the white wire of the next run of cable.   Connect them together, contrary to the general rule but permitted by the Code for switch loops. From the first 3-way switch to the second, 3-wire cable is used.   Since one of these three wires is white, connect it to the white wire of the 2-wire cable and continue it on to the "common" terminal of the second switch.   That leaves one black and one red wire between the two switches; connect them to the remaining terminals of each switch, completing the installation.

In wiring the sequence of $A3$ and $B3$ of Fig. 18-14, as pictured in $C3$, similar problems arise.   There is a 2-wire cable from SOURCE entering the first switch box; from that point a length of 3-wire cable runs to the outlet box in the center, and from there another length of 3-wire cable runs on to the second switch box.   Continue the white wire from SOURCE from the first switch box direct to the fixture, as the Code requires.   Run the black wire from SOURCE direct to the common terminal of the first switch.   Going on to the fixture, the second wire on the fixture may not be white; therefore make it black and continue it onward to the common terminal of the second switch.   That leaves unconnected two terminals of each of the two switches, and your problem is to connect them to each other.   There are yet two unused wires in each run of 3-wire cable, a black and a red in the one, and a white and a red in the other.   Therefore make one continuous red wire out of the two reds, make a continuous white-black out of the other two, and connect the extreme ends to the two remaining terminals on each of the switches, respectively.   This completes the connections.

In the case of the sequence of $A4$ and $B4$ of Fig. 18-14, as pictured in $C4$, the problems are similar, and you should have no difficulty in determining for yourself why each wire is of the color indicated.   The fundamental rule is that the white wire from SOURCE must go to the fixture, and the second wire on the fixture may not be white or green.

Pilot Lights.   In Fig. 18-15 is shown a frequently used combination toggle switch and pilot light, together with its internal wiring diagram. This is nothing more or less than a separate switch and a separate pilot light in a single housing.   The arrangement of the terminals on another brand may be totally different from that shown in the picture. A pilot light is used at a switch which controls a light that is not visible from the switch, for example, at a switch in the house controlling the light in a garage; it is used simply as a reminder that another light is on.

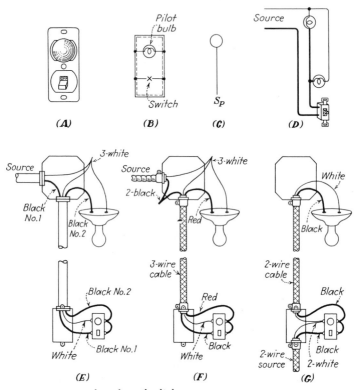

Fig. 18-15   A switch with a pilot light.

In the same figure at C is shown the blueprint symbol for this combination and at D the usual wiring diagram.   Compare this with Fig. 18-13, B; it is merely a case of one switch controlling two different lights, one of which happens to be located at the same point as the switch.

Using the conduit system, the connections are shown at E.   Run the incoming white wire from SOURCE directly to the first fixture and extend it from there to the second fixture, which in this case happens to be the lamp in the combination device.   Run the incoming black wire from SOURCE directly to the switch.   From the other side of the switch, run a black wire to the first lamp and from there on to the second lamp, finishing the job.   Note in the diagram that the two black wires have been labeled "No. 1" and "No. 2"; the two are not interchangeable at the switch end.   This is a good example of a case where it is desirable to make the third wire red for identification purposes, in which case the incoming black wire from SOURCE would run directly to the switch terminal and red would be substituted for the black No. 2.

Wiring the combination with cable, as shown in Fig. 18-15, *F*, presents no problem.   If the wires from SOURCE come in through the switch box, the problem is still simpler, as *G* shows.

Switched Receptacle Outlets.   There is a growing trend toward controlling receptacle outlets with switches, so that all floor and similar lamps can be controlled at one time.   No great problem is involved; simply consider the receptacles as so many fixtures, and connect the devices accordingly.   Go through the process of connecting three receptacle outlets with a pair of 3-way switches.   There can be a great many different sequences of outlets and switches; that shown in Fig. 18-16 at *A* and *B* is perhaps as common as any.   When conduit is used, the problem is simple indeed.   Run the white wire from SOURCE to each of the receptacles, connect the remaining terminal of each of the three receptacles together with black wire, and continue to the common terminal of one of the 3-way switches.   From there run two wires over to the other 3-way switch.   To the common terminal of this second switch, connect the black wire from SOURCE, and the job is finished.   It is pictured in Fig. 18-16, *C*.   All the necessary wires in the boxes housing the receptacles will badly crowd the ordinary switch boxes, so the preferable method is to use 4-in. square boxes with raised covers designed to take a duplex receptacle (see Fig. 10-12, *A*).

To wire this same combination using cable is practically impossible with the particular sequence shown, because it requires four wires at some points and 4-wire cable is not stocked by dealers.   Therefore when using cable, it is best simply to modify the sequence to that shown in Fig. 18-16, *D*, which requires nothing more than 3-wire cable.

Two-circuit Duplex Receptacles.   When receptacles in a living room (or other rooms too in many cases) are *not* controlled by a switch, it is a real nuisance to go around and turn each floor or table lamp off separately, at bedtime.   Moreover, when entering such a room in the dark, it becomes very difficult to find the lamps in the dark to turn them on;

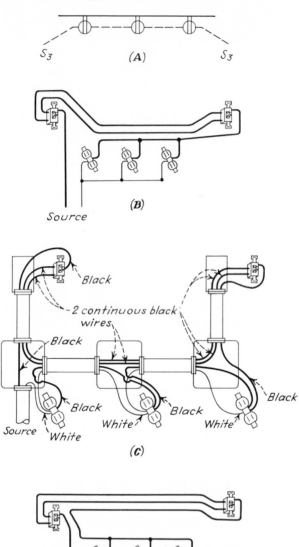

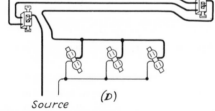

Fig. 18-16 Switched receptacle outlets.

stumbling in the dark can lead to accidents. If they are *all* controlled by a switch, there is the objection that clocks can't be used; neither can a vacuum cleaner be used without first turning on a switch, which then turns on all the lamps. One solution is to have a switch control some of the receptacles in the room, leaving others permanently connected. A better solution is to use "2-circuit" receptacles which are so designed that one of the two openings can be left permanently live for clocks, radio, TV, and so on, and the other opening switched for floor and table lamps.

An ordinary duplex receptacle has a pair of terminal screws common to both halves of the receptacle, and a second pair common to the other side of both halves. The 2-circuit receptacle also has a pair of terminal screws common to both halves of the receptacle, but two separate screws *not* common to the other side of both halves. Each of the second pair of screws serves one of the two halves of the receptacle. Ordinary receptacles of better quality are so constructed that they can be converted in a moment, to the 2-circuit type, by breaking out a small

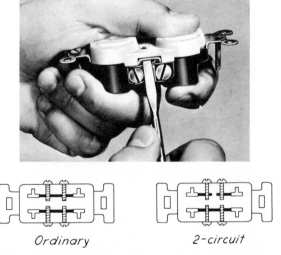

Ordinary   2-circuit

Fig. 18-17 Two-circuit receptacles have many advantages. In better quality receptacles, the ordinary receptacle can be converted to the 2-circuit type by breaking out a small metal strip as shown. (*General Electric Co.*)

part of the metal strap in which the screws are installed. How this is done is shown in Fig. 18-17.

Two-circuit receptacles let you use clocks or other devices which are never turned off, or which you wish to have permanently connected, and yet let you turn off all the floor and table lamps at one time. In installing them, it will be found convenient to have all the switched halves of the receptacles at the bottom, leaving the top halves permanently live, and easy to get at for use of vacuum cleaner and similar appliances.

Another common application for the 2-circuit receptacle is in the two special 20-amp appliance circuits in the kitchen. If these two circuits are merged into one 3-wire circuit (as will be explained in Chap. 20) run the neutral to the common terminal, the red and black wires to the other two terminals. Then, when two different appliances are plugged into the same duplex receptacle, they will automatically be on opposite legs of the 3-wire circuit.

Figure 18-18 shows the blueprint symbol at *A*. You will find this symbol, and many others for various kinds of receptacles other than the most ordinary, in Fig. 16-1.

The wiring presents no great problem. Run the white neutral wire from SOURCE to one of the terminal screws on the metal strap on the

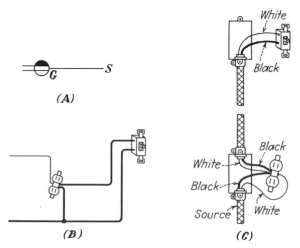

Fig. 18-18 Wiring 2-circuit receptacles using cable.

receptacle that is not broken in the middle. Run the black wire from SOURCE to one of the terminal screws on the opposite side of the receptacle, and continue it to the switch. From the switch run another wire back to the remaining terminal screw on the receptacle, on the side where the strap is broken in the center. All this is shown in *B* of Fig. 18-18, but for more detail, see the additional illustration of Fig. 18-19.

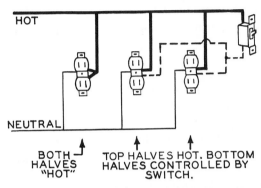

Fig. 18-19 More detail for wiring 2-circuit receptacles.

If you use cable to the switch, again you will have the problem that one of the wires is white where it should be black; take advantage of the Code's special dispensation and connect the white wire in the switch loop to the incoming black wire from SOURCE, as shown in *C* of Fig. 18-18.

If you use 3-way switches, the problem of getting the colors is more difficult, but may be solved exactly as in the case of the outlets shown in Fig. 18-14.

Combining Outlets. Just as in the early part of this chapter the outlets of Figs. 18-2, 18-4, and 18-7 were combined into one three-outlet combination of Fig. 18-8, so outlets may be combined into any desired combination with any desired total number of outlets, which then forms a circuit running back to the fuse cabinet. To add an outlet at any point, connect the white wire of that new outlet to the white wire of the previous wiring (provided it is not a white wire in a switch loop, when using cable), and the black wire of the new outlet to any previous wiring where the black wire can be traced back to the original SOURCE without being interrupted by a switch. Simply connect the new outlet at any point on the previous wiring where a lamp connected to the two points in question would be permanently lighted. Review Chap. 4 if this point is not entire'y clear.

Installing Recessed Fixtures. The temperature inside recessed (flush) fixtures is higher than it is in fixtures mounted in free air. The terminals will be the hottest part, and often that temperature will be higher than the temperature limits of the kinds of wire used in ordinary wiring. Assuming the fixture is listed by Underwriters, it will be marked with the minimum temperature rating of the wire supplying the fixture.

If the fixture is marked "60°" the circuit up to the fixture may be wired with any of the kinds of wire normally used in wiring. Most flush fixtures are marked for a temperature considerably above 60°. In that case you have a choice of two methods of installation. The first is to wire the entire circuit up to the fixture with a wire having a temperature limit, per Table 310-2(a) (see Appendix), at least as high as the temperature shown on the fixture; this is likely to be an expensive wire. Fixture wire may *not* be used.

Alternately, install a junction box not less than 12 in. from the fixture, and run a length of flexible conduit not less than 4 ft or more than 6 ft long, from that junction box to the fixture, looping the conduit if necessary to consume the entire length. See Fig. 18-20. In that length of

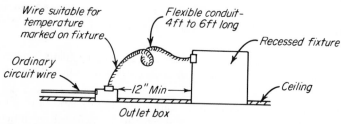

Fig. 18-20 Schematic outline for installing a flush fixture.

flexible conduit, install wire suitable for the temperature; fixture wire may be used, if of a type suitable for the temperature. Install ordinary wire up to the junction box.

A better method is to use fixtures designed to operate within the prescribed temperature limits, which permits using ordinary wire. A fixture so constructed and properly installed is shown in Fig. 18-21.

Testing.   When the roughing-in work has been finished, the installation must be tested.   Then in case of error or mishap, the wiring is still accessible for correction.   With conduit wiring, since the wires are not pulled in until after completion of the building, the test is made at that time.

Fig. 18-21   A well-designed and properly installed flush fixture. (*The Kirlin Co.*)

Usually the test device consists of a doorbell or buzzer in connection with two dry cells connected in series, as shown in Fig. 18-22.   Tape

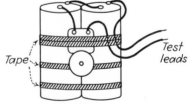

Fig. 18-22   Test outfit consisting of two dry cells and a doorbell.

the bell to the cells; also tape the test leads to the cells so that, if the entire unit is lifted by the test leads, there will be no strain on the bell or on the dry-cell terminals.

Before proceeding to test, go around to each outlet, remove insulation from the ends of the wires, and twist together all those which will ultimately be permanently connected to each other.   Leave those to which a fixture or receptacle is to be attached protruding freely out of the outlet.   At all points where a switch is to be installed, temporarily twist together all the wires that will later be connected to any one switch. Be certain that no exposed bare wire is allowed to touch an outlet box or conduit or the armor of cable.

Testing Conduit Installation. *We must assume that the power has not yet been turned on.* We must assume that the appliances which will be controlled by thermostatic or similar means within the appliance (water heater, motor) have not yet been installed. The transformer for doorbell or chimes must not be connected.

Connect one lead of your test bell to the neutral strap of your service equipment, and touch the other lead to each of the black wires to the individual circuits in turn. The bell should *not* ring. If it does ring, it indicates either a short circuit from the white to the black wire in the circuit under test, or a ground from the black wire to the conduit or a box, possibly at an outlet box where the black skinned wire has been allowed to touch the box.

If this test checks OK, make a different test. If you are using circuit breakers, turn them all on. If you are using fuses, be sure there is one in each fuseholder. Then remove the bell from your batteries, and leave just the leads from the batteries. Connect one of these leads to the neutral strap in the service equipment, the other to *both* of the terminals in the service equipment to which the incoming hot service wires are connected. This will mean a temporary jumper from one terminal to the other which you must later remove—or you will have a direct short circuit at 230 volts when the power is turned on.

Then take the bell to each outlet where a fixture or receptacle is to be installed. Touch it across the black and white wires; in each case it should ring, just as the lamps in your fixture or floor lamp will later light when connected to the same wires. Of course, this assumes that previously all the wires which go to any one switch have been temporarily twisted together, thus duplicating the condition of all switches turned on. After each check across the black and white wire at each outlet, touch the bell across the black wire and the outlet box itself; the bell should again ring because through the conduit all the boxes are connected together and grounded and the white wire in turn is also grounded. The bell will probably not ring loudly because of higher resistance through the conduit than through the wire, but it should ring nevertheless. In making this test, if the conduit and boxes have black-enamel finish, it will be necessary to scrape off some of the enamel before touching the bell to the box, for the enamel is an insulation which might prevent the bell from ringing. If the bell rings feebly, a poor job has been done somewhere; probably the locknut on one or more boxes has not been run down tightly enough. Grounding is a safety precaution and must be properly done. If the bell rings at each point, the wiring is all right.

Testing Armored-cable Installation. Proceed exactly as with conduit.

Testing Nonmetallic-sheathed-cable Installation. If you have used metal boxes and cable with the bare grounding wire, properly installed, test as in the case of a conduit installation.

If you have used nonmetallic boxes (or metal boxes but cable without the bare grounding wire), naturally the bell cannot ring when touched to the black wire and the box. However the remainder of the test is the same as with the conduit system.

If all tests check, proceed to finish the installation as discussed in the next chapter.

19

Finishing: Installation of Switches and Other Devices

In the previous chapter the roughing-in of a considerable assortment of outlets was discussed. Assume now that that work has been finished and that you are ready to install the switches and receptacles, the wall plates, the fixtures, and so on.

Installing Switches, Receptacles, etc. Every device of this kind is provided with a metal strap which has holes in the ends, so spaced as to fit over the holes in the ears of a switch box, on which it is mounted by means of machine screws that come with the device. The wall plate in turn is anchored to the device, not the box (see Fig. 19-1).

For a neat installation, the strap of the device must be flush with the front of the plaster. Since plastering is done after the switch boxes are installed, the front edges of the boxes are not always flush with the plaster; usually they are, and should be, a trifle below the surface of the plaster. Some means must therefore be used to mount the devices flush with the plaster. Most devices come equipped with "plaster ears" on the ends of the strap, as was shown on the switch in Fig. 4-

7. These ears lie on top of the plaster, automatically bringing the device flush with the surface. The metal is scored near the end of the strap so that the ears can easily be broken off if they are not required. If your device does not have plaster ears, you can insert small washers between the box and the strap of the box.

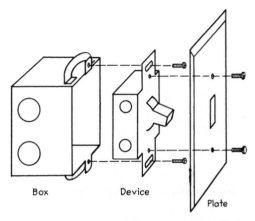

Box   Device

Plate

Fig. 19-1 The switch or other device is installed on the box, and the wall plate is installed on the device.

Neatly installed devices must be absolutely straight up and down; often the boxes are not entirely straight. For this reason mounting holes in the ends of the mounting strap are not round but elongated, so that the device itself can be mounted straight even if the box is not straight. A glance at Fig. 19-2, which exaggerates the usual condition, should make this clear.

Before connecting wires to the terminals of switches and similar devices, cut off some of the excess wire protruding from the box, leaving only enough to make it easy to make connections to the terminals. But do not be too enthusiastic about cutting off the last fraction of an inch, for when much later a switch needs to be replaced, the terminals may be located somewhat differently from those on the original switch, requiring perhaps half an inch more wire. There is a happy medium of length: Cut off enough so that extra wire will not crowd the box, yet leave enough to make installation of the device easy.

Use care in preparing the wires for connection to terminals; skin off only enough insulation to make the connection.   The insulation of the wire should extend up to the terminal.   *There must be no bare wire exposed between the terminal and the end of the insulation.*

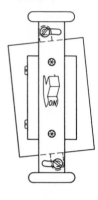

Fig. 19-2   The mounting straps of devices have elongated holes, in the ends, to permit vertical mounting even if the box is not mounted straight.

Installing Wall Plates.   After the device has been installed in a switch box, a wall plate must in turn be installed by means of the screws that come with the plate.   If the device has been properly mounted, the plate will fit snugly against both the device and the wall.   If the device has been mounted slightly below the surface of the plaster, do not pull up too tightly on the screws holding the plate, for this will distort or damage the plates; it is not unusual to crack the bridge in the center of a bakelite duplex receptacle plate by pulling up too tightly.   If the switches, receptacles, or other devices have plaster ears, the problem is automatically solved.   If they do not have such ears, it is best to use spacing washers between the outlet box and the strap of the device in order to bring the device flush with the plate.

If the box is of the 2- or 3-gang type with a number of separate devices, the mounting of the plate will not be entirely simple because of the elongated holes in the mounting straps of the individual devices, as mentioned in an earlier paragraph.   These elongated holes are a tremendous advantage in mounting devices in a single-gang box; they are a nuisance in multigang boxes, because they permit mounting several devices so that they are not entirely parallel with each other, as compared with the absolutely parallel openings in a 2- or 3-gang plate. The only thing that can be said about mounting devices in multigang boxes is that extreme care must be used to see to it that all the devices are absolutely straight up and down, that is, that they are absolutely parallel to each other; then the holes for screws in multigang plates will automatically match up with the tapped holes in the straps of the individual devices mounted in the multigang boxes.   Unless you are very careful in this detail, you will waste a great deal of time in trying to insert screws through holes for them in plates, into corresponding holes in devices under the plate, which holes, however, will be found to be displaced far enough so that the mounting screws will enter with great difficulty (which tends to crack the plates) or not enter at all—there may be no visible holes for them to enter.

If you are not experienced, a knack that will save you much time is shown in Fig. 19-3.   Start with a 3- or 4-gang metal plate, preferably

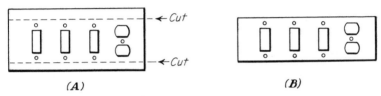

(A)   **(B)**

Fig. 19-3   This gadget simplifies installing several devices in a multigang box.

with openings for switches and a receptacle, as shown at A.   Saw off the edges, very close to the mounting holes that normally are used to install the plate on the switches, all as shown at B of the illustration. After the switches (and receptacle if there is one) have been properly connected, install them in the box, but install the mounting screws loosely; do not tighten them.   Then temporarily install the sawed-off cover on the switches, tightening the mounting screws of the cover, so that all the devices are solidly anchored to the temporary cover.   Then all the devices will be parallel and the right distance apart, in spite of the oval holes in their mounting straps.   Only then tighten the screws in all the individual straps, solidly mounting the devices in the box, then remove the temporary sawed-off cover.   The holes in the final wall plate will now fit perfectly over the various switches and receptacles, and can easily be installed.

It must be noted however that this scheme will work only when using ordinary switches and similar devices, and not when using "interchangeable" devices described in the next chapter.

Wiring Lighting Fixtures.   The fixture may have two terminals for connecting the wires, in which case connect the white wire to the whitish terminal, the black wire to the other terminal. More often the fixture has two wire leads. If these are black and white there is no problem. Usually they consist of fixture wires, as described in Chap. 6: one a solid color, the other of the same color but identified with a colored tracer in the outer cover. The wire identified by the colored tracer is the neutral and corresponds to the white wire in the outlet box. In case the identification is not clear, trace the wires down into the fixture; the one that connects to the outer screw shells of the individual sockets is the neutral.

Hanging Fixtures. There are many ways of mounting lighting fixtures, all dependent on the style, shape, and weight of the fixture, the particular box involved, and method of mounting the box in the ceiling or wall.

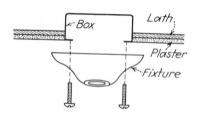

Fig. 19-4 Simple fixtures are mounted directly on the box.

Simple fixtures sometimes mount directly on top of outlet boxes by means of bolts which fit into the ears on the boxes. This method is shown in Fig. 19-4 and requires no further description. At times the fixture is too large to permit direct mounting in this fashion, in which

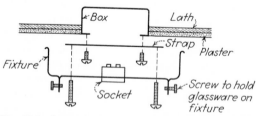

Fig. 19-5 Larger fixtures are mounted on a strap, installed on the box.

case a strap is used, as shown in Fig. 19-5. The strap is first installed on the outlet box, the fixture mounted on the strap. A detailed explanation should not be necessary.

Often a fixture stud, such as is shown in Fig. 19-6, is used in mounting the fixture.   Mount the stud on the bottom of the switch or outlet box by means of bolts, through holes provided for the purpose.   Some boxes

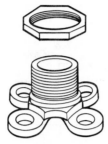

Fig. 19-6   Fixture studs are installed in the bottom of boxes, to support fixtures.

have the stud as an integral part of the box.   If you use a hanger of the types that were shown in Fig. 10-9, the stud that is part of the hanger goes through the center knockout in the box, serves to anchor the box to the hanger, and at the same time permits the stud to be used for support-

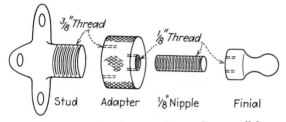

Fig. 19-7   A typical lockup unit for installing small fixtures.

ing the fixture.   The outside of the stud is tapped to fit ⅜-in. trade-size pipe; sometimes there is an inner female thread fitting ⅛-in. trade-size pipe.

Figure 19-7 shows the fittings generally used, called the "lockup unit."   It consists of a reducer fitting over the fixture stud, a length of ⅛-in. running-thread pipe, and a nut to hold the assembly together, the nut usually being of an ornamental nature and then called a "finial."   Figure 19-8 shows the same parts used to hold up a simple fixture.   It is a simple matter to drop down the fixture while making connections, then to mount it on the ceiling.

If the fixture is a larger unit, the mounting is similar, and Fig. 19-9 should make it clear.   The top of the fixture usually consists of a hollow stem with an opening on the side through which the two wires from the fixture emerge.   The top of the stem is threaded to fit on the fixture stud, and the mounting is as shown in the picture.   While the connections are being made, drop the canopy down and then slip it back flush with the ceiling.

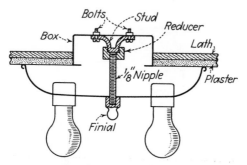

Fig. 19-8   This shows parts of Fig. 19-7, used to install a fixture.

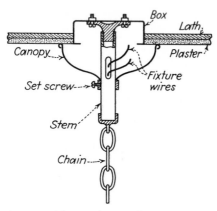

Fig. 19-9   The usual method of installing larger fixtures.

Sometimes the wires from the fixture come out of the end of the stem instead of through an opening in the side.   In that case use a "hickey," shown in Fig. 19-10, between the end of the stem and the fixture stud. Sometimes the stud is too short or the box too deep, in which case use an extension piece, as shown in Fig. 19-11.   Fixtures weighing more than 50 lb must be supported independently of the outlet box (Sec. 410-16 of the Code).

Fig.   19-10 A   hickey, used between the fixture stud and the end of the stem on the fixture.

Fig. 19-11 In deep boxes it may be necessary to use an extension piece over the fixture stud.

Mounting Wall Brackets.   The method of mounting depends to a large degree on the type of box used.   Many brackets are too narrow to cover up a 4- or even a 3¼-in. octagon box, so it has become customary to provide standard switch boxes on which to mount wall brackets. Sometimes a stud is mounted on the bottom of the box; in that case the mounting is completed with the lockup device that was shown in Fig. 19-7, the completed installation having the appearance shown in Fig. 19-13.   More usually a fixture strap, such as shown in Fig. 19-12, is first

Fig. 19-12   Fixture straps are mounted on switch or outlet boxes, and the fixture is then supported by the strap.

mounted on the switch box, the fixture in turn being mounted on the strap, all as shown in Fig. 19-14.

Adjusting Height of Fixtures.   A fixture with chain is adjustable as to the height above the floor and to compensate for ceilings of different heights.   The height is simply controlled by removing as many chain

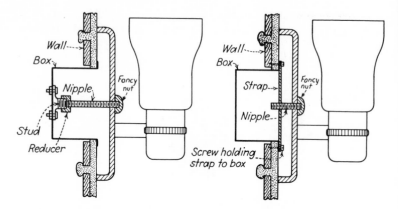

Fig. 19-13 Wall bracket installed on switch box by means of fixture stud in bottom of box.

Fig. 19-14 Wall bracket installed on switch box by a fixture strap mounted on the box.

links as required. The actual height will be governed by personal preference and location. If the fixture hangs above a dining-room table, hang it at least 24 in. and preferably 30 to 36 in. above the table. Again this will be dependent upon the type of fixture, the ceiling height, and similar factors.

20

Miscellaneous Wiring

In the wiring of any house, there are problems and niceties of detail that could not readily be included in the discussions of previous chapters, so they will be grouped here in a separate chapter. Study this chapter well because from it you should get ideas that will help you to make your installation one that is better than just average.

Heavy Appliances. The installation of appliances such as range, water heater, clothes dryer, and so on will be covered in the next chapter.

Quiet Switches. Ordinary switches have an annoying click as they are turned on or off. Usually this click may not even be noticed, but in sickrooms, children's rooms, and similar locations the click can be decidedly annoying. In the dead of night, a switch turned on or off in the hall on the *outside* of a bedroom wall can sound very loud inside that room.

Two kinds of noiseless switches are available and are in common use. The original noiseless switch was the mercury type, in which a pool of mercury inside a glass tube takes the place of mechanical contacts.   Mercury switches must be installed in a vertical position, and right end up, or they will not operate.   The "general-use AC-only" switches described in Chap. 4 are much less expensive, and are very quiet.

Lighted Switches.   Often it is quite a nuisance to find a switch in the dark.   Use switches with a lighted handle.   Such switches have a small neon lamp in the handle, which lights up when the switch is turned *off*.   The lamp consumes only a fraction of a watt and will last almost indefinitely.

Dimmers.   Some people like to be able to control the brightness of the lights in parts of their homes, such as dining room or recreation rooms.   This can now easily be done using special dimming switches, provided the lighting is from incandescent lamps, not fluorescent.   In each case, remove the ordinary switch, in its place install the dimming switch.   Two types are available.

One is a quite inexpensive switch that provides HIGH-OFF-LOW positions, and controls up to 300 watts.   It can be used to replace only a single-pole switch, not a 3-way type.   The somewhat more expensive type shown in Fig. 20-1 controls the brightness continuously from OFF

Fig. 20-1   A dimmer controls the brightness of lights; it also serves as a switch. (*Pass & Seymour, Inc.*)

to full bright, up to 600 watts.   Some models are available in the 3-way type, replacing one of a pair of ordinary 3-way switches.

Interchangeable Devices.   Several switch boxes can be joined together to make one 2-gang or 3-gang or even larger box, permitting two or three or more switches, receptacles, and similar devices to be used side by side.   However, the more devices used side by side, the larger the wall plate becomes; 3- and 4-gang plates are rather unsightly. Sometimes there is not room for a 3-gang plate at a particular point on a wall where three switches are to be located.   Even if there is room, the holes in the mounting straps of the switches are oval, as already explained, making it none too easy to install three switches in a 3-gang box so that a 3-gang plate will later fit easily and neatly.   Therefore it is not surprising that there were developed devices very small in physical size so that two or three can be used in a single-gang switch box.

The basic devices, such as switches, receptacles, pilot lights, and so on, are stocked separately; typical pieces are shown in Fig. 20-2.   They

Fig. 20-2   Separate devices of this kind are assembled on the job into any desired combination.   Three of them occupy no more space than one ordinary device.   (*Pass & Seymour, Inc.*)

are mounted by the user on the skeleton strap of Fig. 20-3 (which comes with the plate), and as shown in successive steps in the same illustration.

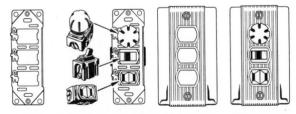

Fig. 20-3   The devices of Fig. 20-2 are assembled on a skeleton strap, as shown in the steps above.   (*Pass & Seymour, Inc.*)

While a single strap with three devices can be mounted in a single-gang switch box, or two straps with six devices in a 2-gang box, this leads to very considerable crowding in the box, and often to more wires than are permitted (per table in Chap. 10) in an ordinary switch box. For that reason, use a deep box which permits more wires; preferably use a 4-in.-square box with a raised cover. If the cover which was shown in *A* of Fig. 10-12 is used, three devices on one strap can be used; if the double type shown in *B* of Fig. 10-12 is used, six devices can be installed.

Plug-in Strip.   No matter how many receptacles that are provided in a home, there always seems to be need for more. At least for better homes, consider the material known as "plug-in strip" shown in Fig. 20-4. This consists of a steel channel with a cover providing outlets at

Fig. 20-4   Plug-in strip provides outlets at very frequent intervals, making a very flexible and adequate installation. (*Wiremold Co.*)

regular intervals. The spacing varies, but for homes 18 in. is popular. Several varieties are available. In some, the receptacles are wired by the installer; in others they are already connected to continuous parallel wires, at proper intervals to fit the openings in the cover.

The material is sometimes installed on top of the baseboard, giving the effect of being part of it. More often it is installed in kitchens some distance above the counter. Connections are made to the back of the channel with conduit or cable, as to an outlet box. More information about similar material will be found in Chap. 29.

Surface Wiring.   Where the wiring is to be permanently exposed as in some basements, attics, or garages, you have a choice of several methods. You may use cable or conduit with surface-type boxes of the type that were shown in Fig. 10-19. An alternate is to use nonmetallic-sheathed cable, preferably Type NMC, and special surface wiring devices which are combinations of box plus device such as switch or receptacle. Since such devices are more frequently used in providing additional outlets in buildings that were previously wired, they will be described at the end of Chap. 22, concerning old work.

Door Switches.   A touch of luxury is a built-in door switch in closets. Opening the door turns on the light; closing it turns the light off. One was shown in Fig. 12-2. Switches of this kind are still somewhat expensive; when they become available at more reasonable prices, they will receive the popularity they deserve.

Telephones. In ordinary residential work, too frequently no attention is paid to telephones, the problem of installation being left strictly up to the telephone people. They do a good job, but still in many cases an exposed run of wire remains in view. Therefore it is suggested that runs of conduit be installed, terminating in switch boxes, at the locations where the instruments are to be installed. Usually the switch box is covered with a special wall plate with a single bushed opening for the telephone cord.

Drop Cords. In making up a drop cord, assemble the parts so that their weight is supported, not by the copper conductor of the lamp cord used, but rather by the entire structure of the lamp cord, including the insulation and the braid. The simplest way of doing this is properly to install an Underwriters' knot at top and bottom. This knot is simply made, as Fig. 20-5 shows. The drop will be supported from the ceiling by a blank cover with a bushed hole in the center, mounted on any type of outlet box.

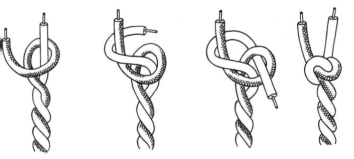

Fig. 20-5 Drop cords are not often used, but if one is installed, be sure to provide an Underwriters' knot at each end of the cord.

Garages. Garage lighting may consist of a single light, or it may incorporate a number of lights with an outlet for a "trouble light" of the type shown in Fig. 20-6, or for a battery charger or similar equipment. If the garage is attached to the house, no particular problems arise; treat the garage as you would another room in the house. If the garage is a separate structure some distance from the house, the following pages will discuss the problem.

If the garage light is to be controlled only at the garage, or only at the house, only two wires are required from the house to the garage. An outlet may also be installed, as shown by dotted lines in Fig. 20-7.

Fig. 20-6 A "trouble light" of this kind is almost a necessity in any garage; provide an outlet for it. (*General Electric Co.*)

If the light is to be controlled by 3-way switches in house and garage, then three wires must be run, as shown in Fig. 20-8. If the wires shown in dotted lines are disregarded, this becomes identical with Fig. 4-22, the basic diagram for 3-way switches. If, however, an outlet is installed as shown in the previous diagram, the outlet will be disconnected with the light when it is turned off at either end. This is undesirable because, for example, the outlet may be used for a charger which is to charge the battery in the car overnight, and the light should not burn all night. Therefore a fourth wire as shown in dotted lines in Fig. 20-8 is necessary, making the outlet strictly independent of the switches

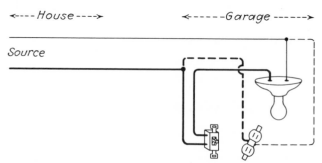

Fig. 20-7 Simple garage circuit. The light is controlled only by the switch in the garage. The receptacle outlet is always on.

and the light controllable from either end.   A "trick" circuit permits
using three wires instead of four, but it definitely violates Code requirements in several important respects and is an unsafe circuit.

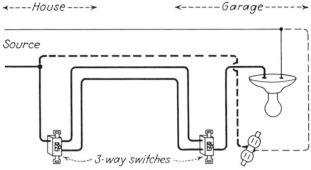

Fig. 20-8   The garage light is now controlled from either house
or garage.   The receptacle outlet is always on.   This requires
four wires between house and garage.

Very desirable also is a pilot light at the switch in the house (see Fig.
20-9).   If the dotted lines are disregarded, the result is the same as the
former circuit of Fig. 20-8.   To add the pilot light, run a fifth wire as
shown in dotted lines.

Whereas the circuit shown in Fig. 20-9 is the usual one when 3-way
switches are used, plus a pilot light at the house end, plus a permanently
live receptacle at the far end, there is another circuit available which
requires only four wires instead of five.   It nevertheless meets Code re-

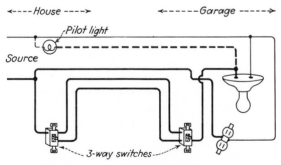

Fig. 20-9   The circuit of Fig. 20-8, with the addition of a
pilot light in the house, to indicate whether the garage
light is on or off.   This requires five wires.

quirements and therefore may be used.   It is shown in Fig. 20-10.   It requires a bit more care in installation to make sure all connections are correct.

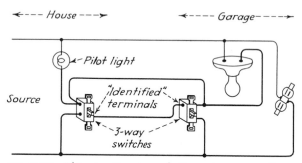

Fig. 20-10   This circuit, using only four wires, serves the same purpose as the circuit of Fig. 20-9 which requires five wires.

If the wires to the garage are to run overhead, they must be securely anchored at each end.   Either of the insulators shown in Fig. 17-3 or 17-4 may be used.   Where the wires enter or leave a building, either of the methods shown in Fig. 20-11 is suitable.   A most convenient fitting to be used at that point is shown in Fig. 20-12 and shown installed in *B* of Fig. 20-11.   Be sure the insulators are mounted at a point lower than the entrance of the wires into the building.

Underground Wires.   Use any of the cables described in Chap. 17.

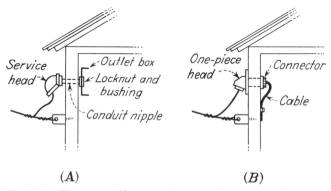

Fig. 20-11   Two ways of having wires enter the garage.   The same methods are used for other wiring, for example, farm buildings.

Outdoor Wiring. In most residential installations the outdoor wiring is limited to garage wiring, already described, and perhaps an outdoor outlet for Christmas-tree lights and similar purposes. A convenient unit for this purpose, shown in Fig. 20-13, is surface-mounted on the

Fig. 20-12 This wall-type entrance fitting is most convenient for bringing wires into outbuildings. (*Killark Electric Mfg. Co.*)

wall and has a flapper-type cover which covers the outlet when not in use. For outdoor switches, use the housing shown in Fig. 20-14. Install any kind of ordinary switch in the housing; it will be automatically

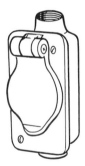

Fig. 20-13 For outdoor use, install an ordinary single receptacle in the housing above. The cover closes automatically when the outlet is not in use.

Fig. 20-14 Install any ordinary toggle switch in this weatherproof housing, for outdoor use. (*Killark Electric Mfg. Co.*)

protected from the weather. Outdoor wiring for farms will be discussed in Chap. 24. Commercial outdoor wiring will be discussed in Chap. 35.

Low-voltage Wiring. In this term is included wiring for doorbells and other signals, thermostats, and similar devices operating at low voltages.   Usually this means 30 volts or less.

The power for operating such circuits is usually derived from small transformers.   Under no circumstances may low-voltage wires be run in the same conduit or armor or cable with other wires carrying full voltage.   They must come no closer than 2 in. to other wires unless such wires are in conduit.   Where the wires come closer than 2 in., use loom or porcelain tubes over the low-voltage wires.   They may never enter an outlet or switch box carrying full-voltage wires unless a metal barrier of the same thickness as the walls of the box separates the two types of wiring.

Transformers.   If only the usual doorbells and buzzers are to be operated, ordinary doorbell transformers are used.   One type is shown in Fig. 20-15.   Mount the transformer near an outlet box.   Connect the

Fig. 20-15 A transformer for operating doorbells and similar equipment.   It delivers about 8 volts.   (*General Electric Co.*)

flexible (primary) leads to the 115-volt wires inside the box.   Similar transformers are available mounted on a box cover, which is then installed on an outlet box.   The screw terminals on the transformer deliver the secondary or low-voltage output of the transformer.   Such transformers have a maximum capacity of about 5 watts and usually deliver somewhere between 6 and 10 volts.   They are suitable only for operating a single device at a time.

Larger transformers are available which give a combination of voltages such as 6, 12, and 18, while again others are available in larger wattage capacity.

Transformers of this type are so designed that, even in case the secondary is short-circuited, the current flowing will be limited to the rating of the transformer.   Such transformers are rated not over 100 volt-amp, and the type used for residential wiring is seldom over 25 volt-amp. Because of this limited current there is no danger of fire, and because of the low voltage there is no danger of shock. Therefore the Code has no limitations on the type wire used or the installation of it, except those points brought out in the previous paragraph.

The wires used for low-voltage work require and have little insulation. Ordinary bell wire or "annunciator wire," as it is formally called, consists merely of the bare copper with a layer of plastic or two layers of cotton, wrapped in opposite directions, then paraffined (see Fig. 20-16). Two or more of these wires are often twisted together, with an over-all braid, forming what is known as "thermostat" cable; it is shown in Fig. 20-17. Each wire in the cable has a different color braid for

Fig. 20-16 Bell wire for doorbells or chimes requires little insulation.

Fig. 20-17 Thermostat cable consists of several wires of the type shown in Fig. 20-16, bundled together.

ease in identification. Use of this cable makes a much neater installation than use of two or more separate wires; there is also less danger of damage to the wires, which would be more a nuisance than a hazard. The usual size of the wire is No. 18, although No. 19 is also used, and the heavier sizes are available. The size must be chosen to match the length of the run, the load, and the voltage available. For ordinary residential use, No. 18 is universally used.

Fig. 20-18 Bell wire and thermostat cable are supported by insulated staples.

In use this wire is merely run over the surface or fished through walls without further protection. It is stapled to the surface over which it runs with insulated staples of the type shown in Fig. 20-18.

Low-voltage Circuits. It is a very simple matter to draw circuits for low-voltage work.

Simply consider the secondary of the transformer as the SOURCE for the circuit, and consider the push buttons as switches, which they are. The basic circuit is shown in Fig. 20-19.   Disregard for the moment the

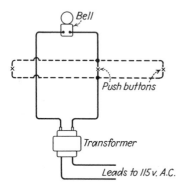

Fig. 20-19 The basic diagram for a doorbell is most simple.

dotted lines, and the circuit becomes most simple.   If the bell is to be controlled from a number of different push buttons, merely add additional buttons as shown in the dotted lines.

Figure 20-20 shows a similar circuit but with both a bell and a buzzer, the former for the front door and the latter for the back door. Figure 20-21 shows the same circuit using a combination bell and buzzer.   This device has three terminals.   One of the three is usually a terminal screw that is fastened directly to the frame of the device, not

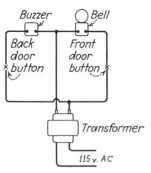

Fig. 20-20 This shows a bell and buzzer operating from one transformer.

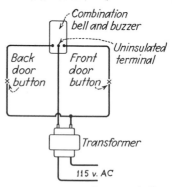

Fig. 20-21 A combination bell and buzzer has been substituted for two separate devices in Fig. 20-20.

insulated from it in any way. This is the terminal that runs direct to the transformer. Sometimes it is the middle terminal, as shown in the drawing; sometimes it is one of the other terminals. The two remaining terminals go directly to the two push buttons.

Additional buzzer circuits may be operated from the same transformer, as required. Mount the buzzer where desired; run a wire from one terminal direct to the transformer or, for that matter, to any other handy nearby wire that runs direct to the secondary of the transformer; from the other terminal of the buzzer run a wire to the push button, and from the push button back to the other side of the transformer or, if handy, to a nearby wire that runs directly to the other side of the transformer.

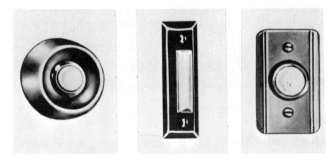

Fig. 20-22 Push buttons are available in many styles. (*NuTone.*)

Types of Push Buttons. Several types of buttons, both surface and flush mounting, are shown in Fig. 20-22; there is a great variety of sizes, shapes, and finishes to suit the user. For apartments, the buttons often are built into the entrance mailboxes.

Chimes. Instead of doorbells and buzzers, chimes are in very common use. They are available in many styles from the very simple to the very elaborate. Several are shown in Fig. 20-23. Most of them are designed to sound two notes (or play a tune) for the front door, one note for the back door.

Wire them as you would a bell or buzzer. Note that most chimes require more power to operate than bell or buzzer. For a good clear signal the transformer should deliver from 15 to 20 volts, as compared with the usual 6 to 8 volts for bells. If you are replacing a bell with chimes, you can use the existing wires, but you will probably have to replace the transformer with one of a higher voltage.

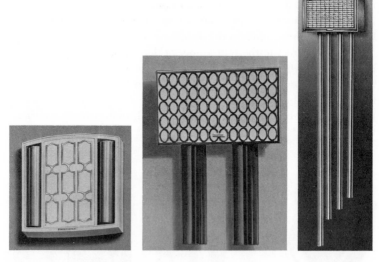

Fig. 20-23   Chimes with musical notes have almost completely displaced doorbells.   (*NuTone.*)

Remote-control Wiring System.   If a light is to be turned on from a single switch, the wiring is very simple indeed.   Even if it is to be turned on and off from two different points, the wiring is not especially complicated but often involves long runs of wire or cable.   If the light is to be controlled from three, four, or more points, the wiring becomes decidedly complicated.   It also becomes expensive because it requires 4-way switches, long runs of wire or cable, and much labor.   Yet lights controllable from many points are very desirable.   The remote-control system makes it possible to control a light from two, three, six, or even a dozen points at very reasonable cost.

In this system the 115-volt wires end at the outlet boxes for fixtures that are to be controlled by switches; they are not run to the switch locations.   An electrically operated switch called a relay is mounted *in the outlet box* for each fixture or receptacle to be controlled separately.   One is shown in Fig. 20-24.   Such relays will handle loads up to 20 amp at any voltage up to 277 volts.   From each relay, three small wires are run to simple switches, located in as many places as you wish.

The power for operating the relays come from a single transformer installed in the basement or any other convenient location. It steps the 115-volt AC down to about 24 volts AC. The transformer is shown in Fig. 20-25 and is the type already described; the voltage is so low that there is no danger of shock, and even if it is short-circuited, it delivers so little power that there is no danger of fire.

Fig. 20-24 The remote-control relay is an electrically operated switch. It is installed in the outlet box on which the fixture is mounted. (*General Electric Co.*)

Fig. 20-25 This transformer delivers about 24 volts. One transformer furnishes all the power for all the relays in an installation. (*General Electric Co.*)

The operation of the relay is shown in Figs. 20-26 and 20-27 which, however, do not pretend to show the exact mechanical arrangement inside the relay, but rather only the principle of operation. Inside the relay are two coils or electromagnets, *A* and *B*, connected for the moment to some dry cells for power, and two push buttons *A* and *B*. When push button *A* is momentarily closed as in Fig. 20-26, current flows through coil *A*, thus making a magnet out of it while the current flows. This attracts the upper end of the armature inside the relay, closing the 115-volt circuit as shown. The circuit stays closed even if coil *A* is no longer energized; a momentary flow of current while the button is closed is sufficient.

If later push button *B* is momentarily closed as in Fig. 20-27, current flows through coil *B*, thus making a magnet out of it while the current flows. This attracts the lower end of the armature, opening the 115-volt circuit as shown. Thus the relay operates like any other switch, except that it is controlled from a distance.

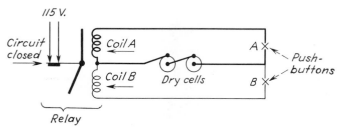

Fig. 20-26   This shows the principle of operation of the remote-control system. Coil *A* in the relay is momentarily energized, which turns the switch on.

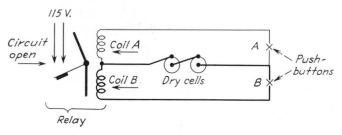

Fig. 20-27   The circuit of Fig. 20-26, but with coil *B* momentarily energized. This turns the switch off.

The relay is made so that the round shank fits into a half-inch knock-out. Push the relay through a knockout from the inside of the outlet box. The 115-volt leads are then inside the box, the three low-voltage leads on the outside of the box. Instead of using dry cells, naturally you will use the transformer already described.

Figure 20-28 shows one of the switches usually used. They resemble ordinary switches but require special plates with the correct size of openings. Switches for this purpose are constructed so that there is a neutral position of the handle. Push the top end momentarily to turn the light on; the handle then returns to the neutral position. Push the bottom end momentarily to turn the light off; again the handle returns to the neutral position. Because the voltage involved is only 24 volts, supplied by a special transformer with little total power, the switches need not be installed in switch boxes, but for the sake of neatness are nevertheless usually installed in boxes. If you want to install them without boxes, use frames for surface mounting, shown in the same Fig. 20-28.

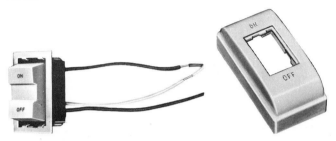

Fig. 20-28 The switch used in remote-control systems is equivalent to two push buttons. If to be surface-mounted, use the cover shown. (*General Electric Co.*)

Because of the low voltage and limited power of the transformer, the wires between relays, switches, and transformer do not have to be the relatively expensive kind used for ordinary 115-volt wiring, nor do they

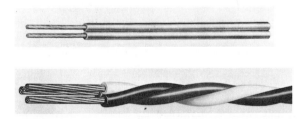

Fig. 20-29 Wire used to connect relays to switches does not need much insulation, because of the low voltage. Special cable shown is convenient. (*General Electric Co.*)

need to be run in conduit or be otherwise protected. Almost any kind of wire No. 20 or heavier may be used. Two kinds of cable specially designed for the purpose are shown in Fig. 20-29; the cable is available in 2-wire and 3-wire type. Run it any way you find convenient. Staple it to the surface over which it runs. In old work, fish it through walls, run it behind baseboards, or run it exposed.

See Fig. 20-30 which shows a complete installation of one outlet controlled by any one of four switches. Use as many as you wish and in any location you wish. Merely connect the white leads of all switches

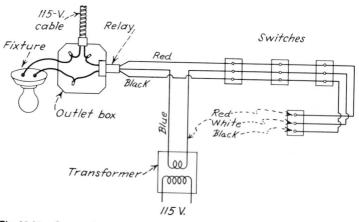

Fig. 20-30   One outlet controlled by any of four switches.

together, all the red, and all the black leads. (All these wires being low-voltage wires, the Code does not require that any particular colors be used. The colors mentioned, and shown in Fig. 20-30, happen to be the colors used on one particular brand of equipment.) If you wish to control two or more outlets at the same time from one switch, merely wire the several outlets together just as if ordinary wiring were being used, and install the relay in the most convenient box.

Numerous diagrams could be furnished for wiring many different combinations, but that should not be necessary. If you have studied the principle carefully, you will be able to make your own diagrams.

Installation Hints.   In new work, two methods of installation are possible. In the first method, all the work is done before the lathing and plastering is done. The relays are installed in the boxes, and the low-voltage wires connected to the relays. In that case be sure to leave at least 6 in. of slack in the low-voltage wires at the relay, so that if a relay at some future time should prove defective, it can be removed from the inside of the box; unless you provide that extra 6 in. of slack on the low-voltage side, it would be impossible to remove the relay. The alternate method is to install the relays after the plastering has been done. Install the low-voltage wires and let them project about 6 in. into the outlet box through the knockout in which the relay will later be mounted. Then, later, connect the low-voltage wires to the proper wire leads on the relay, push them out through the knockout, and install the relay in the knockout.

Use of this remote-control system will make the electrical system of any home many times as flexible as when ordinary switches are used. A switch can be installed in an additional location at a very nominal cost, even if the switch is an afterthought. Use it to approach the ideal —enter a house by any door, go to any place in the house, but never be in darkness, yet be able to turn off lights behind you without retracing your steps.

Using 3- and 4-way switches for basement or garage lights, you never know whether or not the light is on unless you can see the light; with the remote-control system, simply push the off button, and if the light was on, it is now off. Use it on the farm for yard lights, thus avoiding long runs of expensive 115-volt wiring. If you wish, install a master switch in the bedroom and thus be able upon retiring to make sure all lights are off without ever leaving the bedroom.

Three-wire Circuits. Assuming that the building has a 3-wire 115/230-volt service, 3-wire circuits can be used to good advantage as a method of reducing voltage drop. Let us analyze what a 3-wire circuit is.

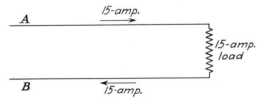

Fig. 20-31 One 2-wire circuit, carrying 15 amp.

See Fig. 20-31, which shows an ordinary 2-wire circuit; assume that it is wired with No. 14 wire, that it is 50 ft long, which means that the current flows through 100 ft of wire. Assume the load is 15 amp, the ampacity of the wire. The voltage drop then is approximately 3.86 volts, or about 3.3%.

Now see Fig. 20-32, which shows two such circuits, one on each leg of the 3-wire service.   The voltage drop on each circuit will still be 3.86 volts.   Note, however, that the two neutrals are *connected to each other* at the service and that they run parallel to each other.   That

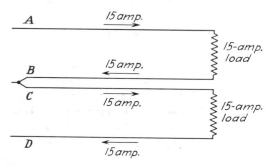

Fig. 20-32   Two 2-wire circuits, each carrying 15 amp and fed by opposite legs of a 115/230-volt 3-wire service.

being the case, why use two wires?   Why not use just one wire, as in Fig. 20-33?   You could easily jump to the conclusion that that might be all right, except that the one neutral wire serving two circuits would have to be twice as big as before to carry 2 × 15, or 30 amp.   That is a wrong conclusion.   In Fig. 20-32, each of the two neutral wires B and C does carry 15 amp, but note the direction of the arrows in the picture. The flow of current in B at any given instant is in a direction opposite

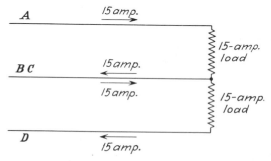

Fig. 20-33   If the two circuits of Fig. 20-32 run to the same location, do not run two neutrals.   Use one and make one 3-wire circuit.   Here each half of the circuit carries 15 amp.

to that in C. So also in Fig. 20-33: at any given instant the single wire BC can be said to carry 15 amp in one direction, also 15 amp in the opposite direction, and the two cancel each other. In other words, wire BC now carries no current at all; the circuit will operate just as well with wire BC missing. Note carefully that this statement is correct only when the amperage in the one circuit is exactly equal to that in the other.

But what about the voltage drop? In circuits of Figs. 20-31 and 20-32, the voltage drop is 3.86 volts, based on 15 amp flowing through 50 ft of wire A, plus 50 ft of wire B, or a total of 100 ft. In Fig. 20-33, however, 15 amp flows only through 50 ft of wire A, for wire BC carries no current. Therefore the voltage drop is only half as great as in the case of Figs. 20-31 and 20-32. In the entire 3-wire circuit there are only 150 ft of wire as compared with 200 ft in two 2-wire circuits. Therefore by using one 3-wire circuit instead of two 2-wire circuits, we save 25% of the copper and still reduce voltage drop by 50%.

If the two halves of the 3-wire circuit are not equally loaded, as, for example, in Fig. 20-34, there is still an advantage. Suppose, as shown

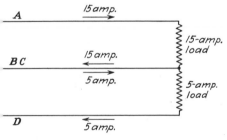

Fig. 20-34 The 3-wire circuit of Fig. 20-33, one half carrying 15 amp, the other half carrying 5 amp.

in that picture, one half of the circuit carries 15 amp, the other half 5 amp, then the neutral carries the difference, or 10 amp. The voltage drop will not be reduced by 50%, as in the case of equally loaded halves, but the total losses in the 3-wire circuit will be less than in two separate 2-wire circuits. If one of the two halves carries no current at all, then the other half functions exactly like any 2-wire circuit. This is also what happens if a fuse blows in one of the two hot wires; what is left is an ordinary 2-wire circuit.

While Figs. 20-33 and 20-34 show 3-wire circuits with a single load at the far end of each line, 3-wire circuits are not limited to such applications.   See Fig. 20-35, which shows a 3-wire circuit with loads connected at various points.   No matter how these loads are spaced, the total losses in such a 3-wire circuit are always lower than in two separate 2-wire circuits.

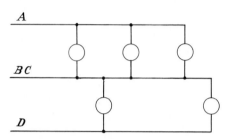

Fig. 20-35   Three-wire circuits need not carry a single load.   Install load where required.

Application of 3-wire Circuits.   If in ordinary wiring (whether in house, barn, or other building) two separate 115-volt circuits would ordinarily be installed more or less parallel to each other, each circuit using two wires, install one 3-wire circuit.   When using cable, remember that 3-wire cable has one white wire; it is the neutral of the 3-wire circuit.

When a 3-wire circuit is installed using cable, the cable will have the usual white neutral wire and one black, one red.   When using conduit, choose these same colors.   Having one hot wire black and the other red makes it easy to know which "leg," or half of the circuit, you are working on.   Be sure that the white wire runs to every outlet, for if you were to connect an outlet between the red and the black wire, it would then operate at 230 instead of 115 volts.   Be sure to distribute the receptacles on the circuit more or less evenly between the two legs.   If two receptacles are located close to each other, so that it is likely both will be used at the same time, be sure that they are on opposite legs of the circuit.

"Safety" Receptacles.   Children automatically have a great deal of curiosity, a tendency to explore.   What is more natural than for a child to push part of a toy, a loose hairpin, or any other loose metallic object into one of the slots of a receptacle?   If the child is on a completely dry floor, probably no harm will be done.   If he is also touching a grounded object such as a radiator, a water pipe, or the framework of a defective lamp, he may be subject to a violent shock. Every year many deaths of children can be traced to this cause.

For that reason "safety" receptacles were developed. One type has a spring-loaded cover over the slots of the receptacle; unless the receptacle is in use with a plug in it, the slots of the receptacle are concealed. Another type is so designed that inserting anything into *one* slot will not make electrical contact; both the prongs of a plug must be inserted before contact is made. Using safety receptacles is sensible procedure.

Swimming Pools. The subject of electrical components in connection with *portable or aboveground* pools is not covered by Code. It is only logical that you use the utmost common sense in such work. Motors as on pumps should be provided with the third grounding wire in the cord; it makes sense to remove such motor-driven pumps entirely from the pool while it is occupied. If there are lights they should be installed so that they (and the wires serving them and the switches controlling them) cannot possibly be touched by anybody in or near the pool.

In Art. 680 the Code does cover the subject of permanent pools, with water level below ground level. The requirements are stringent and severe, as they should be, for nowhere is an individual more subject to shock or injury or even death, than when he is in a pool in which the electrical work is improperly installed.

The student should study Art. 680—and then leave the installation to others who are well versed, and more experienced, in the subject.

21

Wiring of Heavy Appliances

The Code classifies appliances into three groups: portable, stationary, and fixed. Details of installation depend on the class of the appliance, and will be discussed in this chapter.

Portable Appliances.   This group includes those which in normal use are moved about: toasters, irons, percolators, hand tools, and so on. They are equipped with cord and plug, but just because an appliance has a cord and plug does not necessarily mean that it is classed as portable.

Stationary Appliances.   This group comprises those which once installed, are left in the original location, but which can nevertheless be moved fairly easily.   Self-contained ranges and clothes dryers are good examples.   They are connected by cord and plug, and if the owner moves from one location to another, he can easily move them with his other household goods.

Fixed Appliances.   This group consists of appliances which, once installed, cannot readily be moved because of, for example, connections to plumbing or being built into the structure of the building. Examples are water heaters, garbage disposers, oil-burner motors, also *built-in* ovens and counter-installed range tops (but not self-contained ranges). These might occasionally be installed using a cord and plug for convenience, but that does not make them portable or stationary.

Larger Receptacles. The ordinary receptacle installed in a home is rated at 15 amp and 125 volts. Many appliances are rated at more than 15 amp and many operate at 230 volts; in nonresidential work voltages higher than 230 volts are also encountered.

The illustrations of Figs. 21-1 and 21-2 show some of the more common ones, and their usual application. Note that they are identified as 2-pole 2-wire or 2-pole 3-wire; as 3-pole 3-wire or 3-pole 4-wire; and so on. The number of *poles* indicates the number of circuit conductors normally carrying current. If the number of *wires* is one larger than the number of poles, it means that the receptacle has an extra opening, and the plug an extra prong, for connection of a grounding wire, provided only for safety, and never carrying current under normal conditions. Thus a 2-pole 2-wire receptacle is used for an appliance having two circuit wires, but no grounding wire; the 2-pole 3-wire is used for a similar appliance using two circuit wires, plus a grounding wire.

The small figures on the side of each illustration show the diameter of the receptacle in inches. In each diagram, the opening in the receptacle marked "G" is for the grounding wire; the opening marked "W" is for the (white) grounded neutral of the circuit.

The illustrations by no means show all available configurations. Those for higher voltages are not shown. They are available in 2-, 3-, 4-, and even 5-wire types. Besides the ordinary varieties there are others so designed that the plug cannot be inserted or removed without first twisting it to lock or unlock. Other brands are of totally different construction, so that only a plug and receptacle of the same brand will fit each other. However, the types illustrated are the more common varieties.

Receptacles come in a variety of mounting methods to fit various boxes and plates, flush and surface type. The 50-amp 125/250-volt is shown in both the surface-mounting and the flush-mounting type in Fig. 21-3, which also shows a typical plug with "pigtail" cord attached, to fit. These particular receptacles are used mostly in the wiring of electric ranges. A similar receptacle but rated at 30 amp and with different configuration of openings is used mostly for clothes dryers.

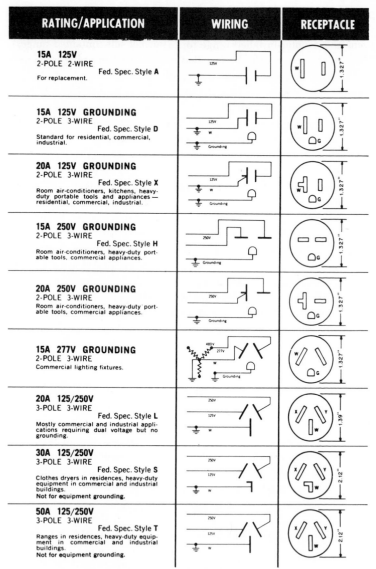

RATING/APPLICATION	WIRING	RECEPTACLE
15A 125V 2-POLE 2-WIRE Fed. Spec. Style **A** For replacement.		
15A 125V GROUNDING 2-POLE 3-WIRE Fed. Spec. Style **D** Standard for residential, commercial, industrial.		
20A 125V GROUNDING 2-POLE 3-WIRE Fed. Spec. Style **X** Room air-conditioners, kitchens, heavy-duty portable tools and appliances — residential, commercial, industrial.		
15A 250V GROUNDING 2-POLE 3-WIRE Fed. Spec. Style **H** Room air-conditioners, heavy-duty portable tools, commercial appliances.		
20A 250V GROUNDING 2-POLE 3-WIRE Room air-conditioners, heavy-duty portable tools, commercial appliances.		
15A 277V GROUNDING 2-POLE 3-WIRE Commercial lighting fixtures.		
20A 125/250V 3-POLE 3-WIRE Fed. Spec. Style **L** Mostly commercial and industrial applications requiring dual voltage but no grounding.		
30A 125/250V 3-POLE 3-WIRE Fed. Spec. Style **S** Clothes dryers in residences, heavy-duty equipment in commercial and industrial buildings. Not for equipment grounding.		
50A 125/250V 3-POLE 3-WIRE Fed. Spec. Style **T** Ranges in residences, heavy-duty equipment in commercial and industrial buildings. Not for equipment grounding.		

All spec. styles shown refer to Federal Specification W-C-596a.

Fig. 21-1 This shows receptacle configurations for many different ampere and voltage ratings. A plug that fits one receptacle will not fit another. (*General Electric Co.*)

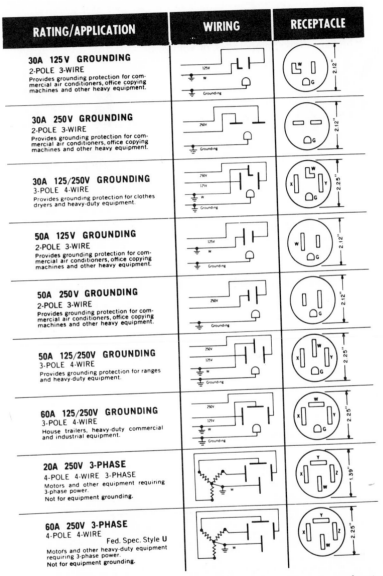

RATING/APPLICATION	WIRING	RECEPTACLE
30A 125V GROUNDING 2-POLE 3-WIRE Provides grounding protection for commercial air conditioners, office copying machines and other heavy equipment.		
30A 250V GROUNDING 2-POLE 3-WIRE Provides grounding protection for commercial air conditioners, office copying machines and other heavy equipment.		
30A 125/250V GROUNDING 3-POLE 4-WIRE Provides grounding protection for clothes dryers and heavy-duty equipment.		
50A 125V GROUNDING 2-POLE 3-WIRE Provides grounding protection for commercial air conditioners, office copying machines and other heavy equipment.		
50A 250V GROUNDING 2-POLE 3-WIRE Provides grounding protection for commercial air conditioners, office copying machines and other heavy equipment.		
50A 125/250V GROUNDING 3-POLE 4-WIRE Provides grounding protection for ranges and heavy-duty equipment.		
60A 125/250V GROUNDING 3-POLE 4-WIRE House trailers, heavy-duty commercial and industrial equipment.		
20A 250V 3-PHASE 4-POLE 4-WIRE 3-PHASE Motors and other equipment requiring 3-phase power. Not for equipment grounding.		
60A 250V 3-PHASE 4-POLE 4-WIRE Fed. Spec. Style **U** Motors and other heavy-duty equipment requiring 3-phase power. Not for equipment grounding.		

Fig. 21-2 Similar to Fig. 21-1, except still more types.   Note the application data at left of each receptacle, also the wiring diagram. (*General Electric Co.*)

All receptacles are so designed that a plug that will fit a receptacle rated at a particular amperage and voltage, will not fit a receptacle rated for higher or lower amperages, or voltage.

Fig. 21-3 Typical 50-amp receptacles, surface and flush, and a "pigtail" cord connector. This is a typical receptacle for a range. (*General Electric Co.*)

Individual Circuits for Appliances. The Code rules as to when an appliance requires an individual branch circuit are not too well defined. In general, you will be meeting Code requirements if you will provide a separate circuit for each of the following:

1. Range (or separate oven or counter units).
2. Water heater.
3. Clothes dryer.
4. Dishwasher.
5. Garbage disposer.
6. Any 115-volt fixed or stationary appliance rated at 12 amp (1,380 watts) or more. This includes motors.
7. Any fixed or stationary 230-volt appliance.
8. Any automatically started motor such as oil or gas burner, furnace fan, pump, and so on.

Appliance Circuits. The Code in Art. 210, and also in Art. 430 in the case of motor-driven appliances, outlines conditions for circuits serving appliances.

If a circuit serves a single appliance and nothing else, per Sec. 422-6 the branch-circuit overcurrent protection may be up to 150% of the ampere rating of the appliance. For appliances smaller than 10 amp the circuit no doubt will be No. 14 wire with 15-amp overcurrent protection, and that is acceptable.

In the case of circuits serving other loads in addition to appliances, no portable appliance may exceed 80% of the ampere rating of the circuit. In the case of 15- and 20-amp circuits there is a further provision that the total rating of *fixed* appliances may not exceed 50% of the rating of the circuit, if that circuit serves also lighting outlets or *portable* appliances.

Disconnecting Means. Every appliance must be provided with some way of totally disconnecting it from the circuit, as is necessary when making repairs, or sometimes while inspecting or cleaning. This is a wise safety precaution. The requirements will be discussed in the following paragraphs.

Disconnecting, Portable Appliances. A plug-and-receptacle arrangement is all that is needed. Of course the plug and receptacle must have a rating in amperes and volts at least as great as that of the appliance.

Disconnecting, Stationary and Fixed Appliances. If the appliance is rated at not more than 300 watts or $\frac{1}{8}$ hp, no special procedure is necessary. If the rating is higher, and the branch circuit to which it is connected is protected by a circuit breaker, no further action is required. If the circuit is protected by a fuse or fuses mounted on a pull-out block in the service equipment, no further action is required. But if the circuit is protected by plug fuses, a separate switch of the general type that was shown in Fig. 13-3, must be installed. If the appliance has a cord and plug (for example, a range or clothes dryer), that is sufficient in place of a switch.

One further exception: If the appliance has built-in switches so that the entire appliance can be disconnected by the switch or switches, no separate disconnecting means is required, for then the service-entrance switch will serve the purpose. However, if the building is an apartment with three or more individual apartments, the service switch for each apartment must be within each apartment, or at least on the same floor.

Wiring 230-volt Appliances with Cable. You have already learned that white wire may be used only as a grounded neutral wire. But the grounded neutral does not run to a load operating at 230 volts. Therefore the wires to a 230-volt load may not be white, but when you are using a 2-wire cable to wire a 230-volt appliance, the cable contains one black wire and one white wire, which may not be used. What to do? Follow Sec. 200-6(b) of the Code: Paint both ends of the white wire black, and the cable will be considered as having two black wires.

Bear in mind that an electric range does not operate at strictly 230 volts. When any burner or the oven is turned to "high" heat, it operates at 230 volts; when turned to "low" heat, it operates at 115 volts. In other words it is a combination 115/230-volt appliance; therefore all three wires including the neutral must run to the range.

Wiring Methods for Heavy Appliances. The Code does not restrict the methods to be used. Use conduit or cable, as you choose. However, the Code in Sec. 338-3(b) establishes one important point. For ranges (including wall-mounted ovens and counter cooking units) and clothes dryers, you may use service-entrance cable with a bare neutral, which otherwise may be used only in the service entrance.[1] There is however one restriction: The cable must run directly from the overcurrent protection in the service entrance. This is practically automatic in ordinary residential work, but in larger installations such as apartments, often there is a feeder from the service equipment to panelboards located at a distance, from which points circuits run to various locations, as required. Such panelboards are *not* part of the service entrance, and cable with a bare neutral is not permitted.

If the appliance is portable or stationary, run your conduit or cable up to the receptacle, which may be either flush-mounted or surface-mounted.

Wiring of Ranges. It is not likely that all the burners of an electric range, and the oven, will ever be turned on to their maximum capacity, all at the same time. For that reason the Code, in Sec. 210-19(c), Exception 1, permits wires in circuits to ranges to be smaller than would be determined by the rating of the range. If the range is rated at not over 12,000 watts, figure your wire size based on 8,000 watts; if it is over 12,000 watts, add 400 watts extra for each 1,000 watts (or fraction thereof) of rating above 12,000 watts. On that basis, No. 6 wires are usually installed; for smaller ranges No. 8 is sometimes used.

Because of the way in which the individual burners and the oven are connected within the range, the neutral cannot be made to carry as many amperes as the hot wires. For that reason, the neutral may be smaller, but never with an ampacity of less than 70% of the ampacity of the hot wires, and in no case smaller than No. 10. The "rule of thumb"

---

[1] Such cable with a bare neutral may be used in any circuit provided the bare wire is used only for grounding; it may not carry current in normal operation, except in the case of ranges or clothes dryers.

is to use a neutral one size smaller than the hot wires: No. 8 neutral with No. 6 hot wires, and so on.

Any wiring method may be used, including service-entrance cable with bare neutral. Run the wires up to a range receptacle of the type shown in Fig. 21-3, flush or surface-mounted, as preferred. The range will be connected to the receptacle using a pigtail cord shown in the same illustration. The plug and receptacle constitute the disconnecting means.

The Code requires that the frames of ranges (and separate ovens and cooking units) be grounded, but does not require a separate grounding wire. The range is so constructed that it is automatically grounded through the neutral wire, when you use one of the pigtail cords and range receptacles just described. The neutral must be No. 10 or larger.

Sectional Ranges. The trend is away from self-contained ranges consisting of oven and burners, toward individual units. The oven is a separate unit, installed in the wall, or in the counter, where wanted. Groups of burners in a single section are installed in the kitchen counter where convenient. All this makes for a flexible arrangement, and permits you to use much imagination in laying out a modern, custom-designed kitchen.

The Code calls such separate ovens "wall-mounted ovens" and the burners "counter-mounted cooking units." Here they will be referred to as merely ovens and cooking units. Self-contained ranges are considered stationary appliances by the Code [Sec. 422-23(c)], but ovens and cooking units are considered fixed appliances [Sec. 422-17(a)].

Two basic methods may be used in the wiring of ovens and cooking units. One way is to supply a separate circuit for the oven, another for the cooking unit. The alternate is to install one 50-amp circuit for the two combined.

Assuming that you install a separate circuit for the oven, proceed as outlined earlier in this chapter for fixed appliances. Use wire of the amperage required by the load. The oven will probably be rated about 4,600 watts, which at 230 volts is equivalent to 20 amp, so No. 12 wire would appear suitable, but the minimum is No. 10 because of the grounding requirement. At the oven the wires may run directly to the oven, but for convenience you will probably install a 30-amp receptacle similar to the 50-amp shown in Fig. 21-3, with a pigtail cord, also shown in Fig. 21-3. This method, using a receptacle and pigtail cord, will make the installation easier, but do note that the receptacle will *not* serve as the disconnecting means as it does when a self-contained range is installed (because a self-contained range is defined as a stationary appliance, and the oven, or cooking unit, as a fixed appliance). Therefore, unless your branch circuit is protected by a circuit breaker, or fuses on a pullout block, you will have to install a separate disconnecting means as outlined earlier in this chapter.

In installing a cooking unit, proceed exactly as with the oven, again using a minimum of No. 10 wire, and preferably a receptacle and pigtail cord. Number 10 wire with an ampacity of 30 amp will provide a maximum of 6,900 watts, which will take care of most cooking units.

If you install a single circuit for the oven and cooking unit combined, follow the circuit of Fig. 21-4. To determine the wire size, add to-

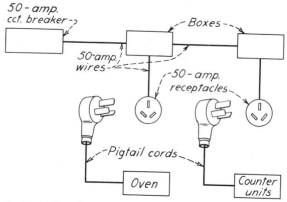

Fig. 21-4. It is best to provide a separate circuit for oven and counter units, but both *may* be connected to one 50-amp circuit.

gether the ratings of the two separate parts, then proceed as if you had a self-contained range of that rating. The same wire size must be used from the starting point of the circuit, up to the oven and up to the cooking unit, or to the receptacles for these components.

Section 210-19(c), Exception 5, of the Code is often misinterpreted to mean that the wire size may be reduced at the point where the circuit splits, with one set of wires to the oven and another to the cooking unit. Not so! It does permit the *leads* from the oven or cooking unit (as in a pigtail cord) to the receptacles, or to the splice with the circuit wires, to be smaller if large enough to carry the load, and if not longer than necessary to service the appliance, and in no case smaller than No. 10.

The receptacles shown in Fig. 21-4 are not required, but will be found convenient in installing and servicing; the oven and cooking unit may be wired directly to the junction boxes in the diagram. The receptacles must be rated not less than the rating of the circuit. If you have used No. 6 wire, the receptacles must be rated at 50 amp.

The neutral of any circuit to a self-contained range, an oven or a cooking unit may be bare only if service-entrance cable with a bare neutral is used. Using any other wiring method, the neutral must be insulated.

Clothes Dryers. Dryers are basically 230-volt appliances, although most have 115-volt motors in them. The Code requires that the frame of the dryer be grounded, but permits it to be grounded by the neutral conductor in the wiring (provided it is No. 10 or heavier wire); therefore if cable is used it must be 3-wire, even if the entire dryer operates at 230 volts. The Code considers dryers to be stationary appliances.

Wire the dryer as you would the range. Use a 30-amp receptacle similar to the 50-amp shown in Fig. 21-3, and a pigtail cord similar to the one used with a range, but smaller. No. 10 wires are generally used. The Code permits you to use service-entrance cable with a bare neutral, if you wish, provided it is No. 10 or heavier. The plug and receptacle will serve as the disconnecting means.

Automatic Washers. Is this a fixed or stationary appliance? The Code isn't clear. If the washer is installed with solid piping to the water system, it certainly should be considered fixed. If it is connected to the water system by flexible hose it could be considered a stationary appliance.

If the washer is equipped with a cord and plug, it will be a 3-conductor cord with 3-prong plug, and the frame of the washer will be automatically grounded, if the receptacle has been properly installed. If it does not have a cord and plug, you must ground the frame of the washer, following the last paragraphs of this chapter.

Water Heaters. The power consumed by a water heater ranges from 2,300 to 5,000 watts. The wiring is simple. Merely run two wires from your service equipment to a disconnecting switch, then to the heater. Number 12 wire is suitable for any heater consuming up to 4,600 watts.

If the service equipment consists of circuit breakers, provide a 20-amp 2-pole breaker. If it is fused equipment of the general type that was shown in Fig. 13-5, you will usually find a pair of special terminals for the water heater, usually not protected by fuses.   In that case, run wires from those terminals to the fused disconnecting switch for the heater.

In most localities power used for heating water is sold at a greatly reduced rate, often in the area of 1 cent per kilowatthour.   Sometimes there is a "catch," inasmuch as the heater is connected to the circuit through a special electrically operated switch furnished by the power supplier, which connects the heater to the power line only during "off-peak" hours.   In other words the heater is *disconnected* from the power line when the power supplier's load is at its peak, when there would not be enough generating capacity if many thousands of water heaters were also connected.   As a result water cannot be heated for several periods of several hours each, on any one day.   If your installation must be of this type, do the wiring as already described, except that the wires must start at the power supplier's time switch, instead of at the usual service equipment.

Grounding.   The grounding of ranges and dryers has already been discussed.   As to any other appliance (fixed or stationary type), if it is equipped with a 3-wire cord and plug, plugged into a properly installed receptacle, the grounding has been taken care of.   If it is permanently installed, and the wiring is in conduit or using armored cable, anchored solidly to a terminal box on the appliance, it is properly grounded.   If however you use nonmetallic-sheathed cable, you must use the kind with the bare grounding wire, which must be connected to the frame of the appliance, and grounded to the box nearest the appliance.

On farms, special care must be used in grounding the water heater. Being connected to water pipes, you can argue that the heater is automatically grounded.   That is so if the service is also grounded to the *same* pipe.   If however the water pipe is very short, and the service is grounded to a driven rod, that is not enough: The short section of pipe must be interconnected with the driven rod.   This is a most important safety measure.

22
Old Work

In old work, or the wiring of buildings completed *before* the wiring is started, there are few *electrical* problems that have not already been covered. Most difficulties can be resolved into problems of carpentry, in other words, how to get wires and cables from one point to another with the least effort and minimum tearing up of the structure of the building.

In new work it is a simple matter to run wires and cables from one point to another in the shortest way possible; in old work considerably more material is used because often it is necessary to lead the cable the long way around through channels that are available, rather than to tear up walls, ceilings, or floors in order to run it the shortest distance.

No book can give all the answers as to how to proceed in old work; here the common problems will be covered, but considerable ingenuity must be exercised in solving actual problems in the field. A study of buildings while they are under construction will help in understanding what is behind the plaster in a finished building.

Wiring Methods in Old Work.   It is impossible to use rigid or thin-wall conduit in old work without practically wrecking the building.   It would be used only when a major rebuilding operation is in process, and installation then would be as in new work.   The usual method is to use nonmetallic-sheathed cable or armored cable.   The material is easily fished through empty wall spaces.   It is sufficiently flexible so that it will go around corners without much difficulty.   In some localities flexible conduit (greenfield) is generally used.   Install it as you would cable, except that the empty conduit is first installed, the wire pulled into place later.

Cutting Openings.   To cut good openings for outlet and switch boxes in walls requires a certain amount of skill and a generous measure of common sense.   The openings must not be oversize and must be neatly made.   Start by marking the approximate location of the box, and, if possible, allow a little leeway so that the opening can be moved a trifle in any direction from the original mark.   First make sure there is not a stud or a joist in the way; usually thumping on the wall or ceiling will disclose the presence of timbers.   Then dig through the plaster at the approximate location and probe until the space between two laths is found; then go through completely.   It would be well to reach through this opening and, with a stiff wire or similar instrument, probe to right and left to confirm that there is no stud or similar obstruction.

The sawing in a lath-and-plaster wall can be done using a keyhole saw; proceed gently so as not to loosen the lath behind the plaster. Some prefer to use a hack-saw blade, one end heavily taped to serve as a handle.   Have the teeth of the blade lie backward, the opposite of the usual position, so that the sawing is done as you pull the blade out of the wall, not as you push the blade away from you as in usual sawing.   This will tend to leave a firm bond between lath and plaster, especially if you hold your hand against the plaster as you do the sawing.   Unless you watch this carefully you may end with a considerable area of plaster unsupported by laths, which have become separated from the plaster during the operation.

Temporary Openings.   The openings discussed in the preceding paragraph are openings into which a box will later be fitted.   In old work it is often necessary to cut temporary openings in odd places to make it possible to pull cable, for example, from the ceiling around the corner into a wall.   The cable does not go through the opening; the opening is merely used to get at the cable during the pulling process to help it along, or to get around obstructions in the wall.   Such openings must, of course, be repaired when the job is finished.

If the room is papered, the paper must be carefully removed in one place and then reinstalled so that the paper will look like the original installation.   This is easily done.   With a razor blade cut the two sides and the bottom of a square, but not the top.   Apply moisture with a rag or sponge, soak the cut portion, and after the paste has softened, lift the cut portion, using the uncut top as a hinge.   Fold it upward, and fasten to the wall with a thumbtack.   These steps are shown in Fig. 22-1.

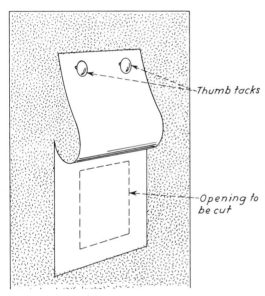

Fig. 22-1.   Sections of wallpaper are easily removed temporarily.   Use the top of the cut section as a hinge; this makes it easy to restore the wallpaper to its original condition, after the installation is completed.

When the opening in the plaster is no longer needed, it is easily patched, using plaster of paris or a ready-mixed plaster, which need only be mixed with water, and set.   The same mixture is used to fill the openings around switch and outlet boxes, for the Code does not permit open spaces; the plaster must come up to the box.   The section of wallpaper is replaced by applying fresh paste and letting down the hinged section which was loosened and pinned up while the opening was being made.

Mounting Outlet Boxes. For new work the Code requires outlet boxes with a minimum depth of 1½ in.; for old work this requirement is waived when use of deeper boxes leads to injury of the building. Boxes ½ in. deep are therefore commonly used. Two of these are shown in Fig. 22-2. If the outlet box is located so that it can be attached to a

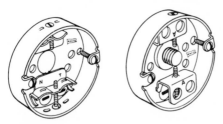

Fig. 22-2. In old work, shallow boxes are permitted if the use of standard boxes 1½ in. deep would result in injury to the building. (*All-Steel Equipment, Inc.*)

joist or similar substantial timber, install as shown in Fig. 22-3. In similar fashion a box may be mounted directly on lath even if it is not backed up by a joist, as shown in Fig. 22-4. However, this method is to be discouraged because, if a fixture of any substantial weight is attached to the box, damage to the ceiling may follow.

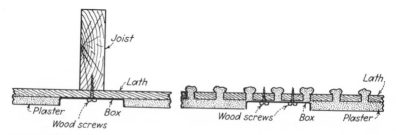

Fig. 22-3. In mounting boxes on the ceiling, install them on a substantial timber where possible.

Fig. 22-4. If it is necessary to install boxes directly on lath, mount them to two different laths, not to one only. This distributes the weight of the fixture.

By far the simplest method is to use one of the old-work hangers shown in Fig. 22-5.   The method of its use is shown in steps in Fig.

Fig. 22-5.   An old-work hanger is very handy for mounting a box on the ceiling.   (*All-Steel Equipment, Inc.*)

22-6.   First make a hole in the ceiling at the proper place.   Then slip the hanger into the hole; note that the hanger has a length of wire attached to the stud so that it is not easily lost inside the ceiling or wall. Then pull back by this wire and allow the stud only to project from the

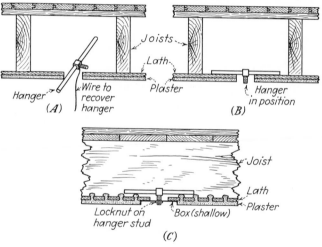

Fig. 22-6.   Installing an old-work hanger.   When the job is finished, the hanger should lie at right angles to the lath.

opening in the ceiling.   Turn the bar crosswise so that it lies at right angles to the lath; this will later distribute the weight of the fixture over a number of laths instead of throwing it all on one or two as is the case when the box is mounted directly on lath.   Remove the locknut from the stud, slip the stud through the center knockout of the box, tighten the locknut on the stud inside the box, and the job is finished.

Regardless of the method of mounting, the cable must be at least partially attached before the box is mounted in place.   This, in the case of ceiling outlets and in similar cases where the cable comes in through the bottom, is no problem.   Be sure that the cable connector is rigidly anchored to the cable and that the locknut on the connector is securely driven home before the box is finally mounted.

Often the flooring above the ceiling in which the box is to be installed can be lifted temporarily (as will be explained later).   This makes possible a simple installation using a straight bar hanger, used as was shown in Fig. 10-11.   Cut an opening in the ceiling for the box; install the box on the hanger, the ends of which have been bent upward; and nail the hanger with the box into place.

Ears on Switch Boxes.   The mounting ears on the ends of switch boxes are adjustable to compensate for various thicknesses of plaster. They are also completely reversible, as Fig. 22-7 shows.   In the posi-

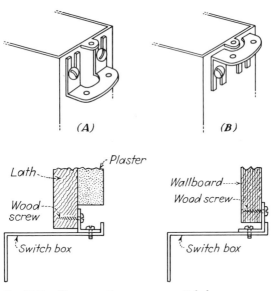

Fig. 22-7.   The mounting ears on switch boxes are reversible.   In the position *A* they are used to install boxes on lath under plaster.   Reverse them as shown at *B* for mounting on wallboard or similar material.

tion in which they come on boxes and as shown at *A*, they are used for mounting such boxes on plastered walls.   The ears are fastened to the lath and are of such proportions as to bring the front edge of the box flush with the plaster surface.

If the box is to be mounted on wallboard or similar material over which there is no plaster, the ears are reversed as shown in *B* of Fig. 22-7, again bringing the front edge of the box flush with the surface.

Openings for Switch Boxes. In cutting the opening for the switch box, take into consideration the dimensions of the box compared with the width of lath. The ordinary switch box is 3 in. long, while two laths plus three spaces between laths measure more than 3 in. If two full laths are cut away, it will be difficult to anchor the switch box by its ears on the next two laths, for the mounting holes on the ears will then come very close to the edges of the laths, which will split when the screws are driven in. Cut one lath completely and remove part of the width of each of the two adjoining laths; this should be clear from Fig. 22-8.

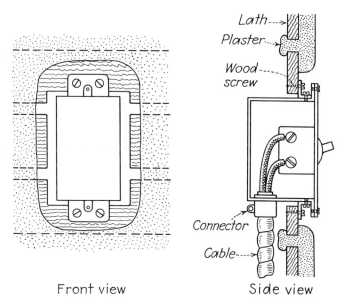

Front view   Side view

Fig. 22-8. In cutting the opening for a switch box, do not cut away two complete laths. Cut away one, and part of each of two others.

Cutting the Opening. Make a mark on the wall, approximately where the switch or receptacle is to be located. Bore a small hole through the mark, with a stiff wire probe to make sure that there is no obstruction, and that there is sufficient space all around. Enlarge the hole a bit, to locate the space between two laths; this will locate the center of your opening, up and down. Then mark the area of your opening, about 2 by 3¼ in.   Drill half-inch holes at opposite corners, and also at the center of top and bottom.   The *centers* of the holes must be on the lines of the outline, so that not over *half* of each hole will be outside the rectangle; unless you watch this carefully, part of the holes may later not be covered by your switch or receptacle plate.   See Fig. 22-9.

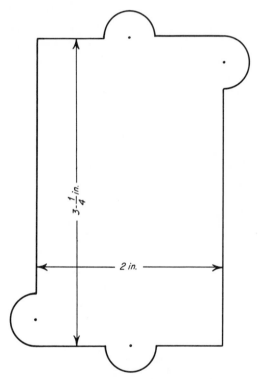

Fig. 22-9.   Size of opening for a switch box.   Make yourself a template of this drawing; it will save you much time.

The holes at the corners are for inserting the saw blade for cutting the opening; the holes at top and bottom are for clearance for the screws for mounting switches or receptacles.

Make yourself a template the size of the illustration.   Lay a piece of stiff cardboard and a sheet of carbon paper under the page with the illustration; trace the outline, cut the cardboard to size.   Trace around the template wherever there is to be an opening.   This will save much time especially if many openings are to be made.

Other Installation Methods.   Instead of mounting the switch box using screws, other methods are available.   They are specially suitable in wallboard jobs because the wallboard is too flimsy for screws to be used.   One method is to use the special metal hanger strips shown in Fig. 22-10.   Place one on each side of the wall opening with the ears projecting into the room.   Slip the switch box into place between these two hangers, then bend the ears of the hanger down inside the box, as shown in the illustration.   Be sure that the ears lie snug against the inside surface of the box; if they are allowed to bend away from the inside surface, they can easily cause grounds from the box to the terminals of receptacles or switches.

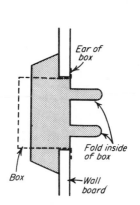

Fig. 22-10. A pair of these hanger strips are convenient for installing a switch box on wallboard.

Fig. 22-11. This box has collapsible straps on each side. Push box into opening, tighten screws on sides, and the box will be clamped into wall. (*All-Steel Equipment Inc.*)

A different kind of box makes the installation even simpler.   As shown in Fig. 22-11, there are two metal strips parallel with the sides of the box.   After the box is pushed into its opening, tighten the two screws on the outside of the box; this collapses the metal strips so that they bulge against the inside of the wall, effectively anchoring the box in the wall.

Installing Box in Wall.   If the cable runs into the bottom of the box, there is no problem involved in cutting the right size of opening.   Usually, however, the cable runs into knockouts in the end of the box, as shown in Fig. 22-12; the installation then is not so simple.   If the cable is rigidly attached to the box with a connector before the box is mounted, then it will no longer be possible to get the box into the opening—the cable is in the way.   If the opening is made big enough so

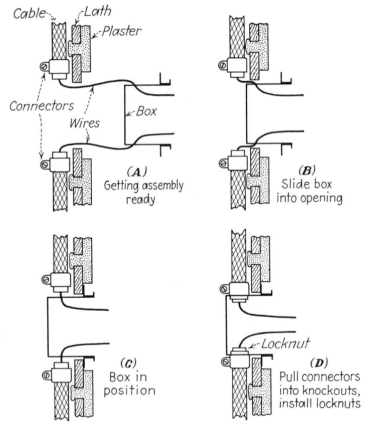

Fig. 22-12.   Installing switch box in opening.   The cable is anchored to the box by locknut of connector, *after* the box has been inserted into the opening.

that a box which has been preassembled with cable and connector will slip through easily, there will be a very sloppy fit that no self-respecting workman will tolerate.   To do a good job, follow the procedure outlined in Fig. 22-12.   Cut the opening only big enough for the box, plus about ⅛ in., or the thickness of the wire inside the cable.   Leave a generous length of wire sticking out of the cable.   Attach the connector to the cable, remove the locknut from the connector.   When ready to install the box, let the wires stick out of the opening in the wall, with the connectors inside the wall.   Push the wires through the knockout into the box and grasp them inside the box.   Push the box into the wall; there will be room at the ends of the opening for the wires to slide through into the wall.   When the box is in its opening, pull on the wires, pulling the connector into the knockout; then slip the locknut over the connector and tighten.

If you are using Type NM or NMC cable, all this will be easier if you use a box with beveled corners, as shown in Fig. 22-13, which has clamps for nonmetallic-sheathed cable.   Such boxes are not available with clamps for armored cable.

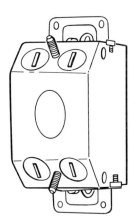

Fig. 22-13. Boxes with beveled corners simplify the procedure of Fig. 22-12.

Cable behind Baseboard.   Assume that there is a wall bracket in the middle of a wall, already wired and controlled by a switch on the fixture, but that now it is to be controlled by a wall switch several feet to the left but on the same wall.   This is a relatively simple job (see Fig. 22-14).   First cut the opening for the switch box at D.   Then remove the baseboard running along the wall at the floor, and cut two holes, B and *C*, behind the baseboard.   Then from *B* to *C*, cut a groove or trough in the plaster between two laths; if the plaster is not very thick, it may be necessary to slice away part of the laths.   In any event, the trough must be big enough to receive the cable, as shown in the cross-sectional view of the same picture.

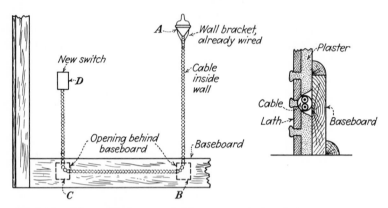

Fig. 22-14.   Often cable is concealed in a groove in the plaster and lath, behind the baseboard.

When this has been done, prepare a piece of cable long enough to extend from *A* to *B* to *C* to *D*, with a connector at each end; remove the locknuts.   Remove the fixture at point *A*; remove a knockout from the bottom of the outlet box on which the fixture is installed.   Then push a fish wire with a hook bent on the end through the knockout in box *A* down toward *B*.   Reach into opening *B* with another piece of fish wire again with a hook on the end.   It will not be difficult to hook the two pieces together so that by pulling at *B* the first piece is pulled in a continuous length from *A* to *B*.   Attach the cable to the fish wire, and pull it into the wall through opening *B* until the end appears at *A*.   Pull it into place so that the connector slides into the knockout in *A*, tighten the locknut, and the job is finished at *A* except for connecting the wires. Next drop the fish wire in opening *D* until it appears at *C*, and fish the cable up inside the wall until it emerges at *D*, in the meantime laying the cable securely into the trough from *B* to *C* to take up slack.   Anchor it at *D* so that it cannot be lost inside the wall; replace the baseboard. Use extreme care that nails are not driven through the cable.   All that remains to be done is properly to attach the cable to the box at *D*, mount the box, and install the switch.

Cable through Attic. In single-story houses or when working on the second floor of two-story houses, it is generally entirely practical to run cable through the attic. It is a simple matter to lift a few boards of the usual rough attic flooring and lead the cable around, in that way avoiding the need for openings in the walls of the living quarters except the openings for boxes. No baseboards need then be lifted. It may require a few feet more of cable, but the saving in labor more than offsets this. Always explore this possibility before proceeding with a more difficult method. For example, in Fig. 22-14 the cable is run from outlet A to attic, under the attic floor over to a point directly above outlet D, and there dropped down to D.

Cable through Basement. In wiring the outlet of Fig. 22-14, often you can run the cable through the basement, going straight down below point B into the basement, then over toward the left, then upward again at point C. More usually there will be obstructions in the walls not making this possible in such simple fashion.

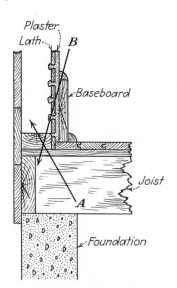

Fig. 22-15. Boring through obstructions. The actual construction found will vary a great deal. Most problems of old work are problems of carpentry.

If the point where the cable is to run down into the basement is on the outer wall, the construction is apt to be something on the order of that shown in Fig. 22-15. In that case bore a hole with a long-shank electrician's bit, of the type shown in Fig. 22-16, either upward as indicated by arrow A or downward from a point behind the baseboard, as indicated by arrow *B*, after removing the baseboard. If the cable is to enter the basement from an interior wall, it is usually possible simply to bore directly upward from the basement, as shown in Fig. 22-17.

Fig. 22-16.   Electrician's bit and extension.   (*Greenlee Tool Co.*)

Cable around Corner Where Wall Meets Ceiling.   Figure 22-18 shows this problem: how to lead cable from outlet *A* in the ceiling to outlet *B* on the wall around the corner at *C*. At first glance this may seem difficult, but it is relatively simple. In houses that are not very well built,

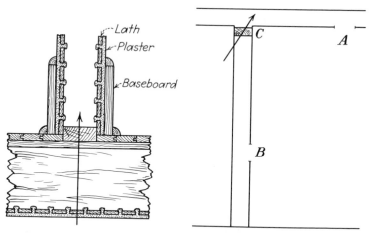

Fig. 22-17.   Sometimes obstructions can be cleared by boring upward.

Fig. 22-18.   To get cable from *A* to *B*, a temporary opening must often be made at *C*.

there may be a clear space at the corner *C*. In that case push a length of fish wire with a hook on the end into the ceiling at *A* until the hook is somewhere around *C*. Then push another length of fish wire upward from *B* until the hook touches the floor above. With one man at *A* and another at *B*, it becomes simply a problem of fishing, jiggling, pulling, and twisting the two lengths of fish wire until the hook on one catches the other. Then pull at *B* until there is a continuous length of fish wire from *A* to *B*, attach the cable to the fish wire at *A*, and pull it into position. It will not come too easily around the corner at *C*, but with help at *A*, it can be pulled through. Much patience is the greatest asset in this work.

If the house is well built, there will be an obstruction at point *C*. Any one of a dozen different types of construction may be used; that shown in Fig. 22-19, which is simply an enlarged view of point *C*, is

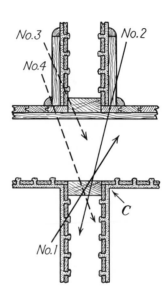

Fig. 22-19. An enlarged view of point *C* of Fig. 22-18.

typical. The usual procedure is to make a temporary opening in the wall at point *C*, but on the opposite side of the wall, away from opening *B*. Bore upward with a long-shank electrician's bit, as shown by arrow 1. Push a length of fish wire into this hole until the end emerges at *A*. If the opening at *C* is large enough, push the other end of the wire downward to *B*, and pull the loop that is formed at *C* into the wall by pulling at either *A* or *B*; there will then be a continuous fish wire from *A* to *B* with which to pull in the cable. More usually the hole at *C* will be small; hence use two lengths of fish wire. Push one through from *C* to *A*, leaving a small hook at *C* just outside the opening. Push another length from *C* to *B*, again leaving the hook just outside the opening at *C*. Hook the two hooks together, pull at *B*, and it is a simple matter then to pull the longer wire from *A* through *C* to *B* and, with this, to pull in the cable.

If there happens to be another wall directly above point *C*, it may be better to bore down from above at a point behind the baseboard, as indicated by arrow 2.   In that case fish wire is pushed down from above through the bored hole to *B*; another length from *A* toward *C*; when the hooks at the end engage, pull down at *B* until a continuous piece of fish wire extends from *A* through *C* to *B*.

In very old homes there may be a molding around the room, at the ceiling; then it is usually better to remove the molding and to chisel a hole in the corner, probably chiseling away a portion of the obstruction, to provide a channel for the cable, as indicated by the arrow in Fig. 22-20.   When the molding is replaced, the cable, if it projects a bit, is concealed.

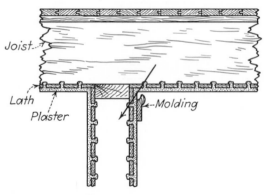

Fig. 22-20.   Occasionally temporary openings can be made behind a molding.

Cable from Second Floor to First.   If the first-floor partition is directly below the second-floor partition, it is usually simple to bring the cable through by boring, as indicated by arrows 2 and 3 (or 3 and 4) in Fig. 22-19.   Use good judgment so that the holes will lie so far as possible in approximately the same plane, thus simplifying the fishing problem. An opening behind the baseboard is usually necessary.

If the first-floor partition is not directly below the second-floor partition, handling as indicated in Fig. 22-21 will usually solve the problem. Bore holes as indicated by the two arrows.

Lifting Floor Boards. In many cases the outlet and switch boxes may be so located with regard to wall and ceiling obstructions that it is necessary to lift hardwood floor boards in the floor above. This should be avoided if possible, but where necessary use extreme care in lifting the

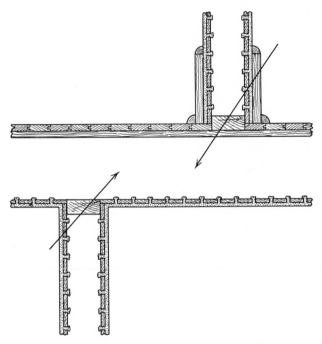

Fig. 22-21. Problem in bringing cable from a second-floor partition into a first-floor partition.

boards so that when replaced there will be no visible damage to the floor. Attic flooring is simply lifted, but the usual hardwood floor with the tongue-and-groove construction presents more of a problem.

It is necessary first to chisel off the tongue on one of the boards. The thinner the chisel used for the purpose, the less the damage that will be done to the flooring. A putty knife with the blade cut off short and sharpened to a chisel edge makes an excellent chisel for the purpose, and a thinner one is not obtainable. Drive it down between two boards and cut off the tongue (see Fig. 22-22). This should be done to the entire length between three joists, although the picture shows only two joists.   Having the cut section extend over a longer space gives the advantage of a better footing when the board is reinstalled.   In cutting off this tongue, the exact location of the floor joists can be determined and in this way points *A* and *B* in the picture located.   Bore a small hole at these two points next to the joists, and with a keyhole saw cut across the boards as shown.

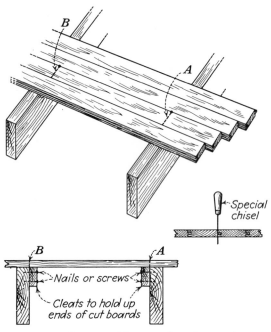

Fig. 22-22. Steps in lifting and replacing floor boards.

The board can then be lifted, the electrical work done, and later the board replaced.   It will be necessary to attach cleats to the joists for the floor board to rest on, at the cut ends.   Anchor these cleats securely so as to give the cut board a really solid footing.   The bored holes are later filled with wooden plugs.

Extension Rings.   In old work it is often desirable to be able to extend a circuit beyond an existing outlet.   If the wiring is entirely flush, it might entail considerable carpentry if the new outlet were also to be made flush, and at least in certain types of work (such as basements) it will be entirely acceptable to have the new outlet of the surface type. In that case an extension ring of the type shown in Fig. 22-23 is used. Extension rings are, to all intents and purposes, outlet boxes without bottoms, and they are available to fit all kinds of outlet boxes.

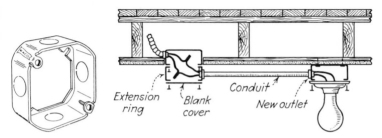

Extension ring   Blank cover   Conduit   New outlet

Fig. 22-23.   If surface wiring is acceptable, an extension ring is handy in installing a new outlet, starting from an existing outlet.

Their use should be made clear by the same picture.   Simply mount the extension ring on top of the existing outlet box and from there proceed as in any exposed wiring.   The extension ring is covered with a blank cover or with the fixture or other device that may have been installed on the original outlet box.

Boxes on Brick Walls.   When a house is of brick construction, a considerable amount of labor is involved in the mounting of the boxes, for the boxes must come flush with the plaster when it is applied.   A space must be chiseled into the brick to receive the box.   Usually 4-in. square boxes are used, together with covers of the type that were shown in Fig. 10-12.   These are available in various depths so that, if the ordinary $\frac{1}{2}$-in. type does not bring the cover flush with the plaster, one of a greater depth, say $\frac{3}{4}$ in., will be found suitable.

The box cannot be secured to the brick directly with screws; conse-

Fig. 22-24.   To drill holes in masonry, use a star drill and a hammer, or carbide-tipped drills with electric drill.   (*The Rawlplug Co., Inc.*)

quently it is necessary to use one of the many types of plugs or anchors available for the purpose.   In any case it will be necessary to drill holes into the brick or masonry, using for the purpose a star drill of the general type shown in Fig. 22-24.   The drill is used by simply pounding on its head with a hammer, rotating the drill a bit after each blow.   Using the drill holder shown in the same illustration makes the operation much simpler, and protects the hand.   Similar drills can be used in power hammers, and a different type in electric drills.

A very common mounting method is that using the well-known lead plugs or anchors which are merely inserted in a hole in the masonry, ordinary wood screws then being used.   The Code prohibits wooden plugs.

The use of lead expansion anchors of the general type shown in Fig. 22-25 provides a mounting which is considerably more secure than

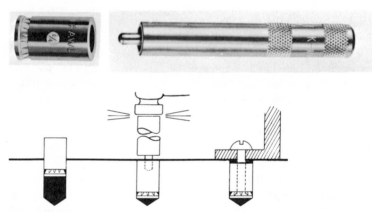

Fig. 22-25. One of many available varieties of screw anchors.   (*The Rawlplug Co., Inc.*)

using just lead plugs.   Drill a hole of the proper size; drop the anchor into it; set it using the special tool shown in the same illustration, and as shown in steps, again in the same illustration.

If the mounting must be over a hollow area such as a tile wall, use toggle bolts.   A typical bolt is shown in Fig. 22-26, and the illustration also shows the installation method.   Merely slip the collapsible wings through the opening in the wall or other surface; a spring opens the wings which then provide anchorage for the bolt.

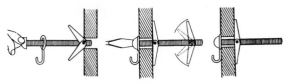

Fig. 22-26. Toggle bolts are convenient and practical in mounting equipment on hollow walls. (*The Rawlplug Co., Inc.*)

Use Common Sense and Patience. No book can outline all the problems in carpentry that will be encountered in old work. The method of construction in houses varies with the age of the house, the general sturdiness of the building, the skill and integrity of the builder, the geographical area, and many other factors. Plenty of patience, coupled with a generous measure of "horse sense," is the greatest asset in old-work wiring.

Surface Wiring. There are available materials which make it quite simple to start from an existing receptacle outlet and add additional outlets, for example in a living room or kitchen. The same materials are equally suitable for more complicated original wiring in areas such as basements, attics, garages, and farm buildings where the wiring is to be permanently exposed.

The basic materials for this type of wiring are combinations consisting of a receptacle, switch, or other device, each combined with a nonmetallic outlet box. Figure 22-27 shows a representative assortment of them; others are available. They are made for use with 2-conductor No. 14 or No. 12 nonmetallic-sheathed cable. They have no terminal screws. Just strip the ends of the cable and push the bare wires into the devices for a good permanent connection; each device has a "strip gauge" molded into it showing how far to strip the cable (this method of connecting without terminal screws was shown in Chap. 8, Fig. 8-5).

The devices are available in either brown or ivory color; the cable likewise is available in both colors. For use in living areas of homes, the cable is available with prepunched nail holes in the cable; it may be nailed to the wall, using the special nails that come with it. The cable

Fig. 22-27. Surface-mounting devices are easy to install. This particular type has no terminal screws. (*General Electric Co.*)

is about $\frac{3}{16}$ by $\frac{1}{2}$ in. in size and can be bent across either the short dimension or the long dimension without buckling, as shown in Fig. 22-28.

Start from an existing receptacle outlet. Use the special attachment plug shown in Fig. 22-29, and connect it to one end of a length of cable,

Fig. 22-28. Where appearance is important, use this special cable. It bends easily in any direction. (*General Electric Co.*)

Fig. 22-29. Start from an existing receptacle, using this special plug. It has a built-in swivel. (*General Electric Co.*)

as shown in Fig. 22-30, by merely pushing the stripped end of the cable into the plug. The plug has a swivel in it so that you can run the cable downward from the starting point or sideways.

Next run the cable to the point where you want the next outlet; in-

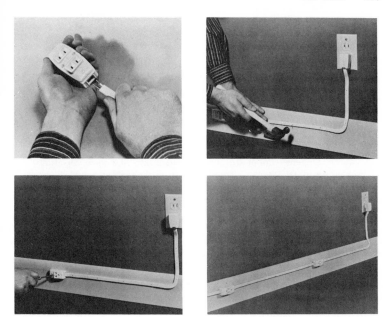

Fig. 22-30. Install the material as shown in the steps above. (*General Electric Co.*)

stall the next outlet and proceed to the one after that. All this is shown in several steps in Fig. 22-30.

Instead of starting from a receptacle outlet, the cable may be run into any existing outlet box containing a grounded wire and a permanently hot wire. Naturally you must use a connector on the cable at such starting point, as in other cable wiring.

The cable described is a special variety of Code Type NMC nonmetallic-sheathed cable, made with particular attention to neat appearance in living areas of homes. In other areas where appearance is not so important, any kind of nonmetallic Type NM or NMC cable may be used, making the system quite suitable for use in basements, attics, garages, farm buildings, and so on.

The devices described up to this point may be used only with 2-wire cable, and only a single length of cable can enter each end. Similar fittings which have screw terminals instead of push-in connections for wires are shown in Fig. 22-31. They can be used with either 2- or 3-

Fig. 22-31. Devices similar to those in Fig. 22-27 but more versatile. They have terminal screws. Two cables can enter either end. (*General Electric Co.*)

wire cable, and two lengths of cable can enter each end. This series includes several devices, including 3-way switches, that are not available in the other kind. All this makes this series a bit more versatile, especially in the wiring of farm buildings where 3-way switches are necessary. However, the two series may be used intermixed on the same job.

In using this material in farm buildings and similar locations, you can, with good planning, run a *continuous* length of cable from one point to another without installing outlets as you go along. At points where an outlet is to be installed, leave a slight bulge away from the surface in the cable. Later cut the cable at that point, and install the receptacle or other device.

In installing these fittings, use the same precaution given later in Chap. 24 concerning nonmetallic outlet boxes. If the fittings are mounted on timbers that may swell with moisture, don't pull the mounting screws too tight, to avoid possible future damage.

23

Modernizing an Installation

The wiring originally installed in what is now an older house, and too often even if the house is only 10 years old, just is not adequate for the job it is called upon to do today. A complete wiring job is in order—or is it? Do not jump to the conclusion that every outlet must necessarily be torn out, every receptacle replaced. Many times a less expensive job may serve the purpose, perhaps not quite as well as a complete rewiring job would, but probably in acceptable fashion.

Is the wiring inadequate because you are using too many lights? Too many floor lamps? Too many radios and TV? That is seldom the case. The wiring is inadequate because of too many electrical appliances that were not allowed for at the time of the original wiring job, probably including some that were not even on the market at that time. The installation does not provide enough circuits to operate a wide assortment of ordinary kitchen appliances, plus range, water heater, clothes dryer, room air conditioners. Some of these appliances operate on 230-volt circuits, which may not be available; others operate on 115 volts but, when plugged into existing circuits, overload those circuits.

Usually, the service-entrance equipment is just too small for the load, just as two-lane highways built years ago are too small for the number of cars they are now called upon to accommodate.

To analyze the problem of your particular house, ask whether *if you disconnect all the appliances* (including motors on oil burner or similar locations), you will have all the *lighting* circuits that you need.   The answer might well be "yes," in which case your rewiring problem is simplified.   You will still have to rewire the house but not so completely as at first may have appeared necessary.   In all likelihood, a large part of the need for wiring will be the fact that the present service entrance is too small.   Installing a larger service entrance alone might go a long way toward solving the problem, but that would result in a "patchwork" job and might cost almost as much as doing the job the right way.

Plan the Job.   Proceed more or less as if you were starting with a house that had never been wired, but leave the existing *lighting* circuits intact.   These lighting circuits will, of course, include many receptacles for small loads like vacuum cleaner, radio, TV, and clock.   However, disconnect all the receptacle outlets in kitchen, pantry, family room, dining room, and breakfast room, for you will want to install the two 20-amp special appliance circuits now required by the Code for those areas.   If the present installation has *individual* circuits to 115-volt motors on the furnace or similar 115-volt loads, leave those circuits also intact.   But if such large loads are connected to general lighting circuits, disconnect them from the existing circuits.   If the present installation has individual circuits to water heater, range, or similar 230-volt loads, leave those circuits intact.

Your rewiring job will consist of installing a new service of at least 100-amp capacity as outlined in other chapters, some new circuit breakers (or fuses), and new circuits for appliances.   You will have to connect the old circuits into the new circuit breaker equipment; two different ways of doing this will be described.

The Code in Sec. 220-8 outlines the method for calculating the service entrance when adding considerable loads to an existing residential installation.   However, if you install a 100-amp service and follow the other suggestions in this chapter, you will automatically meet Code requirements in 99.9 + % of the cases.

Install New Service and Appliance Circuits. First, install a new service of at least 100-amp capacity. Install a circuit breaker with 100-amp main breaker, enough branch-circuit breakers for all the new branch circuits you are going to install (not overlooking the two special appliance circuits and the special laundry circuit), plus a few spares, and one additional 2-pole breaker. This additional breaker may be rated at 30 amp if the circuits you are going to leave untouched serve only lighting loads, but should be rated 50 amp if they also serve water heater, range, or similar loads. That 30- or 50-amp breaker will become the SOURCE for all the old circuits that remain intact.

Install the two 20-amp special appliance circuits as discussed in Chap. 12, or preferably one 3-wire 20-amp circuit described in Chap. 20. Install the 20-amp circuit now required for laundry. Install an individual circuit for each heavy appliance such as range, water heater, clothes dryer, also for each motor on furnace, pump, and so on, as discussed in other chapters (unless they are already served by individual circuits that you are leaving intact). Connect all the *new* circuits to the *new* circuit breakers.

When you have done all this, you will have no power on your new circuits but will still have power on your old circuits. Call your power supplier, have them disconnect the power on the outside of the house, and get along for a day or two without electric power while you connect your existing circuits into your new circuit breaker equipment.

Reconnecting Old Circuits: Method A. You have a choice of two methods in connecting your old circuits into your new equipment. The simplest will be described first: let's call it Method A. Cut off the present incoming service wires where they enter the switch box. Remove the short pieces of wire that remain; observe carefully the terminals to which they were connected. If the original service was 3-wire 30-amp, run a length of 3-conductor No. 10 cable from the extra 2-pole 30-amp breaker in the new circuit breaker cabinet, to the old service equipment, connecting the ends there to the terminals to which the original incoming wires were connected. The white wire of the cable, of course, runs from the neutral strap of the new equipment to the neutral strap of the old equipment. If the original equipment was 3-wire 60-amp, use No. 6 cable instead of No. 10; the breaker in the new equipment then must be the larger 50-amp type.

If the incoming service was of the 2-wire type, not 3-wire, use Method B instead of the Method A just described.

If the present equipment includes a fused main switch, leave it in place.   Neither the switch nor the fuses are required, but there is no reason why they can't be left in place.

Reconnecting Old Circuits: Method B.   Most likely your present circuits terminate in a fuse box with four or more fuses; the fuse box may or may not be in the same cabinet with the main switch and the main fuses.   It is not likely that the existing wires will be long enough to reach the new equipment.

Disconnect the wires of one circuit at a time from the existing equipment.   If the wiring is cable of some kind, remove the locknut from the connector, pull the cable out of the box, put the locknut back on the connector, and let the cable dangle.   Eventually you will have four to six ends of cable, one for each lighting circuit.

When all the circuits have been disconnected, disconnect the main incoming wires from the equipment and tear out the old equipment. Then at the point where the old equipment was removed, install a second circuit breaker cabinet with as many breakers as there are old circuits, with a few spares for future use.   Connect the old circuits to the new breakers.   Run 3-wire No. 10 cable from this new circuit breaker cabinet to the 30-amp breaker in the larger new breaker cabinet, as outlined under Method A.   If there are many circuits, you will use No. 6 cable to the 50-amp breaker in the other cabinet.   All this is shown in Fig. 23-1.

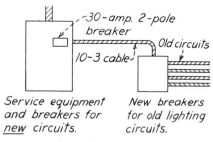

Service equipment and breakers for new circuits.

New breakers for old lighting circuits.

Fig. 23-1.   Circuit for modernizing.

If the location of the old equipment was such that you wish to place the circuit breakers for the lighting circuits in a new location, the original cables of the lighting circuits will certainly be too short to reach the new location.   In that case install a junction box at the point where the old equipment was removed and run cables to the proposed location of the circuit breaker cabinet, as shown in Fig. 23-2. The junction box can be a 4-in. square box if there are just a few circuits but more likely will be a larger cutout box if there are many circuits. This box will contain only splices in the cable, it must be located where it will be permanently accessible, and it must have a metal cover.

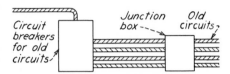

Fig. 23-2. Method of installing junction box.

Additional Outlets. While you are doing this work, you will probably want to add additional receptacle outlets on the old circuits that you are leaving intact. In older homes, there are seldom as many receptacles in living room, bedrooms, and so on, as would be desirable. Why not add such additional outlets at this time? You might also well consider substituting a pair of 3-way switches for a light now controlled by a single switch, or adding a switch for a light now controlled by a pull chain.

Ground. It is not likely that the existing ground wire will be long enough to reach the new service; ground wires must be *continuous*. In any event, the ground wire must run from the *new* service equipment. Tear out the old and install a new ground as explained in Chap. 17; use a size as required for the amperage of the new service.

Call the Power Supplier. Your job is finished. Call your power supplier, have them install the new meter and connect the power to the new service. Then enjoy all the advantages of a house wired in modern, adequate fashion.

24
Farm Wiring

In city homes electrical devices and appliances are purchased primarily for the convenience and utility which their use provides. On the farm the same devices are also used, but in addition many others are found which are used in the *business* of farming—devices which are bought, not so much for their mere convenience, but rather as an investment on which the farmer expects dividends. Into this classification fall such things as milk coolers, milking machines, hammer mills for chopping fodder, silo fillers, hay dryers, water heaters to provide scalding hot water for the dairy, water heaters which during the winter keep the water for chickens at a temperature which experience has shown will promote egg production, and dozens of others.

The wiring of farms involves all the problems so far discussed as well as a considerable number of new ones. The maximum wattage in use at one time is apt to be considerably larger than in city homes because of the liberal use of motor-driven equipment. There is a great deal of outdoor wiring, either overhead or underground, between the various buildings. Substantial distances are involved, which means that wire sizes must be carefully watched, both to avoid voltage drop and for mechanical strength. Relatively poor grounding conditions are usual. These and other factors will be separately considered in this chapter.

Preview. In a typical farm installation, wires from the power line end at a meter pole in the farmyard. From the top of the pole the wires run down to the meter, then back to the top of the pole. Sometimes there is a circuit breaker (or switch with or without fuses) at the bottom of the pole. The wires are always grounded at the pole.

From the top of the pole, a set of wires runs to the house; another set runs to the barn. Often additional sets run to other buildings. Instead of running overhead, underground wires are being used more and more.

At each building, there is a service entrance just as if that building were the only building being wired. This service entrance is installed as was described in Chap. 17, except that the meter is omitted. The Code requires a ground at each building which has more than one circuit and also at each building which houses livestock.

In various buildings, especially barns, there are conditions which require special wiring methods; these will be discussed later in this chapter. Good grounds are difficult to establish, which will also be covered later in this chapter.

Adequacy. In Chap. 12 we studied adequacy of wiring in a house. Everything said there applies to a farm home just as much as to a city home. Indeed, if anything, the farm home deserves more attention and probably needs more circuits, because especially in the case of smaller farms, certain appliances are put into the house, which in the case of larger farms are put into the dairy barn.

In addition to the problems of adequacy in the house, special attention must also be paid to adequacy in and between other buildings and also on the meter pole.

Few farmers whose farms are being wired for the first time can foresee all the different electrical appliances and machines they will use in a few years. Almost always the number of circuits originally provided turns out to be too few; wires between buildings turn out to be too small; wires on the pole should be larger. Always provide more capacity than is needed at the time of installation; doing so will increase the labor hardly at all, the cost of material only a little, but it will do away with the later expensive alterations.

Overhead or Underground?   A farm wired with overhead conductors will have a large number of wires running all over the place.   This gives the farm a very untidy appearance and also invites troubles of various kinds.   Wires to low buildings can be damaged by moving vehicles.   Many overhead wires on an isolated farm invite trouble from lightning.   In northern climates where sleet storms are common, wires can break from the weight of the accumulated ice; a broken wire on the ground is dangerous.   Underground wires cost little more than overhead wires and do away with these problems.

If you are going to use overhead wiring, watch wire size carefully. Make sure the wire size selected is big enough to carry its load in amperes.   Make sure it is strong enough for the length of the span, which means that sometimes you must use wire larger than would otherwise be necessary for the amperage involved.   All this has already been discussed in Chap. 7.

If you are going to use underground wiring, use the materials and follow the methods already discussed in Chap. 17 concerning the service entrance.

If several sets of underground cable are to enter the bottom of the meter socket on the pole, you will have trouble because the threaded hub in the bottom of the socket is too small.   In that case provide a junction box immediately below the meter, inside which quite a num-

Fig.   24-1.   A   weatherproof junction box of this type makes it easy to run several underground runs, starting at the meter on the pole.   (*Hoffman Engineering Co.*)

ber of runs can be terminated, with only a single set of wires from the box into the meter socket.   A typical junction box for such purposes is shown in Fig. 24-1.   A telescopic metal channel to protect a group of underground wires running up the side of the pole is shown in Fig. 24-2, installed in connection with one of the junction boxes shown in Fig. 24-1.

Fig. 24-2. This trough protects the aboveground portion of underground runs, at the pole. (*Hoffman Engineering Co.*)

Made Electrodes. In cities the underground water system provides an excellent ground. On farms you must provide a substitute; how this is done was discussed in Chap. 17. It will probably be well for you to review the subject at this time.

Grounds on Farms. In Chap. 9 we studied the subject of grounds in general. We learned that, for safety, the neutral wire must be connected to the earth through a water pipe or ground rod. A good ground offers protection not only against accidents in the electrical wiring system but also against lightning.

Now, on a farm there is considerably more danger from lightning than in a city. A good ground is many times as hard to obtain on a farm as in a city. All that makes the subject of farm grounds a hundred times more important and difficult than the subject of grounds in a city. It is a most important subject, and too little attention is given to it. Study it well and apply the principles outlined below.

In a city, run a grounding wire to the nearest cold-water pipe, and automatically you have a good ground, ninety-nine times out of a hundred. The underground city water system provides a good, permanent, low-resistance ground. How good a ground is, is determined by the resistance in ohms between the ground rod and the surrounding earth. The Code in Sec. 250-84 says that the maximum permissible resistance is 25 ohms. A 25-ohm ground is passable; it is not a good ground. A good ground should have a resistance considerably under 25 ohms; city grounds are usually under 10 ohms.

A study of over 200 farms was made in Minnesota in a rather dry year. Only 9 out of 215 (1 in 20) farm grounds had a resistance of 25 ohms or less—and remember that a 25-ohm ground is not a good ground. The other 19 out of every 20 grounds had resistance of more than 25 ohms, some well over 100 ohms.

Perhaps one in a hundred farms had a ground that would be considered a good ground in a city. The others had "grounds" in that they consisted of wires leading to earth, but they did not fully serve the true purpose of grounds—safety. They were probably better than no grounds at all, or were they? Perhaps such a ground is like a spare tire in an automobile, which turns out to be without air when needed: it looks like a spare but serves no useful purpose because it is "flat"; the ground looks like a ground but does not really serve the purpose of safety, which is the only basic purpose of a ground.

How does one tell a good ground from a poor ground? We have already discussed the fact that the lower the resistance of a ground, the better the ground is. So quite properly, after the ground is installed, we must measure its actual resistance. If it measures 25 ohms or less, it is a ground that meets Code requirements; it just barely passes. If it measures 10 ohms, it (for the conditions usually prevailing on farms) is better than the average. If it measures under 5 ohms, it is an excellent ground. How are you going to measure that resistance? Unfortunately it is not simple to measure it accurately.

The only instrument on the market for measuring ground resistance simply, accurately, and directly in ohms costs about $300, so few people will be able to afford one of them. So we must use a method which may not give an exact answer but which will nevertheless give an approximate answer, which will be more accurate than a guess.

To test a ground, you will need an alternating-current ammeter reading to at least 25 or 30 amp. Mount this in a convenient box with a fuse of a size that will burn out when the capacity of the meter is exceeded. Provide two test leads. All this is shown in Fig. 24-3.

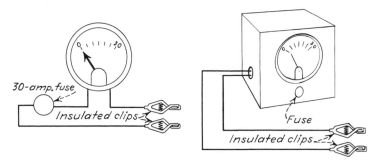

Fig. 24-3. An ordinary alternating-current ammeter lets you determine the approximate resistance of a ground.

See Fig. 24-4, which shows the basic installation of a ground. The only hitch is that we do not know whether or not the ground is a good one. Then see Fig. 24-5 which shows the same installation, except that the ground wire has been disconnected from the neutral wire. You would never do this except for the purpose of the test.

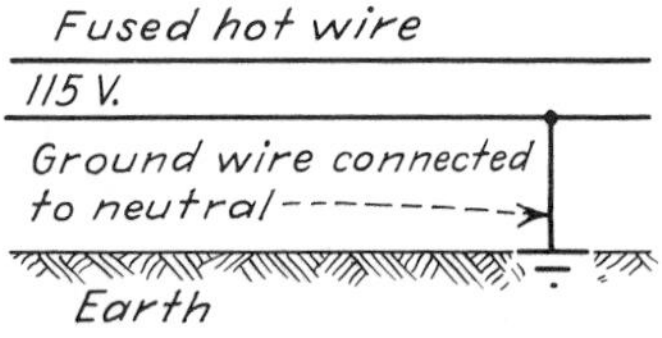

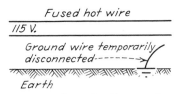

Fig. 24-4. The neutral as grounded in a normal installation.

Fig. 24-5. In testing for ground resistance, temporarily disconnect the ground wire from the neutral.

Now see Fig. 24-6, which shows the circuit used in measuring the approximate resistance of the ground. It boils down to grounding the "hot" wire through the ammeter and fuse. (Be sure all other grounds are connected and that all fuses are in place.)   From the ammeter reading, the approximate resistance of the ground can be determined, using Ohm's law, volts / amperes = ohms.   Assuming that the line voltage is 120 volts, it follows that if 15 amp flows, the resistance is 120/15, or 8 ohms.   Other values are:

Amperes flowing	Approximate resistance	Amperes flowing	Approximate resistance
1	120	10	12
2	60	15	8
3	40	20	6
4	30	30	4
5	24		

In inspecting an 0–30 alternating-current ammeter, you will notice that it has a nonlinear scale.   The scale is very crowded near 0, and it is

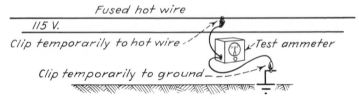

Fig. 24-6.   In determining ground resistance, using an ammeter, temporarily ground the *hot* wire through the ammeter.

almost impossible to tell the difference between 2 and 5 amp.   That, however, is not serious, for if the amperage on test is below 5, you will know that your ground has a resistance above 24 ohms and that it is not a very good ground.

Note that if the hot wire which you have temporarily connected to the ground through the ammeter is wired with No. 14 wire, it will be protected at the service by a 15-amp fuse or circuit breaker.   This fuse will not blow instantly, nor will the breaker trip, if 20 amp flows through it momentarily.   If you get a 20-amp flow you will know that the ground resistance is about 6 ohms.   If the 15-amp fuse or breaker at the beginning of the circuit blows or trips *instantly,* you will know you have an excellent ground, probably in the area of 4 ohms.

After the resistance of any ground has been determined, remove the tester, reconnect the ground wire to the neutral, and proceed to the next ground.

It should be noted that this test will not work if there is only a single ground on a wiring system, for the approximate resistance determined by this system is not really the resistance of the particular ground under test, but rather the resistance of that ground, plus the resistance of all the other grounds on the wiring system in parallel. Since on a farm there are usually from two to six other grounds and since all these are in parallel while the one particular ground is being tested, the total resistance determined by this ammeter method will be higher than the actual resistance. The percentage of error should range from 10 to 25%. The greater the number of grounds connected to the system, the less the percentage of error.

Remember that the resistance of a specific ground connection varies from time to time. The drier the earth, the higher the ground resistance. Therefore it is good practice to try for a ground of, say, 6 ohms or less resistance, one which will pass at least 20 amp when tested as described. But how are you going to make a ground better than it is? There are several methods.

First of all, when possible, install the ground rod where rain off the roof will tend to keep the earth wet. Wet earth makes for low-resistance grounds.

Second, if one rod driven to an 8-ft depth does not provide a ground of sufficiently low resistance, install a longer rod driven 12 or 16 ft into the earth. Alternately, install two or three rods; connect all of them together using ground clamps and wire of the same size as your ground wire. Do make sure that the rods are at least 10 ft apart, for experiment shows that, if they are closer together, extra rods do not greatly reduce the resistance.

Third, salt the ground, pouring salt water around the rod and letting it soak into the ground. This is common practice among the power companies but little used on farms, although it is so simple. Experiments have shown that, if the water in the earth immediately around the rod contains as little as 1% of salt, the resistance of the ground connection may drop as much as 90%; in other words, the 25-ohm ground becomes a 2½-ohm ground. The salting should take place at least once a year, although cases have been found where a single salting was to a degree still effective after two years. Salt the ground when it is first installed; salt it at intervals thereafter.

Meter Pole.   On practically all farms today, the power supplier's wires end on a pole in the farmyard.   On the pole are found the meter and usually a switch to disconnect the entire installation.   The wires are grounded at the pole.   From the top of the pole, sets of wires run to the house, to the barn, and to the other buildings to be served.   At each building there is a service entrance as already described in Chap. 17, except without the meter—more about that later in this chapter.

There is a right and a wrong location for the meter pole.   Why is there a pole in the first place?   Why not run the wires to the house and from there to the other buildings, as was done when farms were first being wired?   That would lead to very large wires to carry the total load involved; a very large main switch in the house; expensive wiring to avoid voltage drop, which is wasted power; a cluttered farmyard; and many other complications.

Locate the pole as close as is possible to the buildings where the greatest amount of power will be used per year; on modern farms, the house rarely consumes the greatest total.   That also means locating the pole so that the largest wires will be the shortest wires.   In that way you will find it relatively simple to solve voltage-drop problems without using wires larger than would otherwise be necessary for the number of amperes to be carried.   The large expensive wires to the building with the big loads will be relatively short, and the smaller less expensive wires to the buildings with the smaller loads will be relatively long.   That keeps the total cost down.

Basic Construction at Pole.   The three wires from your power supplier's transformer end at the top of the pole.   *The neutral wire is usually the top wire.*   In a few localities it is the middle wire; check with your power supplier.   Note that the neutral from the power line is spliced at the top of the pole to the neutral running on to the various buildings, so that in effect the neutral is a continuous wire from the power line direct to every building.   The neutral wire is grounded at the pole, as will be explained later.   The neutral is also continued down to the meter socket, where it ends.

The two hot wires from the power line run down to the meter socket, then back to the top of the pole.   This makes a total of five wires from the top of the pole to the meter socket.   In many localities the usual construction is to run all five of them inside a single conduit to the meter socket, as shown in Fig. 24-7.   In some localities, three wires are run to the socket in one conduit and two run back to the top of the pole in

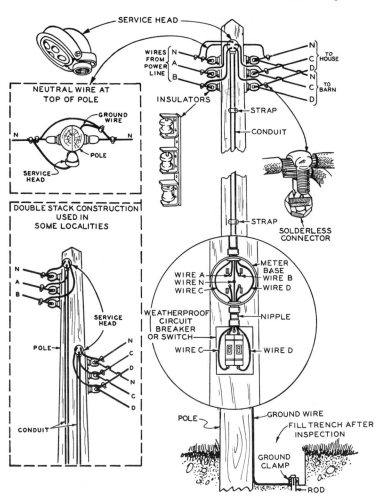

Fig. 24-7.   Typical meter-pole installation on a farm.

another conduit, as shown in the same illustration.   Use whichever method seems to be standard in your locality; observe other farms or consult your power supplier or the local inspector.   Regardless of the details of the installation, leave at least one-third of the circumference of the pole clear, so that linemen and repairmen can climb the pole without difficulty.

Installing the Meter Socket.   The meter socket is sometimes furnished by the power supplier but installed by the contractor.   In other localities the contractor furnishes the socket.   Mount it securely about 5 ft from the bottom of the pole.   If a switch or circuit breaker is also used, mount it about 5 ft from the bottom, the meter socket just above it.

Insulators on Poles.   Near the top of the pole install insulator racks of the general type that were shown in Fig. 17-4.   Provide one rack for the incoming power wires, one for each set of wires running from the pole to various buildings.   Remember that the pull on the wires in a heavy wind or under ice conditions in northern climates is terrific. Anchor the racks with heavy lag screws.   Better yet, use at least one through-bolt, all the way through the pole, for each rack.

Installing the Stack.   Whether you use a single-stack construction (all wires in one conduit) or the double-stack, the general procedure is the same.   In the past the conduit has often ended at a point below the insulators, which then required great care in the installation of the wires to make sure that a drip loop was provided in order to prevent water from following the wires into the service head.   The Code requirement today is to bring the top of the conduit to a point *above* the topmost insulator, as shown in Fig. 24-7, thus automatically solving the water problem.

At the top end of the conduit use a service head of the general type shown in Fig. 17-7, with the right number of holes in the insulator. Run wires through the conduit, white for neutral, black or other color (but not green) for the hot wires.

In practice, the switch, meter socket, the conduit with wires inside, and insulators are usually preassembled on the pole before it is erected. When the pole goes up, it is ready for wires to be installed on the insulators.

Wire Size, Pole to Building.   Wires must run from the pole to individual buildings, and they must be of the proper size.   At each building there must be a service entrance quite similar to that discussed in previous chapters, except without a meter.   The 1965 was the first Code [Secs. 220-4(l) and 220-4(m)] to require a specific method to determine the ampacity of the wire from pole to any building with two or more circuits, and rating of the service equipment at that building.

It is assumed that the building will have a 3-wire 115/230-volt service, so the total amperage *at 230 volts* must be determined.   For motors, use the amperage shown in the table in Chap 13.   For all other loads, start with the wattage. For lights you can determine the total watts from the size of lamps you intend to use. For receptacles, if you allow 200 watts for each, you will probably be on the safe side; they will not all be used at the same time. Then divide the total watts by 230, and you will have the amperage at 230 volts.

Caution: Suppose you have in a building, six 115-volt 15-amp circuits for lights and receptacles. That theoretically makes a total of 6×15 or 90 amp at 115 volts, or 45 amp at 230 volts. But these circuits will not all be loaded to capacity at the same time, so do not use the theoretical 45 amp. Use the total determined by the preceding paragraph.

For each building to which wires run from the pole (except the farm *house;* figure it as outlined in earlier chapters) first determine the amperage at 230 volts of all loads that have any likelihood of operating *at the same time.* Enter the amperage under *a* of the tabulation below. Then proceed through steps *b, c, d, e,* and *f* as outlined in the tabulation.

    a. Amperage at 230 volts of all connected loads that in all likelihood will operate at the same time, including motors if any . . . . _____amp

    b. If *a* includes the *largest* motor in the building, add here 25% of the amperage of that motor (if two motors are the same size, consider one of them the largest) . . . . . _____amp

    c. If *a* does *not* include the largest motor, show here 125% of the amperage of that motor. . . . . _____amp

    d. Total of $a+b+c$ . . . . _____amp

    e. Amperage at 230 volts of all other connected loads in the building . . . . _____amp

    f. Total of $d+e$ . . . . _____amp

Now determine the minimum service for the building by one of the following steps:

    A. If *f* is 30 or less and if there are *not over two* circuits, use a 30-amp switch and No. 8 wire.[1] If there are *three or more* circuits, use a 60-amp switch and No. 6 wire.

    B. If *f* is over 30 but under 60, use a 60-amp switch and No. 6 wire.

[1] The Code in Sec. 230-41 requires a minimum of No. 8 if the building has not over two 2-wire branch circuits, and a minimum of No. 6 in all larger installations.

C. If *d* is less than 60 *and if f* is over 60, start with *f*. Add together 100% of the first 60 amp plus 50% of the next 60 amp plus 25% of the remainder. For example, if *f* is 140, add together 60, plus 30 (50% of the next 60 amp) plus 5 (25% of the remaining 20 amp) for a total of 60+30+5 or 95 amp. Use 100-amp switch and wire with ampacity of 95 amp or more.

D. If *d* is *over 60 amp* start with 100% of *d*. Then add 50% of the first 60 amp of *e*, plus 25% of the remainder of *e*. For example if *d* is 75 amp and *e* is 100 amp start with the 75 of *d*, add 30 (50% of the first 60 of *e*), and add 10 (25% of the remaining 40 of *e*) for a total of 75+30+10 or 115 amp. Use a switch or breaker of not less than 115-amp rating and wire with corresponding ampacity.

The wire sizes determined above will be the minimum permissible by Code. You would be wise to install larger sizes to allow for future expansion. No. 8 wire is acceptable for spans up to 100 ft, and No. 6 for 150 ft. For longer spans, use an extra pole. If the wires are installed in northern areas on a hot summer day, remember that a copper wire 100 ft long will be a couple of inches shorter next winter when the temperature is below zero. Leave a little slack lest insulators be pulled off buildings during winter.

Wires on Pole. The calculations above will determine the size of the service to each individual building. To determine the size of the wires on the pole, before they break down into smaller sizes to separate buildings, proceed as follows using the amperages determined above:

1. Highest of all the amperages for an individual building, as determined above,_____ amp at 100% . . . . . . ._____ amp
2. Second highest amperage,_____amp at 75% . . ._____ amp
3. Third highest amperage,_____amp at 65% . . . ._____ amp
4. Total of all other buildings,_____amp at 50% . ._____ amp
5. Total all above . . . . ._____ amp

In the tabulation above, *include* the farm house, figured as outlined in other chapters. Important: If two or more buildings have the same function, consider them as one building for the purpose above. For example if there are two brooder houses requiring 45 and 60 amp respectively, consider them as a single building requiring 105 amp. If no other building requires more than 105 amp enter 105 amp in *1* above.

The total of 5 above is the minimum rating of the switch or breaker (if used) at the pole, and the minimum ampacity of the wires used. Do note that in listing the amperage for any one building, the amperage to use is the *calculated* amperage, not the rating of the switch used. For example, if for any building you determined a minimum of 35 amp but you use a 60-amp switch (because there is no size between 30 and 60 amp) use 35 amp, not 60 amp.

Connections at Top of Pole. The wires on the pole will be of large size and even on the ground would be difficult to solder. Use solderless connectors of the types that were shown in Figs. 8-22 and 8-23. These connectors being made of metal should be taped after installation.

Connections at Meter. If there is to be neither switch nor fuse (nor circuit breaker) at the pole, the connections are very simple, as shown in Fig. 24-8. The service wires always run to the top terminals of the meter socket; the wires running back up to the top of the pole (and from

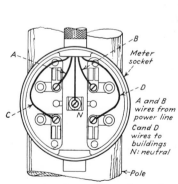

Fig. 24-8. Connections in meter socket on pole, without a switch.

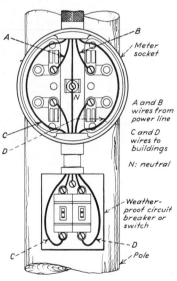

Fig. 24-9. Connections in meter socket on pole, with a switch or circuit breaker.

there to the various buildings) are connected to the bottom terminals of the socket.

If there is to be a circuit breaker or a switch, the connections change only a little, as Fig. 24-9 will show.   Naturally any switch or circuit breaker installed must be the weatherproof type.

If the wires from pole to the various buildings are to be underground, the wires from the bottom terminals of the meter socket will then (instead of running back to the top of the pole) be connected to the various underground wires.

Current Transformers.   As the amperage rating of a service increases, the size of the wires naturally increases too.   If the service is rated at 200 amp or more, as is often the case on farms, it means very large wires running from the top of the pole to the meter, and back to the top again. That is both expensive and clumsy.   It is not necessary to bring the heavy wires down to the meter: Use a current transformer.

Fig.   24-10.   A   current   transformer.   (*General Electric Co.*)

An ordinary transformer changes the *voltage* from one value to another.   The voltage in the secondary as compared with that in the primary, depends on the ratio of the number of turns in the secondary as compared with the number of turns in the primary, and can be either higher or lower than in the primary.   A current transformer is so designed that the *amperage* in the secondary is in proportion to the amperage in the primary, but with a great many *fewer* amperes in the secondary than in the primary. Most current transformers are designed so that not over 5 amp will flow in the secondary, and the total amperage in the secondary is not over 5 volt-amp.

A typical current transformer, shown in Fig. 24-10, has the shape of a doughnut, 4 to 6 in. in diameter. It has only a single winding, the secondary. The wire (the amperage flowing in which is to be measured) is run through the hole of the doughnut, and becomes the primary. Assuming the transformer has a "200 to 5" ratio (for use with a 200-amp service), the amperage flowing in the secondary will be $\frac{5}{200}$th of the amperage in the primary. If the amperage in the primary is 200 amp, 5 amp will flow in the secondary; if it is 100 amp in the primary, $2\frac{1}{2}$ amp will flow in the secondary. If the service were 400 amp, the transformer would have a "400 to 5" ratio.

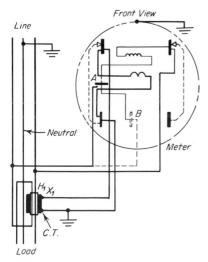

Fig. 24-11. Wiring diagram for one current transformer, on a 115/230-volt service. (*General Electric Co.*)

In an installation of this kind, the heavy service wires do not leave the top of the pole. One current transformer is installed at the top of the pole, with the two hot wires running through the hole. Four small No. 14 wires run from the top of the hole to the meter: two from the secondary of the current transformer, and two from the hot wires, for the voltage.   The meter operates on a total of not over 5 amp, but the dials of the meter will show the actual kilowatthours consumed.   A wiring diagram is shown in Fig. 24-11.

Caution: The secondary winding of a current transformer must always be short-circuited: when connected to a meter, it is effectively short-circuited by the windings in the meter.   If you were to touch the two secondary terminals of such a meter that is *not* short-circuited, you would find a dangerous voltage of thousands of volts.

In an installation using a current transformer at the top of the pole, if you also want a switch at the pole, install it at the top of the pole.   Pole-top switches are operated from the ground by means of a hook or rod, as shown in Fig. 24-12.   Note the current transformer installed in the same cabinet with the switch.   Such switches are also available in double-throw type as required if an emergency standby generating plant is installed for use during periods of power failure.

Ground at Meter Pole.   The neutral wire always runs from the top of the pole, through the conduit, to the center terminal of the meter socket. The neutral is not necessary for proper operation of the meter, but grounds the socket.   At one time it was standard practice to run the ground wire from the meter socket (out of the bottom hub) to ground, but experience has shown that better protection against lightning is obtained if the ground wire is run outside the conduit.   Run it from the neutral at the top of the pole, directly to the ground rod at the bottom of the pole.   It is usually run tucked in alongside the conduit as far as it goes, then to ground.   In some localities it is stapled to the pole on the side opposite the conduit.

In some areas the custom is to let the top of the ground rod project a few inches from the ground.   The ground clamp is permanently exposed.   More often the ground rod is driven about 2 ft from the pole (or building) after a trench has first been dug about a foot deep from rod to pole.   The top of the rod is a few inches above the bottom of the trench. The ground wire runs down the side of the pole (or building) to the bottom of the trench, then to the ground clamp on the rod.   After inspection the trench is filled in and the rod, the clamp, and the bottom end of the ground wire remain buried (see Fig. 24-13).   Use the method favored in your locality.

Grounds in buildings housing livestock should be installed so that seepage from animal manure does not saturate the ground around the rod.   Chemical action in time eats the wire, the clamp, and sometimes

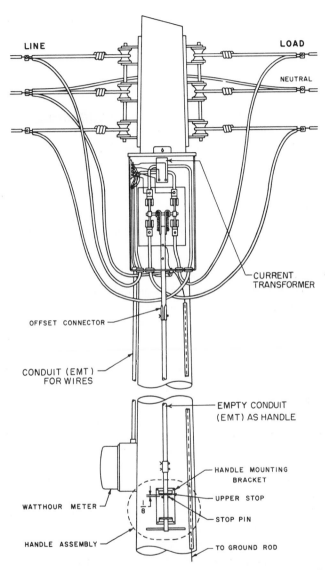

LINE

LOAD

NEUTRAL

CURRENT
TRANSFORMER

OFFSET CONNECTOR

CONDUIT (EMT)
FOR WIRES

EMPTY CONDUIT
(EMT) AS HANDLE

HANDLE MOUNTING
BRACKET

UPPER STOP

WATTHOUR METER

$\frac{1}{8}$

STOP PIN

HANDLE ASSEMBLY

TO GROUND ROD

Fig. 24-12. If a switch is wanted in a high-capacity service, install it at the top of the pole.   Note the current transformer in the cabinet of the switch.   (*Hoffman Engineering Co.*)

even the rod, so that what at one time was a good ground turns out to be no ground at all.

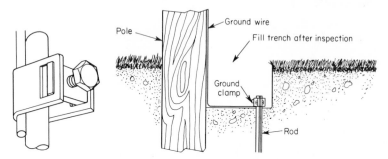

Fig. 24-13.   In some localities, the top of the ground rod is below the surface of the ground.   Fill the trench after inspection.   If the ground rod is copper, use *only* a copper clamp.

Ground at Outbuildings.   The Code in Sec. 250-24 requires a ground at any building with two or more circuits, and also at every building that houses livestock, regardless of the number of circuits.

If a building, regardless of number of circuits, contains equipment that must be grounded in any event, naturally a ground must be provided.   The "equipment" includes disconnecting switches, fuse or circuit breaker cabinets, conduit, and even ordinary switch boxes for a single switch, if made of metal.   But if a building is wired with nonmetallic-sheathed cable and has only a single circuit, no ground is required.

Entrance at Individual Buildings.   The entrance at any one building served directly from the meter pole is made as already outlined in Chap. 17, except that the meter is omitted.   Instead of running the ground wire from the neutral strap of the service equipment, run it from the neutral wire, at the point where it reaches the building, to the ground rod.   Run it along the service cable or service conduit, just as on the pole, for maximum protection against lightning.   If, however, underground wires have been used, ground them in the usual way.

In the house, provide circuit breakers (or fused service switch plus branch-circuit fuses) just as if the house were the only building being wired.   In other buildings you have a considerable choice of equipment, with one restriction: You must be able to disconnect *all* the wiring with one or more switches.   In a very small building with only one or two lights and maybe a receptacle, the switch might be an ordinary toggle switch such as you use to control lights in the house, but it must disconnect the receptacle at the same time.

In the category of buildings with more outlets, there might be at one extreme a building with one circuit, at the other extreme a dairy barn with 20 circuits, and the equipment must be selected according to the load and the number of circuits. For the smaller of the buildings mentioned you might use a small fused switch of the general type that was shown in Fig. 13-3 and in the larger building one of the 100-amp type shown in Fig. 13-5, or even larger. In-between buildings will require in-between switches in proportion to the load and the number of circuits. Circuit breakers are being used more and more in place of fused equipment.

Buildings Fed through Another Building. On farms it often happens that one building is fed by wires from another building. At any such building proceed as if the wires came directly from the pole, except that the switch (or circuit breakers) that controls the wiring may be located either at the building *from* which the wires run or at the building *to* which they run.

Whether this disconnect switch needs to be fused or not depends entirely on the size wires used and the size fuses ahead of the starting point (see Fig. 24-14, which shows several buildings served through the main

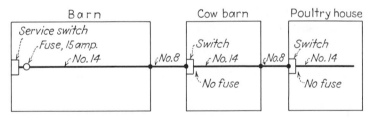

Fig. 24-14. Whether disconnect switches in outbuildings need to be fused or not depends on the size of the fuse at the starting point.

barn). The wires are No. 14 inside the buildings, but, for mechanical strength, No. 8 is used between buildings. Since the fuse in the barn is the 15-amp size, it protects the smallest wire in the entire circuit; therefore the switches in cow barn and poultry house need not be fused.

In Fig. 24-15, No. 8 wires run through the barn, between the barn and cow barn, and through the cow barn, serving a motor there which requires up to 30 amp for starting.   Therefore a 30-amp fuse is used in the barn, and since the No. 8 wire continues straight through to the poultry house, this wire must be fused at not over 15 amp at the point where the change is made to No. 14 wire in the poultry house.   Likewise at the point where No. 14 wire is connected to the No. 8 in the cow barn, a fuse block must be used with 15-amp fuses to protect No. 14 wire.

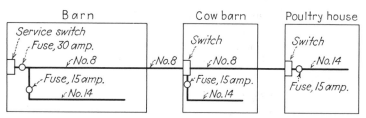

Fig. 24-15.   In this case, fuses are required because the 30-amp fuse in the barn does not protect No. 14 wire.

Farmhouses.   The basic wiring in farmhouses is no different from that in city houses, and the procedure therefore is the same as discussed in other chapters.   But because the ground is usually a rather poor ground, do not use conduit or armored cable: Use nonmetallic-sheathed cable.

Use Nonmetal System.   The Code in Sec. 300-1(c) recommends a nonmetal system, in other words nonmetallic-sheathed cable, for locations where a good permanent ground is not found.   Certainly farms fall into that classification; hence conduit and armored cable are seldom used for farm work.   The reasons for this are not hard to understand. In barns considerable moisture is always present, and in addition there are other corrosive vapors such as, for example, ammonia from the excreta of animals.   This quickly attacks the conduit or metal armor, so that the life of the metal is relatively short.   As the metal raceway rusts out, it may open the continuous ground that otherwise exists from outlet to outlet in the usual metal system.   If an accidental ground occurs in such a system at a point beyond the break and if a person or an animal touches a part of the metal raceway above the break, the circuit is completed through the body with at least an unpleasant, if not a dangerous, shock.   Figure 24-16 should make this clear; the ground through the body is equivalent to touching both wires and may be equally dangerous. Many dairy animals have been killed every year through just such occurrences—metallic systems apparently but not actually grounded. Cattle and other farm animals cannot withstand as severe a shock as human beings, and often are killed by shocks that would be only unpleasant to a man. For these reasons the nonmetal systems are recommended for such locations.

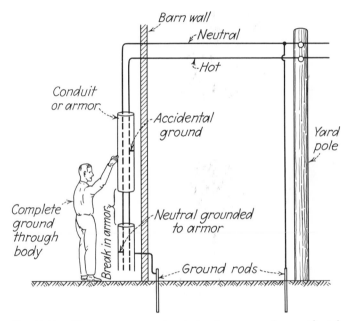

Fig. 24-16. If there were no break in the armor, the accidental ground would blow a fuse. Because the armor has a break, a dangerous condition exists.

The objection can be raised that, even if nonmetallic cable is used, dangerous conditions can and will still arise when an accidental ground happens to occur inside one of the usual metal outlet boxes; anyone touching the outlet box is then still subject to the same shocks encountered with a metal system in which the metal raceway is eaten through as outlined in the previous paragraph. To this objection there is no good answer, with the result that nonmetallic outlet and switch boxes have been developed to make the entire system nonmetallic. Such boxes are made of various plastic materials. Figure 24-17 shows a representative assortment.   Many types are available with mounting brackets or nails, for installing directly on timbers of buildings.   Naturally covers for switches, receptacles, and similar devices are available. Nonmetallic boxes being of one-piece construction, cannot be ganged, but ready-made 2-gang and 3-gang boxes are available.

Fig. 24-17. Boxes made of nonmetallic material are very commonly used on farms.   (*Union Insulating Co.*)

Install such boxes as you would metal boxes, except that per Code Sec. 370-7(c), connectors are not needed at the box, but the cable must be supported within 8 in. of the box (except in the case of "old work" where the cable is fished through wall spaces).   Since there are no cable clamps in the box, one additional wire may be installed in it (see Chap. 10).   These two facts make the use of nonmetallic boxes more and more popular in nonfarm installations, provided nonmetallic-sheathed cable is used.

In the installation of nonmetallic boxes, one precaution is in order. When they are installed in barns or other locations where the humidity is high, wood timbers shrink and swell with the variations in humidity. Steel boxes, if rigidly nailed or screwed to supporting timbers, do not present a mechanical problem as the timbers swell with moisture, for the steel boxes "give" a bit if required. Nonmetallic boxes on the other hand, if screwed down tightly on dry timbers, have been known to break out their bottoms as the timbers swelled with increasing humidity. When mounting such boxes, leave just a little bit of slack and don't drive the mounting screws down completely tight.

Also very popular for barns and similar locations are the combination surface devices shown in Fig. 24-18. Their use and method of installation have already been fully covered in Chap. 22 and will not be repeated here.

Fig. 24-18. These handy devices replace outlet box, cover, and wiring device, all in one piece. (*General Electric Co.*)

Cable for Barn Wiring. In the early period of farm wiring the only kind of nonmetallic-sheathed cable then made (now known as Code Type NM) was used in wiring barns and other farm buildings. It gradually became clear that the usual high humidity in such buildings caused rotting of the cable, leading to danger of shock and fire. It became necessary to rewire many farm buildings because of the short life of the cable.

This led to the development of several types of "barn cable" which have now been standardized into two basic types: Code Types NMC and UF.

In the wiring of barns and other buildings housing livestock, use only one of these two types. There is no good reason why they should not be used throughout all the buildings on a farm.

The physical make-up of the circuits, that is, the combinations of cable and boxes and switches and receptacles, is not different from that already discussed and therefore needs no further amplification. The chief points to observe are points of practicability and common sense. Locate switches and receptacle outlets so that they cannot be bumped by animals in passing. Locate switches at convenient height so that they can be operated by the elbow; farmers' hands are often both full. A great convenience is to have 3-way switches to control at least one light from either end of the barn. Never use brass-shell sockets; use only porcelain or bakelite, for the same reasons that nonmetallic boxes are used.

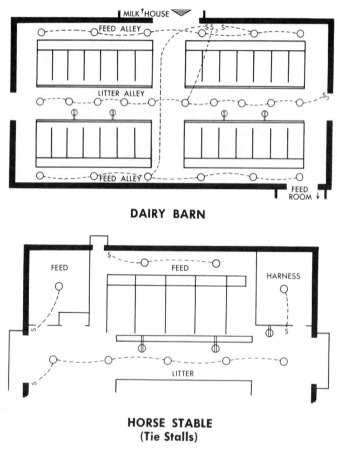

DAIRY BARN

HORSE STABLE
(Tie Stalls)

Fig. 24-19. Suggested wiring diagrams for farm buildings. (*Industry Committee on Interior Wiring Design.*)

Barn Wiring. Barns come in all sizes and descriptions. Provide lighting outlets and receptacles in proportion to the need. In Fig. 24-19 are shown some suggested wiring diagrams, from the "Farmstead Wiring Handbook." It is wise to locate the outlets for lamps between joists, so that the lamp does not project too far down into the aisle between stalls, where it might easily be damaged. Since most barn ceil-

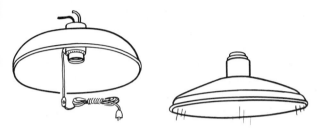

Fig. 24-20. A 60-watt lamp with a reflector is often as effective as a 100-watt without a reflector.

ings are dark and dusty, they reflect practically no light. About half the light from an exposed lamp falls on the ceiling and is lost. Provide a reflector for every socket; the area underneath will be lighted almost twice as well as without a reflector. Several types of suitable reflectors are available at reasonable prices; two are shown in Fig. 24-20. Their use is a good investment. Clean them often.

Fig. 24-21. A typical vaporproof receptacle, often used in haymows. (*Killark Electric Mfg. Co.*)

Provide a light to illuminate the steps to the haymow. In haymows inspectors frequently require vaporproof receptacles of the type shown in Fig. 24-21. A vaporproof receptacle is simply a socket with a tight-fitting glass globe that encloses the lamp. The dust that arises in haymows is inflammable or even explosive, and if an unprotected lamp is accidentally broken, although the lamp burns out instantly, during that instant there is a flash while the filament melts at a temperature of about 4000°F.   This flash may set off an explosion; hence the requirement for the vaporproof receptacles.

Cable should always be installed in a location where it cannot possibly be damaged accidentally.   In haymows, cover it with strips of board at all points where it might be damaged by forks.   Where it passes through a floor, the Code requires that it be protected by conduit or pipe for at least 6 in.; many inspectors sensibly require this protection for about 6 ft, especially where there is danger of forks damaging the cable.   Note, however, that this length of pipe or conduit does not then make the system a conduit system; the pipe is there purely for protection against accidental mechanical damage to a length of approved cable.

Cable in barns and other farm buildings should not be run along or across the bottoms of joists or similar timbers, because this exposes the cable to mechanical injury.   The cable will receive good protection if it is run along the side of a joist or beam.   More cable will be required to run the cable from the side of an aisle out to the middle for an outlet, then back to the side of the aisle, but consider the extra cost as insurance against damage.   Figure 24-22 will show the details of recommended practice.

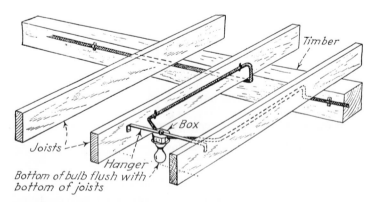

Fig. 24-22.   Keep exposed cables away from centers of aisles.   To avoid damage, it is best to keep lamps from projecting much below the bottom of joists.

In buildings in which the humidity is high, such as buildings with livestock or dairy equipment, it is best not to run cable on the inside of the outer walls; these walls being colder than the other areas, will naturally have condensed moisture on them. The preferred location for cable is where it will stay reasonably dry.

Poultry Houses. One point that should be particularly noted is that special wiring is frequently required for lighting designed to promote egg production. It is well known that hens produce more eggs during winter months if light is provided during part of the time that would otherwise be dark. It is best to provide light at both ends of the day, morning and evening. Opinions vary as to the ideal length of the "day" but a 14- or 15-hr period seems reasonably acceptable. Of course that means that the length of time the lights must be on, varies from season to season; adjusting the period every 2 weeks seems a reasonable procedure.

In the evening, if all the lights are turned off suddenly, the hens can't or won't go to roost, will stay where they are. Therefore it is necessary to change from bright lights to dim lights to dark. All this can be done by manually operated switches, but an automatic time switch of the type shown in Fig. 24-23 costs so little that manually operated switches shouldn't be considered. In the morning, the time switch turns on the bright lights at the time set, then when daylight appears turns them off.

In the evening, again at the time set, the switch turns on the bright

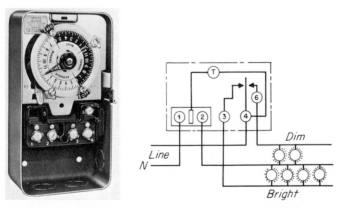

Fig. 24-23. A time switch automatically controls poultry-house lights, dim and bright, for forcing egg production. (*Paragon Electric Co., Inc.*)

lights.   Later again at the time set, the bright lights go out and the dim lights are turned on; during this interval of dim lights the hens go to roost.   Shortly thereafter all lights are automatically turned off.   The wiring for such switches is simple, and wiring diagrams are furnished with the switches.

Be sure that both the bright and the dim light falls on the roosts, for if the roosts remain in darkness when the lights come on, the hens may not leave their roosts.   Neither will they be able to find the roosts in the evening if the roosts are in darkness, while the dim lights illuminate the rest of the pens.   One 40- or 60-watt lamp for every 150 to 200 sq ft of floor area is usually considered sufficient for the "bright" period.   Normal spacing is about 10 ft apart.   For the "dim" period 15-watt lamps are suitable, but only about half as many as the number of "bright" lights are needed.   It is a good idea to provide each light with a shallow reflector; otherwise a good share of the light falls on the ceiling and is lost.

Water Pump.   Every farm will have a water pump.   It not only will serve to provide water for all usual purposes but, in addition, will be a tremendous help in case of fire.   But in case of fire, quite often power lines between buildings fail and fuses blow, so that the pump cannot run just when it is needed most.   That can be avoided by simply considering the pump as a fire pump.   The Code in Sec. 230-2 permits a fire pump to be connected through an independent service.   In practice this simply means that wires are run to the pump house directly from the meter pole, ahead of the main fuses or circuit breaker.   Then, even if the main fuses blow, the pump will still run.   Simply run two wires from the pole, ahead of circuit breaker or main fuses, to the pump. There must be a fused disconnect switch (or a circuit breaker) for the pump only, and if this is mounted on the pole, it must naturally be of the weatherproof type.   Underground wires provide additional insurance against failure.

Other Buildings.   No particular problems are involved in other buildings.   Use only nonmetallic sockets; provide switch control instead of pull chains.   Every building, no matter how small, should have a receptacle outlet, if for no other reason than to provide a connection for a "trouble light" of the type that was shown in Fig. 20-6.

Motors.   Stationary motors should be wired as discussed in Chaps. 15 and 31.   For a portable motor of considerable size it will be necessary to provide a heavy-duty receptacle with a rating in amperes and volts at least equal to the rating of the motor. One of the types shown in Figs. 21-1 and 21-2 will be found suitable. If the receptacle must be outdoors, of course a weatherproof type must be used. Except in the smallest sizes, these are very expensive; one is shown in Fig. 24-24, to-

Fig. 24-24. A heavy-duty weatherproof receptacle and plug, for outdoor use. (*Killark Electric Mfg. Co.*)

gether with a plug to fit. If a weatherproof receptacle is to be mounted on the outside wall of a barn or other farm building, consider installing one of the ordinary nonweatherproof type inside the building, with a door over it that can be opened to make the indoor receptacle accessible from the outside.

Yard Lights. Every farm will have at least one yard light of the general type shown in Fig. 24-25, usually located at the meter pole. In

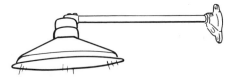

Fig. 24-25. Yard lights should be liberally used in lighting the farmyard.

most cases the light is controlled by two 3-way switches, located at house and barn. A 4-way switch in a weatherproof housing is often located at the pole.

The wiring of a yard light is often haphazard.   It is not uncommon to see it wired by tapping one of the two wires from the light directly to the neutral wire at the pole, running the other to a switch at any convenient location and then on to the nearest "hot" wire.   The scheme works but is contrary to Code, for the wires on the pole constitute a feeder, not a branch circuit.   It is not permissible to tap a feeder, except to form a branch circuit *with overcurrent protection* where the circuit starts.

Today most inspectors will insist on a carefully planned installation. To operate a yard light from one switch, it is necessary to run two wires to it from the nearest fused circuit, and to operate it from two points, it is necessary to run three wires.   Figure 24-26 shows a method that makes a weatherproof and mechanically sturdy job out of the installation; it also shows the wiring diagram.

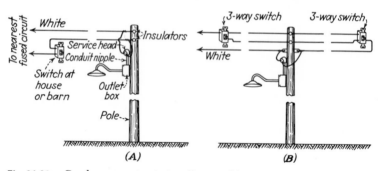

Fig. 24-26.   Good construction in installing yard light.

An alternate scheme that is growing in popularity is the remote-control yard light.   The basic principles involved were discussed in Chap. 20.   Figure 24-27 shows a typical unit which houses the transformer and the relay or remote-controlled switch.   This self-contained unit may be connected directly across the hot wire and the neutral of a feeder on a pole.   That may seem contrary to the prohibition mentioned in a preceding paragraph concerning connections to a feeder, but this self-contained unit contains a fuse; the unit thus becomes a branch circuit where connected to the feeder.   Only low-voltage wires run from this yard-light unit to the locations where the switches are to be installed, so inexpensive 3-wire cable can be used.   The wiring is in accordance with Fig. 24-28, which is substantially that of Fig. 20-30.

The particular unit in Fig. 24-27 shows an exposed lamp with a reflector lamp. The reflector is a silver coating inside the bulb of the lamp, and remains bright throughout its life. Such lamps are available as floodlights or spotlights, depending on whether a large or a small area

Fig. 24-27. A remote-control yard light.

is to be lighted. Be sure the lamp you buy has a "hard-glass" bulb, which will not be damaged if cold rain hits the lighted lamp. Lamps made of ordinary glass shatter when hit by cold rain or snow, therefore can be used only indoors, unless protected by effective reflectors.

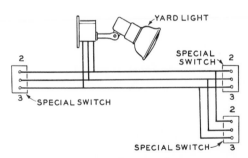

Fig. 24-28. Wiring diagram for remote-control yard light. (Compare with Fig. 20-30.)

A different and very modern method uses mercury-vapor lamps described in Chap. 30. Such lamps produce about $3\frac{1}{2}$ times as much light *per watt* as ordinary lamps, cost about five times as much as ordinary reflector lamps of the hard-glass variety, but last at least 12 times longer. They require special fixtures. Often such lamps used for yard lighting are equipped with photoelectric controls that automatically turn the light on in the evening, off at daylight. Your power supplier can give you information.

Wires Entering Buildings. Use any of the methods described in Chap. 20 in connection with garage wiring.

Tapping Service Wires at Building. Often two buildings are quite close to each other and can then be served by a single set of wires from the pole, which should run to that building with the greater load. Naturally the wires must be heavy enough for the combined load of both buildings. At the service insulators of the first building, make a tap and run the wires on to the second building, all as shown in Fig. 24-29. At the second building, proceed just as if the wires came directly from the pole.

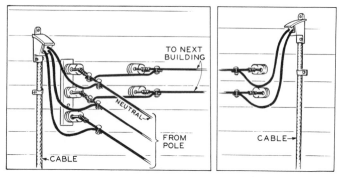

Fig. 24-29. When two buildings are near each other, if the loads are small, tap the service wires from the pole, where they are anchored to the building. Run them to the second building.

If the second building is very small and requires only 115 volts, tap off only two wires including the neutral, as shown in the picture. If the second building has a considerable load so that 115/230 volts is desirable, tap off all three wires. Remember the requirement for a separate service switch and ground at the second building, as discussed in other paragraphs.

If the final building is a small one, the entry to it may also be made as outlined for residential garages in Chap. 20.

Outdoor Switches and Receptacles. On farms it will often be found convenient to install a switch or receptacle outdoors. The simplest way of handling an outdoor switch is to use a cast housing of the type shown in Fig. 24-30.   Inside the housing install any kind of toggle switch whether single-pole or 3- or 4-way.   The housing of Fig. 24-31 will accommodate any single receptacle of the 15- or 20-amp rating.

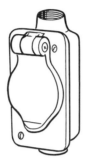

Fig. 24-30 For outdoor use, install any kind of toggle switch in this housing. (*Killark Electric Mfg. Co.*)

Fig. 24-31. For outdoor use, install an ordinary single receptacle in this housing. (*Killark Electric Mfg. Co.*)

When using such housings for switches and receptacles, do not install them on ordinary switch boxes, because they are not watertight.   Instead, use *cast* outlet boxes without openings except for the entry of cable, the "FS" fitting shown in Fig. 29-2 in Chap. 29 is suitable.   Use a watertight connector with a rubber gland, similar to the one shown in Fig. 17-14 for service-entrance cable, but of the proper size for the particular cable you are using.   This then provides a thoroughly weatherproof installation, a safe installation, for the switch or receptacle.

Wiring the Very Small Farm.   On a very few farms, only a little power is used except in the house.   If you are sure this will always be so, the buildings other than the house can be treated like the garage in city wiring.   Bring the main entrance into the house.   Provide one or two circuits for the wiring beyond the house.   From the overcurrent protection in the service, run service-entrance cable to the outside of the house, up the side of the house to an entrance cap and the insulators for the wires that are to run to the other buildings.   That makes a sort of "backward" service entrance at the house.   While service-entrance cable may not be used for ordinary branch-circuit wiring (with the exception of ranges and dryers as discussed in earlier chapters), the Code definitely permits it for the one purpose here described.

Lightning Arresters.   While lightning-caused damage to electrical installation is quite rare in large cities, it is rather frequent in other areas.   The more isolated the location, the more the likelihood of damage.   It occurs frequently on farms, and to a lesser extent in suburban areas and smaller towns.   It is more frequent in southern areas, especially in Florida and other Gulf states.

Fig. 24-32. A lightning arrester should be installed at the pole, sometimes at buildings. (*General Electric Co.*)

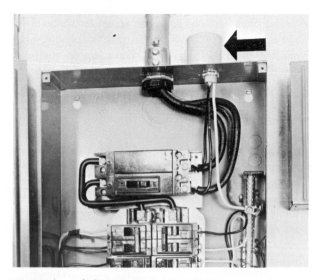

Fig. 24-33. A lightning arrester installed on a service switch. Be sure the white wire is connected to the neutral. (*General Electric Co.*)

Lightning does not have to strike the wires directly; a stroke *near* the wires can induce very high voltages in the wires, damaging appliances and other equipment as well as the wiring. Sometimes the damage is not apparent immediately, but shows up later as mysterious breakdowns.

While proper grounding greatly reduces the likelihood of damage, a lightning arrester properly installed reduces the probability of damage to a very low level indeed. Figure 24-32 shows an inexpensive arrester; it costs about ten dollars and is smaller than a baseball. Three leads come out of it; connect the white wire to the grounded neutral, the other two to the hot wires.

One should be installed at the meter pole. If the feeders from pole to building are quite long, install another at the building. Install it on the service switch, letting the "neck" of the arrester project into the switch through a knockout, as shown in Fig. 24-33.

25

Isolated and Standby Power Plants

An isolated power plant may be defined as any installation by which the owner makes his own electric power. Such installations are used where electric power is not available from commercial power lines. The same generating plants (if they develop 60-cycle AC power) are also commonly used for standby service to provide emergency power when commercial power lines fail.

Battery-type Plants.   The most ordinary type charges a 32-volt battery, thus constituting what has long been known as a "32-volt farm light plant." Others charge 6-, 12-, and 110-volt batteries.   Battery-type plants are little used today.

Alternating-current Plants.   By far the most common type of isolated generating plant today is the type which generates 60-cycle alternating current.   Capacities are available from 400 to 250,000 watts and more. Several are shown in Fig. 25-1.   Among the many advantages of this type of plant are these: Ordinary appliances (motors and so on) are used, costing far less than special 32-volt devices and far more readily obtainable.   Since their voltage is 115 (or 115/230), the wiring is the same as for ordinary electrical work.   Combining these two advantages means that if such a plant is installed in an isolated location and commercial power later becomes available, the wiring and all the appliances can be used without change.   There are no problems of excessive voltage drop as in the case of low-voltage battery plants.   Plants are available in any required wattages, so that there is no limit to the number or size of appliances that can be used.

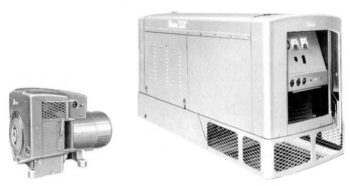

Fig. 25-1.   Typical generating plants developing 5,000 and 35,000 watts of single-phase 115/230-volt power.   (*Onan Division of Studebaker.*)

Since such generators are engine-driven, means must be provided for starting the engine.   The simplest is the hand-starting method, using a rope or crank; this method is practical only with the smallest plants. The usual method uses a 12-volt automobile battery which cranks the engine.   On smaller plants there is a special cranking winding in the AC generator, which also contains a DC winding for charging the battery.   On larger plants the starting system is the same as that of an automobile.

Electrically cranked plants are further divided into those which start by pushing a button on the plant, and remote-control types which can be started from start-stop buttons placed at a distance of up to several hundred feet, and connected by means of inexpensive control wire. The next step is the "full-automatic" type which starts automatically when a light or appliance is turned on, stops when everything is turned off.   The special type of control which starts the plant automatically when commercial power fails will be discussed separately.

In selecting the capacity of the generator needed, naturally you must choose one which has as many watts output as the maximum number of watts to be consumed by all the different loads turned on *at the same time*. Manufacturers of these plants provide helpful literature for that purpose.

Standby Plants. Plants that generate 60-cycle power are very commonly used to provide power during periods when commercial power has failed. In a city home an interruption of a few hours may be inconvenient but not necessarily really important. It is important in a hospital operating room, in a broadcasting station, and on a railway signal system. It is important in a modern windowless office, in a store open during periods of darkness, in a theater, where complete darkness might lead to panic. It is important in a chicken hatchery, where a few hours' interruption may ruin a large load of eggs; in a greenhouse with an electrically operated oil burner, where a few hours' interruption might let the temperature drop to a dangerous point; on a farm, where interruptions are often measured in days rather than in hours—without power, milking machines are out of operation, pumps do not operate, hot water is no longer available in required volume, oil burners do not work. Continuous power is important in many hundreds of different circumstances, and the only guarantee against an interruption is a standby generating plant.

Type of Control for Standby Plants. The user has a choice of two methods for controlling a standby plant. The first and simplest consists merely of a hand-operated double-throw switch. When the power fails, start the plant, then throw the switch. The second is the automatic line-transfer control which, when the power fails, disconnects the load from the commercial line, starts the plant, then transfers the load to the plant. When commercial power is restored, it transfers the load back to the commercial line and stops the plant.

The important part of either control is the double-throw switch, which in one position leaves the load connected to the commercial power line, in the other position it completely disconnects the power line from all wiring and connects the load to the generating plant. (In the case of an automatic control, this double-throw switch is electrically operated.) Do not under any circumstances try to install a standby plant without using such a double-throw switch, for doing so is apt to result in having a dead commercial line connected to the wiring of your establishment at the same time that the standby generator is feeding the power into that wiring system. The standby generator is then *feeding power back* into an otherwise dead line, which can become quite dangerous to linemen working on the supposedly "dead" line. Several deaths on REA lines have been traced to linemen working on a supposedly dead line which was actually energized by a generating plant connected in haphazard fashion without using a double-throw switch. Remember that the same transformer which reduces high-line voltage from 2,300 to 115 volts for use on the farm also steps 115-volt power from a standby plant back to 2,300 volts, if the line is not disconnected while the standby plant is in use.

In selecting such a double-throw switch (whether hand or electrically operated) do not make the mistake of using one which is large enough to carry only the amperage developed by the standby plant. For example, a 5,000-watt 115/230-volt plant will produce about 22 amp, so a 30-amp switch would appear to be large enough. However, the switch is usually connected in the main line, so it will have to carry up to the full capacity of that line, usually 60 amp or more, during normal operation, that is, during periods when the commercial line has *not* failed. The amperage rating of the switch is determined by the maximum number of amperes it must carry during normal periods.

Normally, select a switch with as many poles as are found in the service switch, *not* counting the neutral. Thus for a 3-wire 115/230-volt service you would use a 2-pole switch. Some REA lines, however, require that *all* poles, including the neutral, must be switched, this then requiring a 3-pole switch.

The installation of such double-throw switches is very simple, as Fig. 25-2 will show. Trace the circuit as shown, and you will see

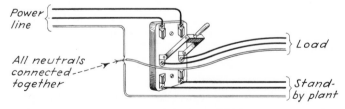

Fig. 25-2. In installing an emergency or standby generating plant, for safety's sake *always* install a double-throw switch. If the controls are automatic, the switch will be electrically operated.

that, when such a switch is properly installed, there is no possible way in which a standby plant can feed back through the transformer into an otherwise dead line.

Advantages of Manual Control.   With the use of a hand-operated switch, part of the load can be turned off by hand before starting the standby generator.   Then a smaller generator can be used.   For example, on a small farm the range and the water heater can be disconnected, making a 3,500- or a 5,000-watt plant big enough.   If this were not done, a 10,000-watt plant might be needed.   In a factory, certain large motors or heating devices might be disconnected if their use is not essential, leaving everything else in operation and thus making practical a much smaller plant than would otherwise be required.   In this way a considerable saving can be made in the original cost.

Advantages of Automatic Control.   For some applications, there is no choice.   If the lights were to go out in a hospital operating room during an operation, a delay might be serious.   If a store is open in the evening and if the lights go out, a delay might lead to panic and injuries.   If installed to provide power for oil burners, the automatic control provides protection even at night and over week ends, when perhaps no one is on the premises.   Automatic installations are frequently made in isolated locations, as, for example, in microwave transmitting stations, where no caretaker is seen for a week at a time.

In addition to the basic advantage of being automatic, such controls usually also provide other conveniences, such as automatic charging of the cranking battery; test switches for checking the operation without taking over the load; "exercisers," if wanted, to run the plant for say 15 min once a day or once a week, to keep the engine always ready to start instantly; many more optional features.

A standby plant with automatic controls does not necessarily have to be big enough to supply power for every load on the premises.   It might serve only specific equipment such as incubators and brooders in a chicken hatchery.   In nonresidential locations, quite frequently special independent circuits are installed for emergency lights, which are never supplied from commercial lines but only from the emergency plant.

Code Requirements.   The Code does not require an emergency plant for any specific location. If, however, any local legislation (city, municipal, federal) does require such a plant, then and only then, Art. 700

of the Code becomes applicable to the installation.   This book will not attempt to discuss all the requirements, but only some of the major ones.

The standby plant must be tested upon installation, and periodically thereafter, as directed by the inspector.   A record must be kept of such tests.   In a hospital, the standby plant must take over the emergency load within 10 sec after power failure.   In any location there must be "where practicable" audible and visual signals to give warning of the failure of the emergency source, to indicate whether the regular power supply or the emergency plant is carrying the load, and to show whether the battery-charging means is functioning properly.

The wiring for separate circuits supplied only by the standby plant may not enter the same conduits, cabinets, boxes, and so on, used for the regular wiring, except in the automatic transfer switch.   There is one exception in the case of exit lights: It is desirable that such lights contain two lamps, one supplied by the regular power, and the second supplied by the standby plant.

There are many other requirements in Art. 700, which you will find in your copy of the Code.

Installation Hints.   Manufacturers' literature provides complete installation instructions, but some general hints will be in order.   The plant must naturally be installed where the engine will cool properly. A small room is not considered a good location unless means are provided for adequate circulation of air to carry away the heat created by the engine.   The fuel-supply system must be considered.   If gasoline is used, an underground tank outside the building provides the best installation.   If you have natural or other gas in your building, use it for fuel, thus doing away with gasoline entirely.   The output of the plant, however, may be reduced somewhat, depending on the richness of the gas in your particular area.   Figure 25-3 shows a properly installed plant.

Tractor-driven Generators.   For emergency use on farms, tractor-driven generators are often used.   They are available in sizes up to 25,000 watts.   Naturally they cost less than generating plants complete with their own engines.   They do have the disadvantage of requiring time to line up the tractor with the generator, after spotting the generator in the right location; they also preclude the use of the tractor for other purposes, during the period of power failure, which on farms sometimes lasts for rather extended periods.

Never install a generator, whether with its own engine or tractor-driven, without installing a double-throw switch.   That is an absolute necessity for safety purposes.

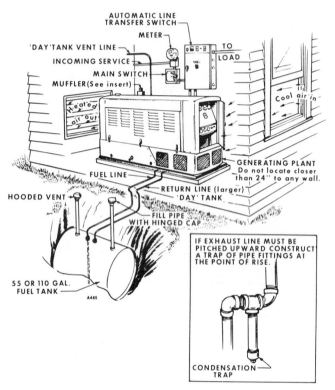

Fig. 25-3. Typical installation of engine-driven generating plant. (*Onan Division of Studebaker.*)

In matching the generator with your tractor, be sure your tractor has sufficient horsepower at the PTO shaft, *at the speed required by the generator*.   It should have at least 2 hp for each kilowatt of generator capacity; $2\frac{1}{2}$ hp for each kilowatt will enable you to better take advantage of the generator's overload capability.   If you don't have enough horsepower in the engine, it will slow down, and so will the generator. The voltage, the frequency, and the capacity of the generator will drop off rapidly.

26

Wiring Apartment Houses

If an apartment is considered merely as that space within a building which is occupied by one family, no new wiring problems of any consequence are encountered. If the apartment house is considered in its entirety as a multifamily dwelling, many new problems arise.

As a matter of fact, the problems in wiring larger apartment buildings can become quite complicated, and the Code requirements can easily be misinterpreted. The "Question and Answer" columns of electrical magazines very often concern such problems. However, you as a student will not be called upon to design a wiring layout for a large apartment building until you have mastered the problems in smaller projects. For this reason detailed discussions of *all* problems in the wiring of larger apartments are considered beyond the scope of this book. The information in this chapter should not be considered complete, but rather looked upon as a foundation, a preview, of the problems you will meet later in your career.

Planning an Individual Apartment. To determine the minimum number of circuits required by Code for any single apartment, proceed as outlined for a single-family dwelling in Chap. 13. For lighting, allow 3 watts per sq ft. For example, a small apartment of 800 sq ft will require 800×3, or 2,400 watts, which means two circuits. To this must be added two separate No. 12 circuits for appliances, just as in the case of the single-family house, thus making four circuits the minimum. Naturally, too, an additional circuit must be provided for each appliance consuming more than 1,650 watts, for example, an electric range.

Service-entrance Problems. In practically all cases there is but a single service drop for the entire building. In most cases each tenant pays for the power he consumes, so there is a separate meter for each tenant, plus usually another meter for "house loads": Hall lights, heating-system motors, water heaters, and similar loads. The service entrance must therefore feed a number of separate meters and disconnecting means. The Code in Sec. 230-70(b) requires that each occupant must have access to his disconnecting means.

Apartment buildings fall into what the Code calls "multiple-occupancy" buildings. Service-entrance problems are discussed in Code Sec. 230-70 (b) which reads as follows:

> **230-70(b). Location:** In a multiple-occupancy building, each occupant shall have access to his disconnecting means. A multiple-occupancy building having individual occupancy above the second floor shall have service equipment grouped in a common accessible place, the disconnecting means consisting of not more than six switches or six circuit breakers. Multiple-occupancy buildings that do not have individual occupancy above the second floor may have service conductors run to each occupancy in accordance with Sec. 230-2, Exception No. 3, and each such service may have not more than six switches or circuit breakers.

The section just cited is usually difficult to understand for people encountering it the first time, for there is no definition of just what is meant by "individual occupancy above the second floor." A building has individual occupancy above the second floor if on the third or a higher floor there is one (or more than one) occupant who does *not* occupy space *below* the third floor.

Do note the distinction in the quoted paragraph between the two types of structures: "A multiple-occupancy building having individual occupancy above the second floor *shall* have . . . ," but "multiple-occupancy buildings that do not have individual occupancy above the second floor *may* have. . . ." In the first case, there is no choice; in the second case, there is a choice.

Service Equipment. The Code in Art. 100 defines the service equipment as "the necessary equipment, usually consisting of a circuit breaker or switch and fuses, and their accessories, located near the point of entrance of supply conductors to a building and intended to constitute the main control and means of cutoff for the supply to that building."

Note that the wires from the power supplier's line to the building, up to the point where they end, are service-drop wires. From the point where they end, up to the meter and service switch, they are service-entrance wires. From the switch to branch-circuit fuses or breakers, they are feeders.

Classes of Buildings. While the Code does not so spell it out, multiple-occupancy buildings can be grouped into three classes so far as the service equipment is concerned:

1. One- and two-story buildings.
2. Three-story or higher buildings, *not* having individual occupancy above the second floor.
3. Three-story or higher buildings that *do have* individual occupancy above the second floor.

In the diagrams that follow, fuses are shown; it must be understood that circuit breakers may be used instead. For the sake of simplicity, only one hot wire is shown, instead of the usual two hot wires plus neutral.

One- and Two-story Buildings. Per the Code, the service equipment *may* be in a common space at all times accessible to all the occupants, but it is not required.

Perhaps the simplest installation is that shown in Fig. 26-1, in which a separate outdoor meter is used for each apartment, regardless of the number of apartments. (Indoor meters are equally acceptable.) From the meter, the wiring to the inside of each individual apartment is exactly the same as if each apartment were a separate house. There is no limit to the number of separate apartments that may be handled in this way.

If preferred, the service wires may be brought into a common space which must at all times be accessible to all occupants, and wires may be run from there to each apartment, as shown in Fig. 26-2.   (If there are not over six individual switches, the main switch may be omitted.) Note that in this scheme all the equipment, including the branch-circuit fuses, is in the common space.   Each set of equipment for one apartment consists of disconnecting means, main fuses, and branch-circuit fuses; all this may be telescoped for each apartment into a single circuit breaker cabinet or fused equipment, the same as you would use in a separate residence.

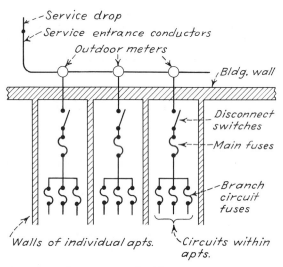

Fig. 26-1.   If an apartment building is one- or two-story, a separate outdoor meter for each apartment may be installed.

That scheme may be changed to the one shown in Fig. 26-3, where the branch-circuit fuses for each apartment are found in that apartment. In that case, the wires from the common space to each apartment become feeders, and must be individually protected where they begin, that is, in the common space.   As a matter of fact, in larger buildings when a common space is used, this is the only logical scheme to avoid waste of materials.

Three-story and Higher Buildings.   The procedure for larger buildings is the same as for one- and two-story buildings, except that the scheme of Fig. 26-1 may not be used *if there is individual occupancy above the second floor.* In that case you must use the scheme of Fig. 26-2 or 26-3 and put all the service equipment into a common space which is accessible to all occupants.

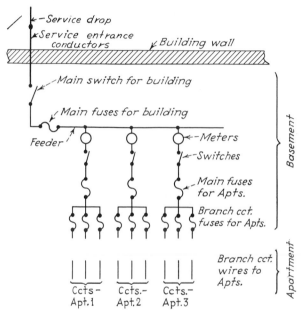

Fig. 26-2. Here all the service equipment is located in a common, accessible space. This scheme *may* be used in any apartment building, and *must* be used in most of them.

A very large apartment may need a service of, say, 750 amp. Since each apartment must have its own disconnecting means (which means many more than six individual switches), you could reach the conclusion that an 800-amp main switch would be required. That need can be avoided by breaking up the total load into not more than six subsections, for example, two fused switches of 200 amp and four of 100 amp. With six such switches installed, no main disconnecting means is required ahead of them all. Each of the six switches then feeds a group of smaller switches, just as a 100-amp switch in a smaller apartment would serve as a disconnecting means for a considerable number of switches each controlling an individual apartment.   See Fig. 26-4.

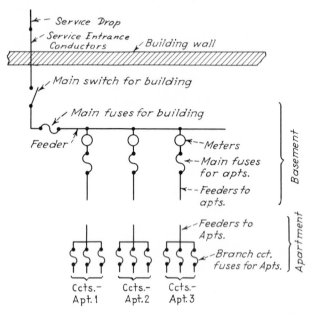

Fig. 26-3. A variation of the scheme in Fig. 26-2. The branch-circuit fuses have been moved to individual apartments.

Determining Feeder to Individual Apartment.   The feeder from the disconnecting means to the overcurrent devices in the individual apartment corresponds to the service-entrance wires in the single-family house (but in the apartment the wires are not service-entrance wires but are feeders).   Wire size is determined in general in the same way that was described in Chap. 13.   Just as in the case of the single-family house, the first 3,000 watts are counted at their full value; in other words, the demand factor is 100%.   For that portion of the load above 3,000 watts the demand factor is 35%.   The range load is added separately, after applying the demand factor to the lighting load.   The 20-amp laundry circuit required for individual dwellings is not required in individual apartments, if separate laundry facilities accessible to all the tenants of the building are provided.

For an apartment of 800 sq ft, figure 3 watts per sq ft, or 2,400 watts, for lighting; add 3,000 watts for the two special appliance circuits, for a total of 5,400 watts.   Count the first 3,000 watts at 100% demand factor.   Count the remaining 2,400 watts at 35% demand factor, or 840 watts.   Add the two, making a total of 3,840 watts.   At 115 volts this is equivalent to 3,840/115, or 33.4 amp.

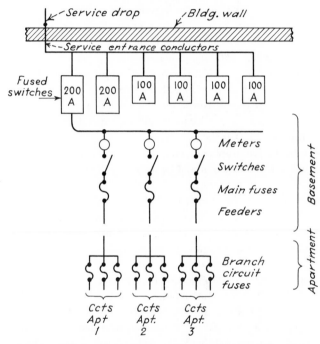

Fig. 26-4.   If there are not over six separate disconnecting means, the main switch may be omitted.

According to Table 310-12 of the Code, No. 8 wire is the smallest that may be used.   Incidentally, the Code in Sec. 215-2 specifies No. 10 as a minimum when a feeder serves two or more 2-wire circuits.

In addition to the two 20-amp special appliance circuits, two general-purpose circuits for all other outlets for lighting, receptacles, and so on will meet the Code requirements for 800 sq ft.   Three instead of two would be desirable, especially if a room air conditioner is to be installed.

All the above assumes you will install one 2-wire 115-volt feeder to the apartment.   If the fuse protecting that feeder blows, the entire apartment will be in darkness, and the tenant will not be very happy. It would be much better to install a 3-wire 115/230-volt feeder.   The 3,840 watts at 230 volts become a load of 16.7 amp, so per Code Table 310-12, No. 12 wire would seem large enough, but again the Code in Sec. 215-2 requires a minimum of No. 10.

If the apartment were larger having, say, 1,200 sq ft, the calculation would be 1,200 × 3, or 3,600 watts for lighting, plus 3,000 watts for the special appliance circuits, or 6,600 watts altogether.   Counting the first 3,000 watts at their full value, and adding 35% of the remainder of 3,600 watts, or a net of 1,260 watts, makes a total of 4,260 watts.   At 115 volts this is equivalent to 4,260/115, or 37 amp, for which No. 8 is the smallest wire permissible.   More likely you would consider the load a 115/230-volt load and run a 3-wire feeder capable of carrying 4,260/230 or 18.2 amp.   No. 12 wire would appear suitable but again Sec. 215-2 requires No. 10 as the minimum.

In addition to the two appliance circuits, three general-purpose circuits are required, and four would not be too many.

In this connection do not overlook Code Sec. 215-3 which requires that the wire in the feeder be of such a size that the voltage drop will not exceed 3%.   See the tables in Chap. 7 which are based on 2% drop; to meet Code requirements you can increase all distances shown by 50%. However, you will be wise to limit the drop to 2% for over-all practicability.

Assume now that a range is to be added to the smaller apartment just calculated.   Follow the recommendation of Code Sec. 220-5, and allow 8,000 watts, assuming the range is rated at not over 12,000 watts.   (If the range is larger, see page 203.)   This 8,000-watt allowance is to be added to the net watts for lighting and small appliances, *after* application of the demand factor.

In the case of the smaller apartment, adding 8,000 watts to the 3,830 watts brings a total of 11,840 watts.   Since a range operates at 115/230 volts, naturally a 3-wire feeder must be installed.   At 230 volts, 11,840 watts is 11,840/230, or 51.5 amp for which No. 6 wire is suitable.

However, you have already learned that the neutral wire serving a range cannot be made to carry as many amperes as the hot wires carry. Therefore the Code in Sec. 220-4(d) permits, for the *neutral* of the feeder, a wire smaller than for the hot wires.   Consider only 70% of the allowance for the range, so far as the neutral is concerned.   In the example above, the ampacity of the neutral only would be figured as follows: For the range count 70% of 8,000 watts, or 5,600 watts. For lighting and small appliances we have already reached an answer of 3,840 watts. The two make a total of 9,440 watts which at 230 volts is equivalent to 41 amp, and thus No. 6 wire is required for the neutral, the same size as for the hot wires. Had the total been under 40 amp, No. 8 would have been acceptable.

Determining Service Entrance. The minimum size of the service-entrance wires for the building as a whole is determined by the probable maximum load of *the entire building* at any given moment. The method of arriving at this probable maximum is very similar to that used for single-family dwellings: The lighting, the small appliances, and heavy loads such as ranges, water heaters, and so on are considered separately.

The greater the number of individual apartments in a building, the less the likelihood that all tenants will at the same moment be consuming power at the maximum rate available to each apartment. Therefore the Code in Sec. 220-4(a) permits a demand factor to be applied to the *lighting and small appliance* loads only. This demand factor is applied against the gross computed watts figured on the basis of 3 watts per sq ft of the total area of all the apartments, plus the minimum of 3,000 watts for small appliances for each apartment, applied as follows:

First 3,000 watts 100%
Next 117,000 watts 35%
All above 120,000 watts 25%

For ranges there is no likelihood whatever that all will ever be used at their maximum capacity at the same moment; the Code in Table 220-5 (see Appendix) establishes demand factors that take this into consideration. Note that the heading of the table specifies that Col. A of the table is always to be used except in the rare cases outlined in Note 4 of the table. Also note that if sectional ranges are installed (separate ovens and counter units, add the ratings of these together, and consider the total as the rating of the range.

The use of electric clothes dryers has become very common. Table 220-6 (see Appendix) shows demand factors that apply, depending on the number of dryers in the building.

Planning Three-apartment Installation. Assume a building containing three apartments of 800 sq ft each or 2,400 sq ft in total, plus the usual basement.   Each apartment will have four circuits, as determined earlier in this chapter.   Each circuit will probably run direct to the overcurrent device for that circuit in the basement.   The service entrance will probably be in accordance with Fig. 26-1 or 26-2.   There is little to calculate except the service entrance.

Service Entrance for Three-apartment Building.   The maximum probable load is very simply calculated in accordance with preceding paragraphs, as shown in the following table:

	Gross computed watts	Demand factor, per cent	Net computed watts
Lighting, 2,400 sq ft at 3 watts . . .	7,200		
Small appliances, 3 apartments at 3,000 watts.	9,000		
Total gross computed watts	16,200		
First 3,000 watts.	. . . . . .	100	3,000
Remaining 13,200 watts	. . . . . .	35	4,620
Total net computed watts	. . . . . .		7,620

The total of 7,620 watts covers only the three apartments proper.   It makes no provision for "house loads," in other words loads connected to the owner's meter, rather than the tenants' meters.   Allow proper wattages for water heaters, furnace motors, basement and hall lights, as you would in a single residence.   Add 1,500 watts for a 20-amp special laundry circuit.   This will come to a significant total; assume 6,000 watts, which added to the 7,620 watts already determined, makes a total of 13,620 watts, which at 230 volts is equivalent to 59.2 amp. Theoretically then according to Table 310-12, No. 3 wire if it is Type T, or No. 4 if it is another type, will be acceptable.   However, the Code in Sec. 230-41 requires wire with 100-amp ampacity for even a single-family residence.   Certainly then nothing smaller should be considered for a building with three apartments.

If an electric range consuming not over 12,000 watts is added to each of the three apartments, Col. A of Table 220-5 (Appendix) shows that 14,000 watts in total must be added for that purpose in calculating the service.   Adding 14,000 watts to the 13,620 watts already determined for lighting and small appliances produces a total of 27,620 watts, which at 230 volts is equivalent to about 120 amp. Install a service of 140-amp capacity (two 70-amp breakers in parallel), or preferably with 200-amp capacity.

Feeders for Apartments, Optional Method. The 1968 Code in Sec. 220-9 introduced an optional method of figuring feeders (both the service entrance, and the feeder from that point to the individual apartment) provided the apartment (a) is electrically heated, (b) has electric range, and (c) has only one feeder from the service equipment to the individual apartment. It may or may not have air conditioning. If all these conditions are met, proceed as follows for the feeder from service to the apartment:

1. For each square foot of area, allow 3 watts for lighting and baseboard receptacles . . . . _____watts
2. Add for two special appliance circuits . . . . 3,000 watts
3. Total of 1 and 2 . . . . _____watts
4. First 3,000 watts of 3, at 100% . . . . 3,000 watts
5. Remaining _____watts at 35% . . . . _____watts
6. Total of 4 and 5 . . . . _____watts
7. Electric range per Table 220-5 . . . . _____watts
8. Other heavy loads (dishwasher, garbage disposer, etc.) at full wattage rating . . . . _____watts
9. Air-conditioning or electric heating equipment (larger of the two loads) at full name-plate rating . . . . _____watts
10. Total of 6, 7, 8, and 9 . . . . _____watts
11. Divide total of 10 by 230, giving you the ampacity required for the hot wires of the feeder. The neutral might be smaller; see discussion in Chap. 28.

For the service-entrance wires for the entire building, proceed as follows:

A. Total 6 in previous tabulation, multiplied by the number of apartments . . . . _____watts
B. Total of all ranges at their full name-plate rating . . . . . _____watts
C. Total of all space-heating equipment, or air-conditioning equipment, at full name-plate rating. If both are installed, use the larger of the two . . . . _____watts
D. Total of all other heavy appliances (dishwashers, garbage disposers, etc.) in the building, at full name-plate rating . . _____watts
E. All "house loads"—equipment not connected to tenants' meters . . . . _____watts
F. Total of A to E above . . . . _____watts

G. Refer to Code Table 220-9 (see Appendix) and apply the demand factor shown there, for the number of apartments in the building.   For example, if there are five apartments, use 45% of the total *F*.   If there are 12 apartments, use 41%.   Enter here, total *F*,_____watts, times_____% demand factor . . . ._____watts

H. Divide G by 230, giving you the ampacity required for the hot wires of the service.   The neutral may be smaller; see discussion in Chap. 28.

See Code Example 4 (a) in the Appendix for a more detailed explanation.

Planning a 40-apartment Installation.   Assume a larger building with 40 apartments is to be wired.   Assume each apartment has 800 sq ft of area, and that half of the apartments are to have electric ranges.   Since all apartments are of the same size, all will be wired in the same fashion, and the calculations for the individual apartments will be as discussed in previous paragraphs.

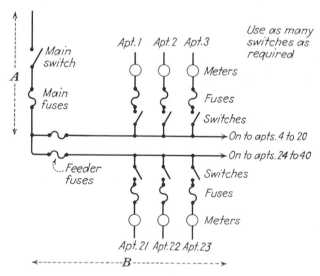

Fig. 26-5.   Feeder diagram for larger apartment building.

Since 40 meters and 40 disconnecting means are to be installed, it is likely that they will be arranged in several banks: two banks of 20 or four banks of 10.   Since there will be more than six disconnecting means a main disconnecting means will be required ahead of all.   The hookup will probably be similar to that shown in Fig. 26-5, which shows two banks of 20.

The final installation will resemble that shown in Fig. 26-6, which shows only 12 meters; similar equipment is available for any number of

Fig. 26-6. A typical group metering panel for use in larger apartments. (*Square D Co.*)

meters needed.   The main circuit breaker is at left.   At the top are individual breakers, each protecting the feeder to one apartment.   In each apartment, install a small fuse cabinet or circuit breaker cabinet, as you would in an individual residence.

Similar equipment using fuses instead of circuit breakers is available for those who prefer fused equipment.

The calculation of the various feeders and subfeeders is covered by the Code in Example No. 4, which will be found in the Appendix of this book and need not be repeated here.

Low-voltage Wiring.   The usual doorbell and buzzer system will be installed in accordance with the principles already outlined in other chapters.   However, it is necessary to take into consideration the prob-

lem created by the longer lengths of wire involved in buildings of considerable size.

In the wiring of fairly large apartments, it is not uncommon for the
doorbell and buzzer system to be installed in accordance with the diagram of Fig. 26-7. Theoretically this diagram is correct, but if ordi-

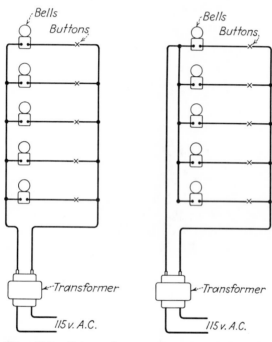

Fig. 26-7. Using   this
wiring diagram, the dis-
tant bells ring too faintly,
the   nearby   ones   too
loudly.

Fig. 26-8. Using   this
diagram, all bells ring
with equal volume.   It
requires only a little
more material than the
other diagram.

nary No. 18 wire is used, as in residential work, it will be found that the
more distant bells ring rather faintly because of the substantial voltage
drop in the long run of the small wire.   A drop of 2 volts in a 115-volt
circuit is not so serious, but when the starting voltage is only 10 volts or
thereabouts, a 2-volt drop is of the order of 20%.   If the transformer
voltage is stepped up sufficiently so that the more distant bells ring

properly, the nearby bells ring too loudly.   A larger size of wire helps, but the distant bells still ring less loudly than the ones nearer the transformer.

The solution is to use the circuit of Fig. 26-8.   A little more wire is involved, but if this circuit is analyzed, it will be found that the number of feet of wire involved for any one bell is exactly the same as for any other bell.   Accordingly all will ring with equal volume; a transformer voltage that is correct for one is correct for all the others.

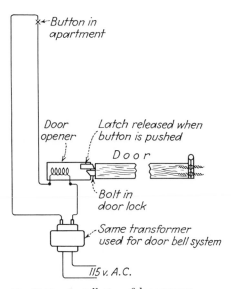

Fig. 26-9.   A typical door opener. (*Edwards & Co.*)

Fig. 26-10.   Installation of door opener.

Door Openers.   Many apartments have the door leading into the inside hall equipped with a door opener so that intruders cannot enter. Pushing a button in any apartment releases a latch which permits the door to be opened.   A typical door opener is shown in Fig. 26-9.   It consists simply of an electromagnet, similar to that found in a doorbell, which releases a latch when the button is pushed, permitting the door to be swung open.   The opener is mortised into the door frame opposite the lock in the door.   The diagram is most simple, as shown in Fig. 26-10.

Actual Wiring: Nonresidential Projects

Part 3 of this book covers the same subject matter as Part 2 but applies to nonresidential structures, such as factories, stores, schools, and similar projects.

Buildings of this type involve greater amounts of power, higher amperages, heavier wires, different kinds of wire, and other problems not met in residential work. Nevertheless the method of installation is in most respects greatly similar to that employed for residential work, and accordingly only new problems will be here explained.

Only the smaller projects are included. The very large projects, such as skyscrapers and steel mills, have been deliberately avoided as being beyond the scope of this book. This is also the case with all wiring at voltages in excess of 600, transformer vaults, and similar subjects which are encountered only in the larger projects.

27

Additional Information about Wire

Chapter 6 described the kinds of wires used in the wiring of residential and farm buildings. In this chapter additional kinds of wire used mostly in nonresidential and nonfarm wiring will be discussed. In addition, some of the fundamental facts concerning insulations, constructions, ampacities, and installation restrictions will be discussed in more detail than in Chap. 6.

What Determines Ampacity of Wire. Before entering into a discussion of the Code and its provisions concerning wire, it will be well to analyze the reasons why the ampacity of any given size of wire varies with the kind of insulation on the wire and the method of installation.

Copper is not harmed by heat; insulation is harmed by heat. If insulation is overheated, it is harmed in various ways, dependent on the degree of overheating and the kind of insulation. Some kinds melt, some harden, some burn. In any event insulation loses its usefulness if overheated, leading to breakdowns and fires.

The ampacities specified in various tables for any particular kind and size of wire is the amperage that it can carry *without increasing the temperature of its insulation beyond the danger point.* The insulation on Types T and TW stands the least heat; consequently these types have the lowest ampacity of all the different kinds. Asbestos will stand the most heat; consequently asbestos-insulated wire has the highest ampacity. The temperature of the asbestos-insulated wire when carrying its rated amperage will be higher than the temperature of the plastic-insulated wire carrying its rated amperage, but its insulation, being designed for it, will not be harmed by the higher temperature.

The rated ampacity of each kind and size of wire is based on the assumption that the wire is installed where the temperature[1] is 30°C or 86°F. The Code in Table 310-2(a) (see Appendix) shows the maximum temperature that the insulation of each kind of wire is permitted to reach. That temperature will be reached when the wire is carrying its full rated ampacity in a room where the temperature is 30°C or 86°F.

It will be worth repeating: When a wire is listed for 60°C or 140°F, it does not mean that the wire may be used in a room with temperature of 140°F. It means that the maximum operating temperature of the wire itself may not exceed 140°F. If the room temperature is above the assumed standard of 86°F and the wire is carrying its full rated ampacity, its operating temperature will exceed 140°F. Therefore if any kind of wire is installed in hot locations, its ampacity is reduced from that shown in Code tables. How to apply the proper correction factors will be explained later.

The ampacities of various kinds and sizes of copper wire are shown in Code Tables 310-12 and 310-13. For aluminum wires see Tables 310-14 and 310-15. All four are shown in the Appendix. The footnotes following the tables are most important.

Ampacity vs. Diameter. From an earlier chapter you will remember that of two wires six sizes different (for example, No. 6 compared with No. 12), the larger will have *twice the diameter,* twice the circumference and *surface* area, but *four times the cross-sectional area in circular mils,* of the smaller wire (the diameters and areas of all sizes can be found in Code Table 8 in the Appendix). It would seem logical then that the larger wire would have four times the ampacity of the smaller

---

[1] The temperature referred to is the "ambient" temperature: the normal temperature in an area while there is no current flowing in the wires. When current does flow, heat is created, and the temperature will increase above the ambient.

wire, but study of Table 310-12 shows that is not the case. Consider the facts in the following tabulation, ampacities being for Type THW per Table 310-12:

Wire size	No. 12	No. 6	No. 1/0	2,000MCM
Area, circular mils	6,530	26,240	105,600	2,000,000
Ampacity	20	65	150	665
Ampacity per thousand circular mils, approx.	3	2.5	1.4	0.30

The No. 12 has an ampacity of 20 amp, and No. 6 with four times the cross-sectional area might then be expected to have four times the ampacity or 80 amp, but the actual ampacity is only 65 amp; No. 1/0 with four times the cross-sectional area of No. 6 might be expected to have four times its ampacity or 260 amp, but it is only 150 amp. Study the last line above, which shows the ampacity of the various sizes in terms of ampacity *per thousand circular mils* of cross-sectional area; the larger the wire, the smaller that amperage.

Why? Assume several sizes of wire, *all* carrying the same number of amperes *per thousand circular mils.* The heat developed in the wires will then be exactly the same in all sizes, again per thousand circular mils. So the ampacity might be expected to be in proportion to the circular-mil area, but it isn't. The answer lies in the fact that one wire that has *four* times the circular-mil area of another wire is only *twice* the diameter, and has only *twice* the circumference and *surface* area. But the heat developed in a wire can be dissipated only from its surface. Compare No. 1/0 with No. 12. The No. 1/0 has 16 times the cross-sectional area, but only four times the diameter and surface area, from which heat can be dissipated. The ampacity of a wire, while not directly in proportion to its surface area, is more nearly related to that surface area than to its cross-sectional area.

Types -W and -H. Ordinary Type T or Type R wire may be used only in dry locations, with a temperature limit of 60°C or 140°F. But wires must often be installed in wet locations, and often must be capable of operating in high-temperature locations.

If the Type designation includes a "W," the wire may be used in dry or wet locations, but the temperature limits are still 60°C or 140°F (unless the Type designation also includes an "H").

If the Type designation includes an "H," its temperature limit is 75°C or 167°F. If it includes "HH" its temperature limit is 90°C or 194°F. But such wires may be used in dry locations only, unless the Type designation includes a "W" in addition to "H" or "HH."

Various "Type R" or "Type T" wires recognized by the Code are:

Type R: Once the most common kind of wire used in wiring, it is no longer being made.

Type RH: No longer made.

Type RHH: Dry locations only; 90°C or 194°F.

Type RW: No longer made.

Type RH-RW: No longer made.

Type RHW: Dry or wet locations; 75°C or 167°F.

Type T: Dry locations; 60°C or 140°F.

Type TW: Dry or wet locations; 60°C or 140°F.

Type THW: Dry or wet locations; 75°C or 167°F.

Type THWN: Dry or wet locations; 75°C or 167°F.

Type THHN: Dry locations; 90°C or 194°F.

Practical Advantages of -H or -HH Wires. Let us assume the wires are being used in a room where the normal temperature is not above the basic Code assumption of 30°C or 86°F. The more amperes being carried, the greater the temperature rise in the wire. But for -H or -HH wires, the Code permits a higher rise, since the insulation is designed to withstand a higher temperature. Therefore the Code assigns a higher ampacity to such wires than it does to ordinary wires. As already explained, Code Tables 310-12 and 310-13 specify the ampacity of each size, depending on the type. For ordinary Type T, use the 60° column; for Types RH, RHW, THW, and THWN, use the 75° column; for Types RHH and THHN use the 90° column (except in Nos. 14, 12, and 10 use the 75° column). Examine the tables; you will find that the differences in ampacities for the various types become more important in the larger sizes, more so than in the smaller sizes.

The more expensive Type THW often becomes less expensive for a particular amperage, than the ordinary Type T. This is especially true when a smaller size of Type THW also permits a smaller size of conduit to be used. For example, to carry a 200-amp 3-wire 3-phase load in conduit, if Type T is used, three 250MCM cables will be required, which in turn means that 2½-in. conduit must be used. If Type THW is used, No. 3/0 is sufficient, and requires only 2-in. conduit.

One word of caution may be in order; while the use of Type THW permits a smaller size of wire, the smaller wire leads to a greater voltage drop.   If the runs are long enough for voltage drop to become a factor, the advantage of Type THW may disappear.

Type -L.   If the last letter in a Type designation is an "L," it shows that the wire or cable is covered with a final seamless layer of lead; the wire may then be used in a wet location.   Lead-covered wires are not used a great deal, for many insulations are now available that may be used in wet locations, without lead covering.

Type RR.   Although you see "Type RR" mentioned in magazine articles and advertising, and see wire so marked on its cartons, no such type is recognized by the Code or by the Underwriters.   It is a wire with the usual rubber insulation of varying grades, but instead of having a final *fabric* cover, it has a second layer of tough rubber or neoprene. That in a general way also describes Type USE service-entrance cable (described in Chap. 20).   But if a wire is tagged only "Type RR" it is not a listed wire.   If it is labeled "Type USE," some manufacturers may add their own private designation "Style RR."

Types THHN and THWN.   These are relatively new types of wire. They are essentially the standard Types THH or THW, plus a final extruded layer of nylon which is exceedingly tough mechanically, besides having excellent insulating properties.   The construction leads to an over-all diameter and over-all cross-sectional area (including the insulation) much smaller than that of ordinary wires, especially in the smaller sizes.   In larger sizes they have a higher ampacity than ordinary wires. These are expensive wires, but their use is often justified because for a given amperage capacity of a circuit, smaller conduit may be used than when using the more ordinary kinds of wire.

Type XHHW.   This type of wire was first recognized by the 1968 Code.   Its insulation is "cross-linked thermosetting polyethylene."   It has no outer braid and in appearance is substantially like Type T.   The insulation is mechanically tough, is moisture- and heat-resistant, and has exceptional insulating qualities.   These facts combined lead to an over-all outer diameter that is considerably smaller than the various types of Types T and R (although not as small as Types THHN, THWN, FEP, etc.).   It would be no surprise if in due course of time this *one* type would take the place of the many types of Type R and Type T now in use, thus simplifying life for manufacturers, distributors, and users alike.

Types FEP and FEPB.   These are new types of wire with insulation of fluorinated ethylene propylene.   If there is no outer braid it is Type FEP which has a temperature limit of 90°C or 194°F.   If it has an outer braid (glass in Nos. 14 to 8, asbestos in No. 6 and heavier), it becomes Type FEPB; it normally has a temperature limit the same as Type FEP, but, on special applications determined by the inspector, may be used up to 200°C or 392°F.   These types may be used only in dry locations.

Type MI Cable.   The letters "MI" stand for mineral-insulated.   This cable consists basically of solid bare wire, one or more conductors properly spaced, with a seamless copper sheath around the outside.   The wires are separated from each other and from the outer copper sheath by a tightly packed mineral insulation (magnesium oxide), which is both a good insulator and also unaffected by heat.   See Fig. 27-1 which shows both the material and the special connectors used with it.

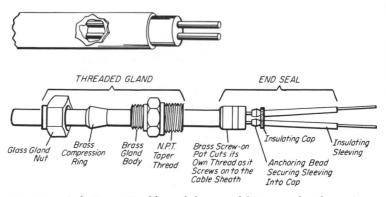

Fig. 27-1.   Code Type MI cable, and the special fittings used with it.   (*General Cable Co.*)

It is installed like other cables, requires no protection such as conduit, and being of relatively small diameter can be easily installed even in crowded quarters.   It must be supported every 6 ft unless fished through wall spaces.   The outer copper sheath of course is waterproof so the material may be installed in any location.   However, immediately after cutting the copper sheath, a seal must be installed to prevent the entrance of moisture.   In bending, the radius of any bend must be at least six times the diameter of the cable.   It is not an inexpensive material, yet, because no conduit is required, it can become an economical type of cable so far as total installed cost is concerned.

Other Types of Wire and Cable. The types discussed here are the more common varieties. There are many others that cannot be described here. A listing of all types together with their maximum operating temperatures and limitations of use can be found in Table 310-2(a) (see Appendix). Incidentally, the listing contains some types no longer made. Further details of construction can be found in Sec. 310-2(b) in your copy of the Code. Their ampacities can be found in Tables 310-12 and 310-13 of the Code (see Appendix).

Identification of Different Kinds of Wire. In most cases it is impossible, just by looking at the wire, to determine what type it is. For example, there is no difference in appearance between Types T, TH, THW, and several others. Wisely, the Underwriters require that all wire be plainly identified by printing the type designation on it at frequent intervals. Other markings not required but usually shown on the wire include the size, the voltage limitation, the manufacturer's name and trademark, the month and year of manufacture. The abbreviation "UL" (for "Underwriters' Laboratories") may be shown. If the conductor is aluminum *and* if the manufacturer elects to print the size on the wire, he must also show the word "Aluminum" or its abbreviation "AL" on the wire.

All this has to do with the markings on the wire itself. The tag on the coil or reel must show the type, size, voltage limitation, manufacturer's name, the month and year of manufacture, the designation for aluminum if the conductor is aluminum, the words "National Electric Code Standard" (or an abbreviation), and any other marking "if it does not confuse or mislead."

On cables Types USE, NM, NMC, and UF the wire size in the cable must be imprinted on the cable itself, as must the voltage limitation.

Using Code Table 310-12. At first this table may seem very complicated, but for installations within the scope of this book, it is quite simple. The "Notes" following the table are very important. Remember that the ampacities shown for different sizes apply only if all the following conditions are met:

1. The wires are used inside a raceway (conduit) or in the form of cable or buried in the ground. If the wires are suspended in free air, use the ampacities of Table 310-13.

2. Not more than three wires are installed in the same conduit or cable. The neutral wire is not counted except in the case of a 3-wire circuit consisting of two phase wires and the neutral of a 4-wire 3-phase system (Note 10, Tables 310-12 and 310-13). But if the circuit supplies mostly fluorescent or HID lighting, the neutral must be counted. The reason for this is a bit complicated and its explanation is beyond the scope of this book. If the group of wires includes a grounding wire, which normally carries no current, its status is not defined by the Code. Most inspectors would probably rule that it need not be counted. If 4 to 6 wires are used, reduce the ampacities to 80%; if 7 to 24 wires are used, reduce to 70% of the values shown.

3. The room temperature is not over 30°C or 86°F. If higher room temperatures are encountered, apply the correction factors shown in Note 15 following the tables. For example, if No. 8 Type RHW or THW with a normal ampacity of 45 amp is to be used in a room temperature of 40°C or 104°F, the corrected ampacity is 45×0.88 or 39.6 amp.

4. If a bare, uninsulated wire is used in conjunction with insulated wires (for example, as the neutral in a service), the ampacity of the bare wire is considered the same as that of the insulated wires with which it is used.

5. The wire is copper. If it is aluminum, use Tables 310-14 and 310-15, also found in the Appendix of this book.

Using Code Table 310-13. When wires are installed in raceways such as conduit, or in the form of cables, the heat that develops in the wires does not dissipate very fast. The ampacity under those conditions is established by Table 310-12. But if the individual wires are installed in free air, the heat is dissipated much faster, and any given size of wire can carry more amperes without overheating. For wires installed in free air, use Table 310-13 (see Appendix).

For weatherproof wire, use the final column of Table 310-13, headed "Bare and Covered Conductors."

"Simplified Wiring Table." The 1968 Code added Secs. 310-20 and 310-21 covering the Simplified Wiring Table. Instead of following the procedures outlined up to this point, the new procedure may be used if there are six or fewer wires in a conduit or cable, for either a feeder or branch circuit, but only if a demand factor of 80% or less exists for the load. To use this short-cut method, follow these steps:

1. Determine the load in amperes, and whether the load is continuous or not.   Continuous loads are those expected to continue for 3 hr or more.   Noncontinuous loads are those where 67% or less of the total is expected to be continuous.

2. Select wire size from Table 310-21 (see Appendix).   Use the proper column, depending on whether the load is continuous or noncontinuous.

3. Determine ambient or room temperature from Table 310-20(c) (see Appendix).

4. Select a wire type with insulation suitable for the temperature. This can be done using Code Table 310-2(a) in your copy of the Code, or from the headings of Table 310-12 (see Appendix).

It seems likely that many people will question whether the new method is really "simplified," and just what the various requirements for its use mean in practice.   Discussions in inspectors' and contractors' meetings, and in electrical magazines, will no doubt bring clarification over the years.

Number of Wires in Conduit.   In defining the number of wires of any given size permitted in a specific size of conduit, Codes earlier than the 1968 did not differentiate between wires of different types.   But for a *given size of wire,* the over-all outside diameter varies a great deal, depending on the type of insulation.   It would seem logical that for a given size of conduit, a larger number of wires of a small diameter should be permitted, than when using wires of a larger diameter.

Beginning with the 1968 Code, this factor was given due weight, and the Code in Table 1 (see Appendix) permits a greater number of the small-diameter wires, than it does for the more ordinary or large-diameter wires of the same gauge.   Col. A of the table shows the number permitted for the more ordinary wires listed at the beginning of the table, and Col. B shows the number for the smaller-diameter wires also listed at the beginning of the table.

The advantages will be most apparent in the smaller sizes of wire. Assume you want to install six No. 14 wires.   Using ordinary wires, per Col. A you will need ¾-in conduit; if you use the smaller-diameter wires, per Col. B you will need only ½-in.   Another example: You want to install six No. 6.   Per Col. A, you must use 1½-in. conduit; per Col. B, 1¼-in. is adequate.   The small-diameter wires that must be used when following Col. B cost considerably more than ordinary wires, but the saving in cost by using smaller conduit and fittings will offset at least some of the difference.

In the larger sizes, wires such as THWN and THNN have a much higher ampacity than the same size of ordinary wire such as Type T. Assume you need a 120-amp circuit. If you use Type T you will need No. 3/0. If you use Type THWN (or any other type listed in the 75°C column of Table 310-12) No. 1/0 is adequate. Another example: For four No. 2/0 Type T with ampacity of 145 amp you will per Col. A need 2½-in. conduit; for four No. 1/0 Type THWN with ampacity of 150 amp you will need only 2-in. conduit.

Replacing Wires in Existing Conduit. If you are replacing wires in an existing conduit, the Code permits a greater "fill" than in the original installation. The answers appear in Code Table 1A (Appendix), and as in the case of Table 1 (which covers original installations) the number of wires permitted depends on the outside diameter of the wire and the type of insulation.

Use Col. A of the table only when using the types of wire listed under A at the beginning of the table, and Col. B only when using the types listed under B at the beginning of the table.

As an example of the advantages gained when replacing wires in an existing conduit, assume you now have a run of 1½-in. conduit containing three No. 1 Type T wires with an ampacity of 110 amp, which is then the capacity of the circuit or feeder. Using Col. A of Table 1A, you will find that you can replace these No. 1 Type T with No. 2/0 Type T with ampacity of 145 amp, or (using Col. B) with three No. 3/0 Type TWHN with ampacity of 200 amp, or with three No. 3/0 Type THHN with ampacity of 210 amp (all ampacities being determined by Code Table 310-12).

In other words, without going to the expense of tearing out the old conduit and installing a larger size, you merely pull out the old wires, and install the new, thereby increasing the capacity of the circuit from 110 amp to as much as 210 amp.

Wires in Parallel. Instead of using one large wire, the Code in Sec. 310-10 permits two or more smaller wires in parallel or multiple. There are half a dozen conditions, *all* of which must be met. Each of the smaller wires that are to be paralleled to form the equivalent of one larger wire must: (a) be No. 1/0 or larger, (b) be of the same material (all copper or all aluminum), (c) have precisely the same length, (d) have the same cross-sectional area in circular mils, (e) have the same type of insulation, and (f) be terminated in the same manner (same type of lug or terminal on all).

For example, assume you have to provide a 3-wire circuit carrying 490 amp. Per Table 310-12 (75°C column) you will need 800MCM cable. In place of each 800MCM cable you can use two smaller conductors each having half of its ampacity, or 245 amp. Per Table 310-12 two 250MCM cables each with ampacity of 255 amp would appear to be suitable. But if you plan to run all six cables through a single conduit, 250MCM cable is not acceptable, because per Note 8 to Table 310-12, when you have four to six conductors, you must derate the ampacity to 80% of what the table shows. The ampacity of the 250MCM cable then becomes 80% of 255, or only 204 amp. Use 350MCM cable with ampacity of 80% of 310, or 248 amp.

But instead of six cables through one conduit, you may use two separate conduits, with three cables in each; derating then is not required and 250MCM cables are suitable. But when running parallel wires for a circuit or a feeder through two separate conduits, special care must be used to meet Code Sec. 300-20. One of each pair (or larger number) of wires must run through one conduit, the other through the second conduit. See Fig. 27-2, which shows the six wires, *a* and *A* paralleled with

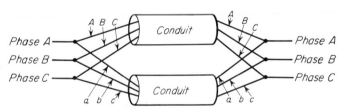

Fig. 27-2. When paralleled wires are run through conduit, each conduit must contain one wire from *each* phase.

each other, also *b* and *B*, and *c* and *C*. Each conduit must contain *a* (or *A*), *b* (or *B*), and *c* (or *C*), in other words *one* wire from *each* phase. If you installed *a*, *b*, and *B* in one conduit, *c*, *A*, and *C* in the other, each conduit would contain wires from only two of the three phases. In that case, very considerable induced and eddy currents would be set up in the metal conduit, resulting in very heavy heating and power losses.

If you are installing a feeder or circuit with a neutral wire, often the neutral can be smaller than the hot wires. If that is the case in your installation, the two wires paralleled for the neutral may be smaller than those paralleled for the hot wires, just so that their combined ampacity is equal to that which would be required if you were not paralleling.

Conductors in parallel are quite common if the use of a single conductor would require very large sizes of cable. The smaller conductors have much higher ampacity, *per thousand circular mils*, than the larger sizes. Thus much less copper by weight is used when two smaller conductors are paralleled to replace one larger one. The total installation cost is thus reduced. However, a word of caution is in order. When you use two smaller cables in place of one larger one, the total circular-mil area of the two smaller cables together will be less than that of one larger cable. That of course will lead to a higher voltage drop than you would have using one cable of the larger size. If the feeder or circuit is a long one, this may become an important factor that must not be overlooked.

28

Planning Nonresidential Installations

The wiring of nonresidential projects follows the same basic principles covered in Parts 1 and 2 of this book, but many new problems arise. Much larger amperages must be handled than is customary in ordinary residential work. Sometimes the voltages are higher. Usually 3-phase power is involved in addition to the usual single-phase power. Devices and materials are used which are not common in residential work. Some points, which in residential work are left to the discretion of the contractor, are covered specifically by the Code for nonresidential work.

This book cannot possibly cover all details of wiring projects of the largest kind, but must be limited to projects of moderate size. As a matter of fact, much of the information in chapters from here to the end of the book should be considered more or less a sort of preview of the kinds of problems you, as a student, will encounter later in your career.

Heavy-duty Lampholders. A "lampholder," which term is seldom met outside of the Code, is simply a socket or similar device by which current is carried to a lamp. Heavy-duty lampholders as defined in Code Sec. 210-8 are those rated at 750 watts or more. They include, but are not limited to, mogul sockets as used to fit mogul bases on incandescent lamps larger than 300 watts. The Code requires mogul bases on all lamps over 300 watts but not over 1,500 watts. Lamps larger than 1,500 watts have special bases, not screw-shell type; they are nevertheless heavy-duty type.

Three-phase Lighting. In larger buildings the service is usually 3-phase, but the two wires that run to any one lighting fixture are necessarily single-phase. How then is the lighting to be handled? There are several ways.

Sometimes a separate single-phase 115/230-volt service is installed for the lighting, so there is no further problem. At other times a transformer is installed on the 3-phase system, furnishing single-phase 115/230-volt power for the lighting. This is the method usually used when a separate single-phase service is not installed, and the 3-phase voltage is too high for lighting.

If, however, the 3-phase service is 4-wire 120/208-volt, the usual procedure is to run all four wires (neutral and three hot wires) to the fuse or breaker cabinets for the lighting circuits. See Fig. 28-1. The

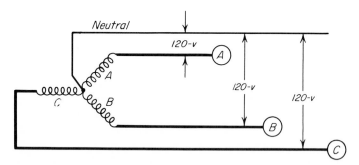

Fig. 28-1. Three separate 120-volt single-phase feeders for lighting are derived from a 3-phase 120/208-volt 3-phase service.

voltage between the neutral and any one of the hot wires *A*, *B*, or *C* is 120 volts. In such installations balance the lighting or other 120-volt circuits so that the load is distributed as nearly as possible equally among the three hot wires. Once the wires leave the fuse or breaker cabinet, each pair is a single-phase circuit, and the wiring for any single circuit is exactly as it would be if the service were 115/230-volt, single-phase.

The same scheme is used in buildings if the service is 4-wire 277/480-volt, the lighting then being at 277 volts, as will be discussed later.

Circuits in Nonresidential Work. If the load on any circuit is likely to continue for 3 hr or more, it is considered to be continuous. In that case only 80% of the nominal capacity of the circuit may be used: A 15-amp circuit with 15-amp overcurrent protection must be considered a 12-amp circuit; a 50-amp circuit must be considered a 40-amp circuit, and so on. The reason for this derating is twofold. The wires in the conduit generate heat, and if the load is continuous, the heat dissipates more slowly than if the load is not continuous. A very considerable amount of heat is also generated by the fuses or breakers in their cabinet; if the load is continuous, it is not unusual for fuses to blow or breakers to trip, even if carrying a bit *less* than their rated capacity. The greater the number of fuses or breakers in a single cabinet, the greater the likelihood of this happening.

But note that if the ampacity of the wires has already been derated in accordance with Note 8 to Tables 310-12 to 310-15, they do not have to be derated to 80% of an already derated value.

Note that per Code Sec. 210-21(a), fluorescent lighting may be installed only on 15- and 20-amp circuits. Lampholders installed on circuits of 30-amp or larger rating must be the heavy-duty type; on 15- and 20-amp circuits they may be any type.

Branch Circuits, Lighting. The basic requirement is that each circuit must have the capacity to handle the load intended to be carried on that circuit. The Code requires that you provide enough circuits so that there will be available the number of watts per square foot prescribed in Table 220-2(a) (see Appendix) for the kind of occupancy involved. In doing so you will be fulfilling the Code requirements, but you will probably not have the amount of light that will be needed, considering the footcandles of lighting that are recommended.

The logical procedure is for you to make your plans showing what lighting outlets are to be installed to provide an adequate level of light, and then provide the number of circuits required, not overlooking the derating to 80% if the loads will be continuous. Since each lighting outlet will consume more watts than in residential work, the number of outlets per circuit will be limited by the rating of the circuit and the load per outlet. The Code in Sec. 220-2(b) requires a *minimum* of 1½ amp per outlet, and 5 amp if heavy-duty lampholders are to be installed.

Branch Circuits, Nonlighting. The basic requirement again is that each outlet be figured on the basis of the actual load to be connected to it. If motors are involved, see Chap. 31. For specific appliances and other loads, count the outlet at the ampere rating of the appliance or other load. Miscellaneous outlets may be counted at 1½ amp per outlet,[1] unless the load is known to be greater.

Branch Circuits, General Information. The essential facts concerning circuits serving *two or more* outlets are recapped in Code Table 210–25, and will be discussed here. There are no restrictions on circuits serving a single outlet. Since the largest circuit for general use is the 50-amp, it follows that any single load of more than 50 amp must be served by a circuit serving only that load.

Fifteen-ampere Branch Circuit. As in residential work, this circuit is wired with No. 14 wire and with 15-amp overcurrent protection; if used for continuous load it must be derated to 12 amp. All receptacles installed must be the grounding type, rated at 15 amp. Any type of lampholder may be used on the circuit.

No portable appliance on the circuit may be rated at more than 12 amp; the total rating of fixed appliances on the circuit may not exceed 7½ amp if the circuit also serves lighting outlets or portable appliances.

Twenty-ampere Circuit. As in residential work, this circuit is wired with No. 12 wire and with 20-amp overcurrent protection. If used for a continuous load it must be derated to 16 amp. Receptacles may be either 15- or 20-amp, and must be the grounding type. Any kind of lampholders may be used on the circuit.

No portable appliance on the circuit may be rated at more than 16 amp. The total rating of all fixed appliances on the circuit may not be more than 10 amp if the circuit also serves lighting units or portable appliances.

A 20-amp circuit has a capacity of 115×20, or 2,300 watts. If the

---

[1] When multioutlet assemblies of the type shown in Fig. 29-19 (in the next chapter) are used, the Code specifies in Sec. 220-2(c) Exc. 3 that "each 5 ft or fraction thereof of each separate and continuous length shall be considered as one outlet of not less than 1½ amp capacity, except in locations where a number of appliances are likely to be used simultaneously, when each 1 ft or fraction thereof shall be considered an outlet of not less than 1½ amp. The requirements of this section are not applicable to dwellings or the guest rooms of hotels."

common four-lamp 40-watt (160-watt total) fluorescent unit is used for lighting, the error may be made of dividing 2,300 by 160 and arriving at an answer of 14 such units per circuit.   However, as pointed out in another chapter, the 40-watt rating of a fluorescent lamp is that of the lamp itself; its ballast requires additional power; the lamp does not have 100% power factor.   The over-all wattage of a four-lamp 40-watt unit is more nearly 200 watts, and considering power factor, each unit consumes about 2 amp.   The maximum capacity then is 10 units per circuit or 8 units if the lighting is continuous.   Moreover, since in nonresidential work the runs are often quite long, loading the circuit to its full capacity may easily lead to excessive voltage drop.   It is wise to limit carefully the number of units per circuit.

If the circuit serves heavy-duty sockets, the Code requires a minimum of 5 amp per outlet, which means a maximum of four per 20-amp circuit, reduced to three if the lighting is continuous.

Thirty-, Forty-, and Fifty-ampere Branch Circuits.   These are wired with Nos. 10, 8, and 6 wire respectively, with 30-, 40-, and 50-amp overcurrent protection.   Fluorescent lighting fixtures may *not* be installed on them; incandescent may be used but the sockets must be the heavy-duty type.   On the 30-amp circuit only 30-amp receptacles may be used; on the 40-amp they may be either 40- or 50-amp type; on the 50-amp they must be the 50-amp type.

Appliances connected to the 30-amp circuit may be of any type but if portable may not exceed 24 amp.   On the 40- and 50-amp circuits only fixed cooking appliances or infrared heating units may be installed.

Which Circuit to Use.   Many factors will influence the choice among the 15-, 20-, 30-, 40-, and 50-amp circuits.   If each fixture is on a separate circuit controlled by a wall switch, the smallest wire permissible may be used, considering the amperage, and not overlooking voltage drop and derating if the load is continuous.

If a number of outlets are to be on one circuit and controlled simultaneously by one switch, the heavier circuits will be automatically required.   If a number of outlets are to be on one circuit, but individually controlled by switches, the heavier circuits will probably be found more economical.

Taps in Branch Circuits.   On branch circuits serving two or more outlets, taps sometimes may be smaller than the circuit wires.   Code Sec. 210-19(c) covers the subject.   On 15- 20-, and 30-amp circuits the tap wire may be as small as No. 14, and on 40- and 50-amp circuits may be as small as No. 12.   All this is contingent on the tap wire's having sufficient ampacity for the load to be served, that it serves either a single receptacle or a single lighting fixture, and is not over 18 in. long.

If the tap instead of being concealed in conduit or other raceway, is exposed as in the case of lighting fixtures suspended by chain with the wires interlaced in the chain, it may be fixture wire as small as No. 18 on 15- and 20-amp circuits.   On 30-amp circuits it must have ampacity of at least 10 amp, and on 40- and 50-amp circuits must have ampacity of at least 20 amp.   There is no limit to the length.

277-volt Lighting.   In general, the lighting in offices, stores, factories, and so on has been at 115 volts.   The trend has been and is still continuing toward better and better lighting, consuming more and more watts per square foot of area.   That in turn requires more and more circuits or larger wire to take care of the increased amperage.   Instead of using 115 volts, why not use a higher voltage, thus reducing the amperage?   Then any given size of wire would carry a higher wattage.   For example, No. 14 wire with ampacity of 15 amp will at 115 volts carry 15×115 or 1,725 watts; at 230 volts, 15×230 or 3,450 watts; at 277 volts, 15×277 or 4,155 watts.

The Code in Sec. 210-6 permits lighting at voltages above 150 volts but not over 300 volts *to ground,* under certain conditions, in certain locations.   Today such installations are common, operating at 277 volts.   If 277 volts seems a peculiar voltage, note that if an establishment is served by a 480-volt 3-phase system, and if the transformers are Y-connected with a neutral wire, the voltage between the neutral and any one phase conductor is 277 volts.   The 3-phase 480-volt wires are still available for power loads.   (If the over-all voltage of the basic system is 460 volts, the voltage from the neutral to the phase wires is 265 volts.)

Under the Code, lighting at voltages above 150 but below 300 volts to ground, is permitted by Sec. 210-6 only under specified conditions. If the lighting is *not* fluorescent, it may be used in industrial establishments or stores if all the lampholders are heavy-duty type or other types approved for the purpose; they must not have switches as part of the fixture; they must be installed at least 8 ft above floor level; last but not least, only competent individuals may service the fixtures.

If the lighting is fluorescent, then "in industrial establishments, office buildings, schools, stores and public and commercial areas of other buildings, such as hotels or transportation terminals" 277-volt lighting may be installed, provided the fixtures are permanently installed, and do not have switches as part of the fixture.   Additionally, "electric-discharge lamps" (mercury-vapor or similar types described in Chap. 30) *if using screw-shell lampholders* may be installed under the same conditions but must be installed at least 8 ft above the floor.

This 277-volt system will permit the installation of a large load for lighting, using smaller wires and usually smaller conduit than would be required for the same wattage load at 115 volts.   This in turn leads to a very considerable saving in the installation cost.   Install it as you would a 115-volt system, but be sure to use switches and branch-circuit circuit breakers that are listed for use at the higher voltage.   Since the voltage to ground is over 150 volts, fuses if used must be of the cartridge type; plug fuses may not be used.

In all 277-volt installations, a word of caution is in order.   All 277-volt circuits of course consist of the neutral wire and one of the three hot or "phase" wires of a 277/480-volt system.   In an installation of any significant size, some of the 277-volt circuits consist of the neutral and phase wire *A*, others of the neutral and phase wire *B*, and the remainder of the neutral and phase wire *C*.   Thus it is quite likely that in a ganged box containing a number of switches to control various groups of lights, the box could contain two (or even all three) of the phase wires.   The voltage between such wires of different phases is not 277 volts, but 480 volts.   The Code in Sec. 380-8 requires that switches grouped in such boxes must be arranged so that the voltage between *exposed* live parts (such as terminals) of adjacent switches will not exceed 300 volts.   If all the circuits controlled by the switches in any one box are connected to the same phase wire, the maximum voltage between adjacent switches is only 277 volts, and the Code conditions are met.

But if the voltage between adjacent switches is over 300 volts, as when two adjacent switches control fixtures connected to two different legs of the 3-phase installation, you must install a metallic divider in the box, between the switches.   A simpler and much less expensive procedure is to use switches that do not have exposed terminals.   Use switches that have no terminal screws, but only the push-in type of connection shown in Fig. 8-5 in Chap. 5.   This scheme will save much labor.   Switches that have *both* terminal screws *and* push-in connections are not suitable.

Appliances.   The requirements for nonresidential use are the same as for residential use; therefore this subject need not be covered again (see Chap. 13).

Motors.   The wiring of electric motors is a sufficiently complex subject to warrant an entirely separate chapter, which will follow later.

Service-entrance Problems.   Before going into this subject, it will be well to review a few definitions that were originally covered in Chap. 13.   Service conductors are the wires that extend from the power supplier's distribution system up to the service switch.   The service drop is that portion of the service conductors which runs overhead; the service drop ends where the wires are anchored to the building.   The service-entrance conductors consist of that portion of the service conductors from the point where the service drop ends up to the service switch.   If the service is underground, there is no service drop, and the entire underground run of wire from the power supplier's wires up to the service switch becomes what the Code calls a "service lateral."

If a building is served by a 3-phase 4-wire *delta-connected* system which supplies both 3-phase power and 115/230-volt (or more usually 120/240-volt) power for lighting, a service-entrance problem exists. This system is discussed in more detail toward the end of this chapter. If a single circuit breaker is used for both the 3-phase and the single phase power, a rather special kind is needed; some power suppliers prohibit it.   The simplest method is to run a separate 3-phase 3-wire feeder from the meter to the 3-phase panelboard, and a separate 3-wire feeder (two hot wires plus neutral) from the meter to the single-phase panelboard.

A building is defined by Code as "a structure which stands alone or which is cut off from adjoining structures by fire walls with all openings therein protected by approved fire doors."

Several Service Drops per Building.   This subject is thoroughly covered by the Code in Sec. 230-2 which reads as follows:

230-2.   Number of Services to a Building or Other Premises Served. In general, a building or other premises served shall be supplied through only one set of service conductors, except as follows:

Exception No. 1.   Fire Pumps: Where a separate service is required for fire pumps.

Exception No. 2.   Emergency Lighting: Where a separate service is required for emergency lighting and power purposes.

Exception No. 3.   Multiple-occupancy Buildings:

(a) By special permission, in multiple-occupancy buildings where there is no available space for service equipment accessible to all the occupants.

(b) Buildings of multiple occupancy may have two or more separate sets of service-entrance conductors which are tapped from one service drop or lateral, or two or more sub-sets of service-entrance conductors may be tapped from a single set of main service-entrance conductors.

DEFINITION: Sub-sets of service-entrance conductors are taps from main service conductors run to service equipment.

Exception No. 4.   Capacity Requirements: Where capacity requirements make multiple services desirable.

Exception No. 5.   Buildings of Large Area: By special permission, where more than one service is necessary due to the area over which a single building extends.

Exception No. 6.   Different Characteristics or Classes of Use: Where additional services are required for different voltages, frequency, or phase, or different classes of use.   Different classes of use could be because of needs for different characteristics, or because of rate schedule as in the case of controlled water heater service.

The exceptions noted above should require no great amount of explanation.   Obviously it is desirable to have a separate source of power available for fire pumps and for emergency lighting.   Likewise, where a large building is occupied by a number of tenants, it would be objectionable to have one service entrance controllable by one tenant, with the other tenants not able to get at the service switch at all.   In that case a number of service drops may be used.   Another choice is one drop, with several subsets of service-entrance conductors.

It should be noted that a single-phase distribution system and a 3-phase distribution system are considered separate systems, so that it is entirely in order to serve a building with a single-phase drop for lighting and similar purposes and also a 3-phase drop for the power requirements.

Several Sets of Service-entrance Conductors per Building.   Generally speaking, only one set of service-entrance conductors is permitted per building, but there are a number of exceptions.   Obviously, where more than one service drop is permitted, as discussed above, each drop will require its own set of service-entrance wires.

In a multiple-occupancy building it is usually necessary to provide each tenant with a separate meter with its disconnecting means and overcurrent protection.   The several methods that may be used are the same as already covered in connection with apartments in Chap. 26.

Skeleton of Nonresidential Installation.   In a house the service-entrance wires run direct to the service switch or main breaker.   There are no feeders between the service and the branch circuits.   The individual branch circuits begin at the service.

In larger installations the distances involved and the number of branch circuits make it totally impractical to lead all branch circuits back to a common starting point.   In that case the service-entrance wires end at the service switch with a feeder to the main switchboard;

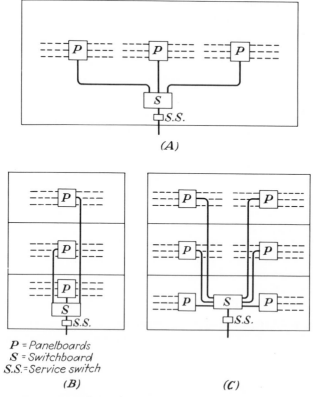

P = Panelboards
S = Switchboard
S.S. = Service switch

(B)   *(C)*

Fig. 28-2.   Typical riser diagrams for larger buildings.

from the switchboard feeders run to panelboards located where needed and placed so that the individual branch circuits (which begin at a panelboard) will be reasonably short.

The riser diagrams of Fig. 28-2 show several types of buildings. A typical single-story building which might be a factory or office building or a store is shown at *A*, a three-story building is shown at *B*, and a larger three-story building at *C*.

Switchboards and Panelboards. Switchboards and panelboards are nothing but convenient distribution points where one incoming set of wires is broken up into more and more individual runs of wire. All necessary switches, overcurrent devices, instruments, and similar accessories are located at such points. Look at such a system as you would a big oak tree: The trunk is the service-entrance wire, the point where the trunk breaks up into half a dozen large branches is the switchboard, and the points where the large branches in turn split up into smaller branches are the panelboards.

There is nothing in the Code which requires that an installation must have one switchboard plus a number of panelboards. These devices are installed to suit the convenience of the user. Often the service wires run from the service switch to a main panelboard and from there to other panelboards. As a matter of fact, it is difficult to define when a panelboard ceases to be a panelboard and becomes a switchboard. The Code, however, makes some distinctions and in Art. 100 defines the two as follows:

> **Switchboard.** A large single panel, frame, or assembly of panels, on which are mounted, on the face or back or both, switches, overcurrent and other protective devices, buses, and usually instruments. Switchboards are generally accessible from the rear as well as from the front and are not intended to be installed in cabinets.

> **Panelboard.** A single panel, or a group of panel units designed for assembly in the form of a single panel; including buses, and with or without switches and/or automatic overcurrent protective devices for the control of light, heat, or power circuits of small individual as well as aggregate capacity; designed to be placed in a cabinet or cutout box placed in or against a wall or partition and accessible only from the front.

Generally speaking, then, if the device is enclosed in a cabinet with a master door, it is a panelboard; if mounted away from a wall and if accessible from back as well as front, it is a switchboard. There is no precise line of demarcation between the two.

A simple fuse cabinet of the type shown in Fig. 28-3 is a panelboard. A larger panelboard is shown in Fig. 28-4. The fuses may be ordi-

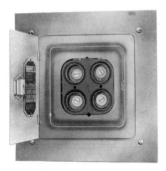

Fig. 28-3. This simple fuse cabinet is a small panelboard. (*Square D Co.*)

Fig. 28-4. A much larger panelboard, containing many circuit breakers. (*Square D Co.*)

nary plug fuses, or the largest type of cartridge fuses, depending on the ampere and voltage ratings of the circuits to be protected. Naturally circuit breakers may be used in place of fuses.

No more than 42 overcurrent devices may be used in any one cabinet. Each separate fuse is one overcurrent device. If circuit breakers are used, each 2-pole breaker is counted as two overcurrent devices, each 3-pole breaker as three. If the panelboard contains one or more *main* circuit breakers, or sets of *main* fuses, these are *not* counted as part of the 42 limit.

Lighting and Appliance Branch-circuit Panelboard. The Code in Sec. 384-14 defines this as a panelboard in which more than 10% of the overcurrent devices are rated at 30 amp or less, *and* feeding circuits for which a neutral is provided.

Rating of Panelboards. Every panelboard must be clearly and permanently marked with its rating in amperes, volts, and number of phases. If the board includes individual switches rated at 30 amp or less, so that any individual circuit can be turned on and off at the board, then the maximum permissible rating is 200 amp.

Overcurrent Protection. If the panelboard is fed by a feeder which itself has overcurrent protection no higher than the rating of the board, no main overcurrent protection is required on the board. If, however, it is fed by a feeder which has overcurrent protection higher than the rating of the board (for example, a 100-amp board fed from a feeder with 200-amp protection), then overcurrent protection must be provided on the board. This may consist of one or two (but never more than two) main circuit breakers or sets of fuses; the total amperage rating of the two combined must not exceed the rating of the board.

It should be worth repeating: Do not overlook the requirement of Sec. 384-16(c) which specifies that if in normal operation the load will continue for 3 hr or more, each overcurrent device in the board must control a load not over 80% of its rating. This ties in with the explanation earlier in this chapter that under such conditions no circuit may be loaded beyond 80% of its normal rating.

Neutral Bars in Panelboards. Check the neutral bar in the panelboard, both for the incoming wire, and the neutrals for the individual circuits. This neutral bar may *not* be grounded to the cabinet. The grounded neutral wire *must* be grounded at the service, but *may not* be grounded elsewhere (except in farm wiring, as was explained in Chap. 24). Nevertheless, the *cabinet* of the panelboard must be grounded. If the wiring is in conduit or metal-covered cable, usually the cabinet is then automatically grounded as explained elsewhere. But if the wiring is by nonmetallic-sheathed cable, use the kind with the grounding wire, and ground that to the cabinet.

Feeders. A branch circuit begins where the overcurrent devices for the circuit[2] are installed. All wires up to that point are feeders. The

---

[2] Note that the overcurrent device mentioned is the one that protects the *circuit,* and not one that is installed to protect the *load.* For example, in a circuit serving a motor, an overcurrent device may be installed near the motor to protect the motor against overloads; that is not the branch-circuit overcurrent device, for it does not protect the circuit against short circuits, grounds, or similar faults.

wires from the service switch to the main switchboard constitute a feeder. Wires from the switchboard to panelboards are feeders. When a large panelboard serves other smaller panelboards, the wires are feeders, sometimes called subfeeders.

Feeder Sizes.   Into the feeder classification fall the service-entrance wires also, as far as method of calculation of the size is concerned.   This calculation is not a difficult procedure and is substantially the same as outlined in Chap. 13 in connection with ordinary residential work, except that the heavier loads as well as the varying Code requirements of watts per square foot must be taken into consideration.   Total the watts required for lighting, plus appliances, plus motors; then divide by the voltage to arrive at the amperage, which, in turn, establishes the minimum-size wire required to meet Code requirements.

Feeders—Lighting Load.   The watts per square foot required by the Code for the purpose of such calculations vary considerably with the type of occupancy of the building and are found in Table 220-2(a) of the Code (see Appendix).   Reference to this table shows that the requirements vary from $\frac{1}{4}$ watt per sq ft in storage warehouses, to 3 watts per sq ft in schools and stores and some other occupancies, with 5 watts per sq ft in offices.   Demand factors, shown in Table 220-4(a) of the Code (see Appendix) reduce this in many cases.   The demand factors apply only to *lighting* loads, including receptacles normally installed on lighting circuits.

Feeders—Appliance.   Handle as in residential work.

Feeders—Motor.   See Chap. 31.

Determining Size of Service Entrance.   Proceed as in residential wiring.   Assume that the total calculated load for lighting and appliances, after application of the demand factors, is 26,800 watts.   The amperage will then be 26,800/230, or 117 amp.   Assume the load for motors or other heavy loads is calculated as 120 amp; this makes a total of 237 amp.   Table 310-12 (see Appendix) shows that if Type T or TW wire is used, 300MCM cable is the smallest that may be used.   If Type RHW or THW is used, 250MCM cable is sufficient.   Again note this is the Code minimum, and it is wise to install a larger size to provide for future loads.

Neutrals of Feeders, Single-phase.   In a 2-wire feeder, both wires must be the same size.   If a 3-wire feeder serves only 115-volt loads, all wires must be the same size.   But if a 3-wire feeder serves both 115- and 230-volt loads, the Code in Sec. 230-4(d) permits the neutral in

some circumstances to be smaller than the hot wires. The feeder
under discussion might be the service-entrance conductors serving the
entire installation, or a feeder serving part of the load.

See Fig. 28-5 which shows a feeder serving both 115- and 230-volt
loads. Assume now that the 115-volt loads are disconnected; in your

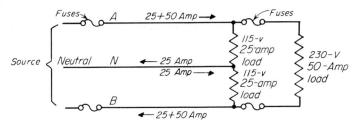

Fig. 28-5. A feeder serving both 115-volt loads and 230-volt loads.

mind erase them from the circuit. The neutral does not run to the 230-
volt load; it is not necessary to the operation of the 230-volt load.
Therefore the 230-volt load does not need to be considered so far as the
size of the neutral is concerned. The neutral needs to be large enough
only to carry the "maximum unbalance" which in Sec. 220-4(d) is defined thus: "The maximum unbalance shall be the maximum connected
load between the neutral and any ungrounded wire."

Reverting now to Fig. 28-5, if the 115-volt loads are 25 amp each, and
the 230-volt load is 50 amp, each hot wire must carry 25 + 50 or 75 amp.
If the two 115-volt loads are identical, for example 25 amp as in the
illustration, the neutral carries no current at all, but if one of the 115-
volt loads is turned off, the neutral must carry the current for the other
115-volt load, in this case 25 amp, and must be sized accordingly.

In ignoring the 230-volt load, an electric range will give you a problem, for it is not a 230-volt appliance, but a 115/230-volt appliance.
At low heat it operates at 115 volts, at high heat at 230 volts. For the
purpose of determining feeder size, consider the range as 30% 230-volt,
and 70% 115-volt. To calculate the feeder, proceed as follows:

A. Total load in watts (including the range at the value
normally required by Code) . . . . _____watts
B. Deduct all loads operating at 230 volts, but not including the range . . . . _____watts

C. Deduct 30% of the watts that you have included
   in A for the range . . . . _____watts
D. Remainder (A minus B minus C) . . . . _____watts

The total wattage of A, divided by 230, gives you the ampacity of the
minimum size of wire acceptable for the hot wires.   The remaining
watts of D, divided by 230, gives you the required ampacity for the neutral, assuming that the 115-volt loads are evenly balanced between the
two hot wires.   If for some unusual reason it is impossible to divide the
115-volt loads into two substantially equal portions, determine the
wattage of the larger of the two groups, divide by 115, which gives you
the maximum unbalance and determines the minimum ampacity of the
neutral.

If the maximum unbalance is over 200 amp, then per Sec. 220-4(d),
count only 70% of the amperage above 200.   For example, if the total
unbalance is 300 amp, the total to use is 200 amp, plus 70% of the remaining 100 amp or 70 amp, for a total of 270 amp.   If, however, part of
the load consists of fluorescent lighting, the 70% factor may *not* be applied to that portion of the load consisting of electric-discharge lamps
(fluorescent, or HID lamps discussed in Chap. 30).   Follow the example of later paragraphs concerning 3-phase feeders.

It must be noted that all the above is correct *only* if the service wires
come from a single transformer (as in B of Fig. 3-13 in Chap. 3), or from
one leg of a 3-phase *delta-connected* transformer bank (as in Fig. 3-16).
Occasionally the premises are served by the neutral and two hot
wires of a 3-phase 4-wire *Y-connected* system (as in Fig. 3-15); the resultant voltage will then be 120/208 volts.   In that case the neutral
must *always* be the same size as the hot wires.

Neutrals of Feeders, 3-phase.   Now assume a 3-phase 4-wire feeder
of 120/208 or 277/480 volts, serving both 3-phase loads connected to
the three hot wires, and also one or more single-phase lighting load.
Any one single-phase load will of course be connected to the neutral
and one of the hot or phase wires.   For the purpose of determining the
neutral, ignore the 3-phase loads such as motors.

Now consider each of the three single-phase loads (lighting, appliances, single-phase motors, etc.) separately.   Determine the amperage
of each, but exclude the amperage for any electric-discharge lighting
load.   Select the amperage of the largest of these three loads; assume it
is 350 amp, excluding the electric-discharge lighting.

If the unbalance is over 200 amp, the Code in Sec. 220-4(d) permits a demand factor of 70% for the unbalance over 200 amp, whether the feeder is single-phase or 3-phase, but *not* for any portion of the load that consists of electric-discharge lighting. (The reason for excluding electric-discharge lighting is rather involved technically, and its explanation is beyond the scope of this book.) Assume that the unbalance for the electric-discharge lighting load determined separately is 90 amp. To determine minimum neutral size, proceed as follows:

First 200 amp of load other than electric-discharge lighting, demand factor 100% . . . . 200 amp

Remaining 150 amp of load other than electric-discharge lighting, demand factor 70% . . . . 105 amp

Maximum unbalance, excluding electric-discharge lighting 305 amp

Unbalance for electric-discharge lighting only, demand factor 100% in every case . . . . 90 amp

Final maximum unbalance . . . . 395 amp

The neutral must have an ampacity of at least 395 amp.

Taps in Feeders. As explained in Chap. 5, normally overcurrent protection is required where a smaller wire is tapped to a larger one. In addition to exceptions already discussed, there are two more important exceptions.

Section 240-15, Exception 5, permits a tap *not over 10 ft long* provided it meets all three of these conditions: (a) it must not extend beyond the switchboard, panelboard, or control device that it serves; beyond that device, say a panelboard, there may be any number of *properly protected* additional circuits; (b) the tap wire must have an ampacity equal to the sum of the ampacities of all the wires in the circuits served; (c) the tap wires must be protected by conduit or similar metallic protection.

Section 240-15, Exception 6, permits a tap not over 25 ft long provided it meets all three of these conditions: (a) the tap wire must have an ampacity at least one-third that of the feeder; (b) it must be protected against physical damage; (c) it must terminate in a single circuit breaker or set of fuses with an ampere rating not greater than the ampacity of the tap; beyond this circuit breaker or set of fuses there may be any number of circuits each protected by its own circuit breaker or fuses.

Calculating Different Occupancies. With the above general discussions it should be possible to calculate almost any type of building, taking into consideration the requirements of Sec. 220-4(a) of the Code (see Appendix).   However, some types of buildings will be separately covered in some detail in later chapters.

Grounding.   The theory and importance of grounding was discussed in Chap. 9, and the actual method of grounding small projects in Chap. 17.   It will be well for you to review the subject there before proceeding with this chapter.   You must consider separately (a) system grounding, and (b) equipment grounding.   *System* grounding refers to grounding one of the current-carrying wires of an installation.   *Equipment* grounding refers to grounding exposed, non-current-carrying components of the system.

System Grounding.   This subject is covered by Code Sec. 250-5. Alternating-current systems *must* be grounded if this can be done so that the voltage *to ground* [3] is not over 150 volts.   In practice this includes single-phase installations at 115 volts, 115/230 volts (also 230-volt installations if the service is derived from a transformer that is grounded, as is usually the case), also 3-phase installations at 120/208 volts.

The Code does not require but does recommend grounding if the voltage to ground is over 150 volts but not over 300 volts (as in the case of 277/480-volt systems).   Systems operating at a voltage of more than 300 volts to ground *may* be grounded.

Occasionally there is a building which has only 230-volt single-phase loads, or only 3-phase loads.   Such loads do not require a neutral for proper operation.   Nevertheless the Code in Sec. 250-23(b) requires that if there is a ground at the transformer or transformers serving the building (as there usually is), the neutral must be brought into the building.   It ends at the service switch, where it is grounded in the usual way.   The reasons for this requirement are too involved for explanation in this book, but are in the interests of greater protection of the system, in case of accidental grounds in the wiring, to the conduit or metal armor of cable.   The neutral in the service, ending at the service switch, need not be larger than the size of ground wire required for the system.

---

[3] "Voltage to ground" in the case of grounded systems means the maximum voltage between the grounded neutral wire and any other wire in the system; in the case of ungrounded systems, the maximum voltage between *any* two wires.

"Separately Derived" Systems. Mention has been made of installations in which the service is at a voltage too high for lighting. It might be a 480-volt service; it might be 277/480 volts. In this latter case 277-volt circuits would be acceptable for much of the lighting. But in either case, 115-volt or 115/230-volt circuits would be necessary for operating small lamps, as well as a multitude of items such as office machines, small appliances, and so on.

The lower 115/230 voltage is obtained by installing one transformer with a 480-volt primary, and a 115/230-volt secondary. In other cases three transformers might be installed, fed by three 480-volt 3-phase primary wires, and with the secondaries connected in wye-fashion to deliver 120/208-volt 3-phase power (as in Fig. 3-15 of Chap. 3), the fourth wire being the neutral. In either case the Code stipulates that the secondary or secondaries constitute a "separately derived" system, and discusses the subject in Sec. 250-26. The Code in turn requires that the transformer secondary or secondaries be considered as a separate wiring system, and that this starting point must be handled as a new service entrance: You must provide a disconnecting means and ground the new system, just as if it were a separate service entrance in a separate building. Much confusion exists as to the proper grounding methods.

If a single transformer supplies single-phase 115/230-volt power, install a service switch with an ampere rating not less than the ampere output of the secondary of the transformer; run the wires from the secondary of the transformer to this switch. Ground the neutral of the transformer secondary to the metal case of the transformer, to the neutral grounding bar of the switch, to the cabinet of the switch, and also to the actual ground. This latter actual ground may be accomplished in several different ways, as shown in Fig. 28-6. If the neutral of the basic system that feeds the primary of the transformer is present at the point where connection is made to the primary of the transformer, ground the neutral of the new 115/230-volt system to the neutral of the original higher-voltage system. That completes the grounding of the new 115/230-volt system.

If the neutral of the original system is not available where connection is made to the primary of the transformer, you have two choices. Run the neutral of the new 115/230-volt system all the way back to the original ground in the building. This can be accomplished by running a separate grounding wire from the original ground at the service switch in the building, to the transformer location.   This grounding wire may be run in the same conduit with the current-carrying wires that run to the primary of the transformer (this wire would not be a neutral, but only a grounding conductor).   In some cases this might involve very long runs back to the original ground, and then not entirely practical. In that case, run the grounding wire of the new 115/230-volt system to the nearest cold water pipe, or ground it in any of the ways that would be acceptable if the derived 115/230-volt system were a completely independent system in a building and served by wires from the power supplier's line.

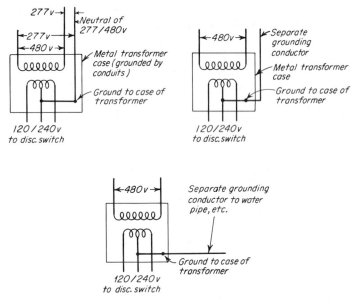

Fig. 28-6.   Three ways of grounding the neutral of a "derived system."

Often, a building may contain a considerable number of separate step-down transformers of this type; each one must be treated as outlined.

Do note that the conduit through which the wires run to the primary of the transformer, even if properly grounded, may *not* be used as the ground for the new 115/230-volt system.   Raceways such as conduit may be used to ground *equipment*, but not to ground *neutrals* of a wiring system, at a point distant from the service—and the secondary of the transformer does constitute a service.

If the derived system delivers 3-phase power, it will often be at 120/208 volts, and will then be 4-wire including the neutral. Ground this neutral exactly as in the case of single-phase 115/230-volt single-phase system.

The derived system might deliver power, either single- or 3-phase, without a neutral wire. The power might be at 230 volts, or at 480 volts (from a primary of still higher voltage). Run a grounding wire from the cabinet of the switch, to the case of the transformer, then ground in any of the ways already outlined.

As already explained, a 4-wire 3-phase Y-*connected* system automatically includes a grounded neutral wire. But what about a 3-wire 3-phase system? A 3-wire 3-phase *delta* system often is not grounded; sometimes one of the wires is grounded. That does not make a *neutral* wire out of that grounded wire; it is a grounded *hot* wire. However, quite often the service is basically a 3-wire delta system at 230 volts, but with the midpoint of one of the three legs grounded (as was shown in Fig. 3-16), thus producing a 4-wire *delta* system. The center-tapped leg then delivers 115/230-volt single-phase power; the grounded wire then is a grounded neutral wire. Do note that in such an installation, the voltage between the neutral wire, and the junction of the other two legs of the transformers, is about 199 volts (or 208 volts if the basic voltage is 240), as the diagram of Fig. 28-7 shows. You must use extreme

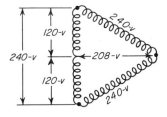

Fig. 28-7. In a delta-connected 240-volt system with one phase center-tapped, the voltage from neutral to the "wild" leg is 208 volts.

caution that nothing is ever connected between the neutral and this "high" or "wild" leg of the system. As a matter of fact, the Code in Sec. 200-6(c) requires that this wire must be tagged or otherwise identified at any point where a connection could be made, if the neutral is also available at that point.

The actual installation of the ground is substantially as in residential work. The ground wire may be bare or insulated. The size of the ground wire depends on the size of the service, and is defined in Code Table 250–94(a) as follows:

TABLE 250-94(a)   Service and Common Main Grounding Conductor for Grounded Systems

Size of largest service conductor or equivalent for multiple conductors		Size of grounding conductor	
Copper	Aluminum	Copper	Aluminum °
2 or smaller	0 or smaller	8	6
1 or 0	2/0 or 3/0	6	4
2/0 or 3/0	4/0 or 250 MCM	4	2
Over 3/0 to 350 MCM	Over 250 MCM to 500 MCM	2	0
Over 350 MCM to 600 MCM	Over 500 MCM to 900 MCM	0	3/0
Over 600 MCM to 1100 MCM	Over 900 MCM to 1750 MCM	2/0	4/0
Over 1100 MCM	Over 1750 MCM	3/0	250 MCM

° See installation restrictions in Section 250-92(a).

Grounding bushings must be installed; also install grounding jumpers if any part of the concentric knockout is left in a cabinet.   In larger installations however, there is some confusion as to the number of points where bonding or jumpering is required.

If there is a single main switch or breaker, proceed as in residential work.   Bonding is not required at the points where conduits leave the cabinet of the service switch, running on to feeders, circuits, or panelboards.   If however a separate metering cabinet is installed for the current transformers and similar equipment, you must bond the conduit running to such cabinet.

If there is *not* a single main switch or breaker, as for example in Figs. 26-1 or 26-4 in Chap. 26, each of the separate switches is considered part of the service entrance, and bonding is required at each separate switch, where the wires enter the switch; each neutral must be grounded.   Moreover, if in an installation similar to that shown in Fig. 26-4, you have used a wire trough of the kind discussed in Chap. 29 and shown in Fig. 29-21, that trough becomes part of the service entrance.

You must bond where the wires enter the trough, and at each point where they leave the trough to run to the individual switches. Better yet, bond at each switch where they enter the cabinet.

If steel conduit is used as protection for the ground wire (or if the ground wire is inside armor), it must be securely and permanently connected to the ground wire, or the cabinet that the ground wire enters, at both ends. Unless you do this, for example if one or both ends are allowed to float without being tied together, the total ground will be much less effective, and afford much less protection, than a ground wire alone without conduit or armor.

Equipment Grounds. "Equipment" here means any exposed non-current-carrying metal parts of fixed equipment that are liable to become energized. Such parts include conduit or armor of cable, boxes and cabinets, metal parts of fixtures, the frames of motors and appliances, and so on. It must be grounded even if there is no *system* ground, under the conditions of Code Secs. 250-42 and 250-43. Basically this includes: (1) all equipment if served by a metal-clad system such as conduit or armored cable; (2) regardless of wiring method, any equipment in a wet location, or located so that it can be touched by a person standing on the ground, or where he can touch a grounded surface or object, as, for example, plumbing; (3) all equipment in "hazardous locations," which will be discussed in Chap. 35; (4) all equipment in contact with metal or metal lath; (5) any equipment operating with any terminal at more than 150 volts to ground (with a few exceptions); (6) most motors; (7) many other items detailed in Sec. 250-43 (see your copy of the Code).

If you have installed conduit, or used cable with a metal armor, you will have automatically met the requirement if the conduit or armor is solidly anchored to the motor or appliance, and to every box or cabinet all the way back to the service entrance. If you have used flexible metal conduit, install a grounding wire inside of it as explained in Chap. 11. If you have used nonmetallic-sheathed cable, you must use the cable with the bare grounding wire in addition to the insulated wires, as already explained in an earlier chapter.

If an appliance or motor that must be grounded is connected to the system using a cord and plug, the cord of course must contain a grounding wire in addition to the current-carrying wires, and the receptacle naturally must be of the grounding type.

If conduit is used, in very large buildings there are cases where a separate grounding conductor must be used, in addition to the circuit wires, per Table 250-94(b) in your copy of the Code. Such cases will not be encountered by the student in the kinds of installations he will meet during his training period.

Install Only *One* Ground Wire. Regardless of all other considerations, one point is important: Use only *one* ground wire in one installation (with the exception of separate buildings on farms, as was discussed in an earlier chapter). This ground wire must begin inside the service equipment.

Size of Ground Wire with Ground Rods. If there is not a good underground water system available for grounding, and you must use a driven ground rod or rods, the ground wire never need be larger than No. 6.

29

Miscellaneous Problems in Nonresidential Wiring

In nonresidential installations the conduit system is used in practically all cases. This automatically provides a really good continuous ground. It provides a certain amount of flexibility in that circuits may be changed, wires added, and breakdowns repaired by merely pulling in new wires, with a fair degree of ease and not too much mechanical change in the actual conduit. Where exposed, the conduit is reasonably neat in appearance and certainly affords ample protection for the wires.

Conduit Fittings. For exposed runs of conduit, it is customary to use, instead of ordinary outlet boxes, cast fittings of the type shown in Fig. 29-1. They are known by trade names such as Electrolets, Condulets, Unilets, etc. These devices are merely specialized forms of outlet boxes, but instead of being provided with knockouts which can be removed to form openings, they have one or more ready-made openings. Accordingly, with a few basic body shapes, dozens of different combinations are available. Each basic type is available for each size of conduit.   Each opening is threaded to fit the size of conduit for which it is designed.   These fittings are also available with threadless openings but with clamping devices for thin-wall conduit.

E   LB   LL   LR   TA

Fig. 29-1.   For exposed runs of conduit, fittings of the type shown here are commonly used.   There are dozens of different shapes or types.   (*Killark Electric Mfg. Co.*)

A few of the more common types are shown in Fig. 29-1.   The Type E with the cover shown is frequently used at the end of a run to a motor or similar device.   The LB is commonly used at a point where a run of conduit comes along and then must go at right angles through a wall or ceiling; it is equally useful in going around a beam or similar obstruction.   Fittings of this kind avoid awkward bends in conduit.   The Types LL and LR are handy for 90-deg turns on a straight run.   There are all kinds of combinations, some as complicated, for example, as the Type TA, which obviously is not used very frequently.

In a different style of body there are available many types similar to the Type FS shown in Fig. 29-2, which is used mostly for the mounting of switches and similar devices.   In the same illustration is shown a weatherproof cover with a spring-loaded flap for a receptacle, and another which will operate a toggle switch mounted in a Type E fitting—a handy combination for outdoor switches.   On exposed runs of conduit, lighting fixtures are mounted on round Type P fittings.

Other Wiring Systems.   The Code in Arts. 318 to 364 defines the many wiring systems that are acceptable.   In addition to the systems described in this book, there are, for example, those involving Aluminum rigid conduit, Nonmetallic rigid conduit, Liquid-tight flexible metal conduit, MC cable, ALS cable, Underfloor raceways, Busways, and others. The student will not need information about them until after he has mastered the more ordinary systems described in this book; consequently these additional systems will not be described here.

Fig. 29-2. Larger fittings are used to house switches, receptacles, and similar devices. (*Killark Electric Mfg. Co.*)

Pull Boxes. Wires of ordinary sizes as used in residential work are sufficiently flexible so that they can be pulled through long lengths of conduit even if there are offsets and bends. The heavier the wire, the more difficult it becomes. In really heavy sizes, such as the circular-mil cables, it becomes more and more necessary to install pull boxes at strategic locations; it is customary in many cases to use them instead of conduit bends. Such a pull box, as the name implies, is nothing but a steel box located where the wires can be helped along as they are pulled into the conduit. Pull boxes may be used only where they will be permanently accessible. A single pull box is often used for a number of runs of conduit as shown in Fig. 29-3.

The Code covers the subject of pull boxes in Sec. 370-18. There are specific requirements only if the conduits entering the box are 1-in. or larger, *and* if the wires are No. 6 or heavier. For straight pulls, the length of the box must be at least 8 times the trade diameter (not actual diameter) of the largest conduit entering the box.

If angle or "U" pulls are involved, the calculation of size becomes a bit more involved. If only one conduit enters any one wall of the box, the distance to the opposite wall must be at least 6 times the diameter of the conduit.

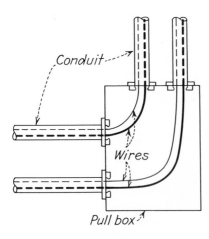

Fig. 29-3. Pull boxes are used with conduit to make it easier to pull wires through long lengths of conduit, and for other purposes.

If several conduits enter the same wall, the distance to the opposite wall must be at least 6 times the diameter of the largest conduit, plus the diameters of all the other conduits entering that wall. The distance from the conduit by which a set of wires enters the box, to the conduit by which it leaves the box, must be at least 6 times the diameter of the conduit; if there are two different sizes, use the larger of the two.

These are minimum Code figures. Good practice often suggests using boxes larger than the minimum (especially if a large number of conduits enter the box), to avoid crowding which increases labor and detracts from neatness.

Concrete Boxes. In nonresidential buildings of all types, walls and ceilings frequently are of reinforced-concrete construction. The conduit and the boxes must be embedded in the concrete if the devices later are to be flush with the surface of the wall. Ordinary outlet boxes may be used, but special concrete boxes are also available. One of these is shown in Fig. 29-4. These boxes have special ears by which they are nailed to the wooden forms for the concrete. Stuff the boxes full of paper before installing; this will prevent concrete from seeping in. The conduit and the boxes must be in position before the concrete is poured. When the forms are removed, the conduit and the boxes are solidly embedded; the interior of the box is clean and ready for use. These boxes come in a variety of depths up to 6 in. Figure 29-5 shows an installed view.

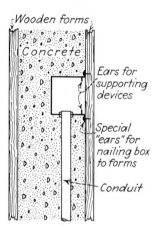

Fig. 29-4. A concrete box, designed to be embedded in concrete as it is poured.

Fig. 29-5. Concrete boxes are nailed to the wooden forms before the concrete is poured.

Deflection of Wires. Small wires are so flexible that it is not likely that they will be damaged even by sharp bends. In the larger sizes, on the other hand, it is conceivable that, if bent too sharply where they emerge from a run of conduit, the insulation might become damaged to the point where grounds might be caused. The Code in Sec. 373-6(b) requires that ungrounded wires No. 4 and heavier must be further protected "by a substantial bushing providing a smoothly rounded *insulating* surface unless the conductors are separated from the raceway fitting by substantial insulating material securely fastened in place."

Several methods of meeting this requirement are available. The simplest is to use metal conduit bushings with a molded insulating material on the inside, as shown in Fig. 29-6. Such bushings are also available with a lug and connector for a grounding wire, as shown in the illustration. Alternately, use a bushing made entirely of insulating material, as shown in Fig. 29-7; in that case two locknuts must be used, one on the outside and one on the inside of the cabinet, before the bushing is installed.

Fig. 29-6. Grounding bushing with insulated throat, and terminal for grounding jumper. (*Union Insulating Co.*)

Fig. 29-7. Conduit bushing made entirely of insulating material. (*Union Insulating Co.*)

Also acceptable is the use of flexible fiber bushings similar to those used with armored cable but of much larger size; they must be constructed so that they snap into place and cannot easily be displaced. They are used in addition to the usual metal bushing.

Moreover, heavier wires are rather stiff, and it is difficult to bend them sharply. Even if no damage to the insulation occurs from bending sharply, efforts to do so often lead to untidy, overcrowded conditions within a cabinet. All this has led to the provisions contained in Code Sec. 373-6(a) which provides minimum gutter widths for each size of wire. The gutter is the space between the wall of a cabinet, and the mechanism contained in the cabinet. The requirements are spelled out in Table 373-6(a) which you will find in your copy of the Code. This requirement is of greater concern to manufacturers than to contractors, and panelboards and similar equipment automatically incorporate the proper gutter widths. But note that wires may *not* be spliced in gutters unless special cabinets with wider-than-usual gutter widths are used.

Locate cabinets so that the incoming runs of conduit will be so placed as to require a minimum deflection of wires where they emerge from the conduit, and so that in vertical runs the weight of the wire will not be supported by the bend at the end of the run. Figure 29-8 shows the wrong and Fig. 29-9 the right method. Let the bends in the wire be sweeping and gentle rather than abrupt.

Supporting Vertical Runs of Wire. Terminals on panelboards and similar equipment are not designed to support any substantial weight. When there is a vertical run of wire, the weight of the wire itself is quite considerable, especially in the larger sizes. If such runs of wire

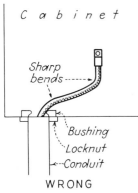

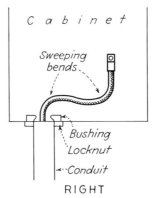

WRONG

Fig. 29-8. The bends shown here are too abrupt. The wire rests on the conduit bushing, which could lead to damage to the insulation.

RIGHT

Fig. 29-9. The bends in the wires should be gentle and sweeping. This tends to prevent grounds where the wires emerge from the conduit.

are connected directly to terminals, damage may result. The Code in Sec. 300-19(a) therefore requires that in such vertical runs the wire be independently supported at intervals as follows:

Nos. 18 to 1/0	At least every 100 ft
Nos. 2/0 to 4/0	At least every 80 ft
250,000 to 350,000 cm	At least every 60 ft
350,001 to 500,000 cm	At least every 50 ft
500,001 to 750,000 cm	At least every 40 ft
750,001 cm and larger	At least every 35 ft

There are several ways of accomplishing the required support. The usual way is to use a clamping device of the general type shown in Fig. 29-10. Another method is to install boxes or cabinets at the intervals shown in the table, and clamp each wire in an insulating support, as shown in Fig. 29-11. A third method is similar and is shown in Fig. 29-12. Each wire must be deflected at least 90° from the vertical and supported by at least two insulating supports; horizontal deflection must be at least twice the diameter of the wire. But this last method has its drawbacks: the wire must be supported *five* times as often as the table shows; the distance between supports must be no more than 20% of the figure in the table.

Fig. 29-10. This fitting is very effective in supporting vertical runs of wire. (*O. Z. Electrical Mfg. Co.*)

Grounded Neutral. The Code in Sec. 200-6 requires that the grounded neutral wire must be white.   But No. 4 and heavier wires are usually not available in white.   The Code permits you to use any color, but at every terminal that neutral wire must be marked; paint the insu-

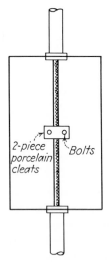

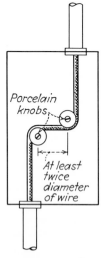

Fig. 29-11. Another way of supporting a vertical run of wire.

Fig. 29-12. Still another way of supporting a vertical run of wire.

lation white, or wrap white tape around it. When the wire is No. 6 or smaller, the grounded neutral wire must be white throughout its entire length, unless it is bare wire, which is acceptable under some circumstances explained elsewhere.

Continuity of Ground. Previous chapters showed how the various runs of conduit or metal-covered cables tie together outlet boxes and other equipment into one continuously grounded system. If, however, in any such system, one or more of the wires have a voltage above 250 volts *to ground*, the usual methods are no longer acceptable. Instead, the Code in Sec. 250-76 gives a choice of several methods.

If conduit fittings of the type shown in Fig. 29-1 and 29-2 are used, that is sufficient for either rigid or thin-wall conduit. The same fittings may be used for armored cable or flexible conduit, for the connectors that are used will fit directly into the threaded openings of the fittings.

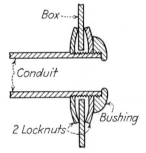

Fig. 29-13. When the voltage *to ground* is over 150, use the double-locknut construction shown here, instead of the ordinary construction shown in Fig. 10-6.

Fig. 29-14 A grounding locknut.

If ordinary outlet boxes or cabinets of drawn steel are used, several methods are available for use. The most common so far as rigid conduit is concerned is the double-locknut system shown in cross section in Fig. 29-13. This involves simply one locknut on the outside of the box or cabinet, another inside, plus the usual bushing. This however is not required in conduits containing only service-entrance wires.

With thin-wall conduit, armored cable, or flexible conduit, the simplest method is to use on the connector involved, inside the box, a grounding locknut shown in Fig. 29-14, in place of the usual locknut that is used on such connectors.   It is not necessary to use jumper wires from one such locknut to another inside the same box or cabinet; the screws on the locknuts are designed to bite into the metal of the box, and will stay in place where vibration or other causes might loosen the ordinary locknut.

In case the run of conduit in question is the service conduit, any of the schemes discussed may be used, except the double-locknut method.

Made Electrodes.   If an underground water system is not available for grounding, you must use a "made electrode," as was explained in Chap. 17.   In addition to the methods discussed there, a new method was first recognized by the 1968 Code, in Sec. 250-83(a), for new installations.   A bare copper wire is installed near the bottom of the poured concrete footing of a building, as the footing is poured.   The wire must *not* be in direct contact with the earth, but rather inside the footing within a couple of inches of the earth; the footing must be in contact with the earth.   The wire must be of the size required for grounding, in Table 250-94(a), at least 20 ft long, and in no case smaller than No. 4.   Aluminum wire may not be used.

Double-pole Switches.   When lighting fixtures are supplied by two *ungrounded* wires, only double-pole switches may be used to control them; such switches open both wires.   This is merely part of the general requirement that switches must open all ungrounded wires, although there are some exceptions in connection with motors, as will be explained in a separate chapter.

Surface Metal Raceway.   In nonresidential buildings the wiring is usually in concealed conduit, embedded in the concrete.   It is a horrendous job to make later changes, or additions, if any new conduit is also to be concealed.   Ordinary conduit installed on the surface is most unsightly.   Yet changes are often necessary especially in offices, stores, and other locations where layout changes are frequent.

Instead of conduit, use surface metal raceway which has a cross section that is attractive and blends well with the general appearance of the area.   The material is used not only for power wires, but for telephone and similar circuits as well; however, both may not be installed in the same channel.

Two styles are available.   One is called the one-piece type (although it actually consists of two pieces, preassembled at the factory), and the other the two-piece type.   The one-piece is installed empty, the wires then pulled into it just as in conduit wiring.   In the two-piece type, the base member is first installed, the wires then laid into place, the cover then installed.   The Code does not publish the number of wires that may be installed in any one size, but in Sec. 352-3 merely states that no conductor may be larger that that for which the raceway was designed, and in Sec. 352-4 that the number of wires installed may not exceed the number for which the raceway was designed.   All that sounds a bit vague, but the answer is simple: Consult the tables furnished by manufacturers.

The Code in Art. 352 limits both styles to *exposed* runs in dry locations, although the material may be run through (but not inside of) dry walls, partitions, and floors; the material must be in continuous lengths where it passes through.   The voltage between any two wires may not exceed 300 volts, unless the material of the raceway is at least 0.040 in. thick.

The one-piece type, in which the wires are pulled in after the installation of the raceway, is shown in Fig. 29-15.   This also shows the di-

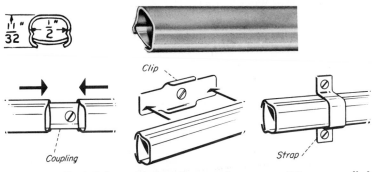

Fig. 29-15. Typical "one-piece" surface metal raceway.   Wires are pulled into place after the raceway is installed.   (*The Wiremold Co.*)

mensions of the smallest size; three larger sizes are available.   It is held in place by several methods, using clips, couplings, or straps shown in the same illustration.   There are available many appropriate kinds of elbows, adapters, switches, and receptacles to fit each size of material.   Some of these are shown in Fig. 29-16.

The two-piece type is shown in Fig. 29-17, which also shows the dimensions of the smallest size; three larger sizes are available up to $4\frac{3}{4}$ in. wide by $3\frac{9}{16}$ in. deep.   The base is first installed, wires then laid in place, the cover then installed. Many fittings are available, similar to those shown in Fig. 29-16 except designed for this type of

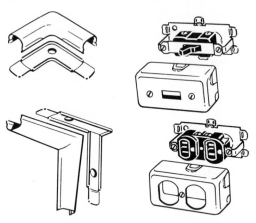

Fig. 29-16. Fittings used with the material shown in Fig. 29-15. (*The Wiremold Co.*)

raceway. One of the larger sizes is available with a divider in the base, thus providing two separate channels. Use one channel for the power wires, the other for telephone or other nonpower purposes. This provides a neat installation and meets the Code requirement that power wires, and other wires, may not occupy the same channel.

Fig. 29-17. Typical "two-piece" surface metal raceway. Install the chan nel, insert the wires, then snap cover into place. (*The Wiremold Co.*)

The Code requires that surface metal raceway may not be installed where "subject to severe physical damage, unless approved for the purpose." There are many cases where the material must be run on the floor, which certainly is a location where physical damage might be expected. For this purpose there is available a "pancake" material shown in Fig. 29-18, which also shows the dimensions of the smallest of three sizes available.   As in the other two-piece type, the base is installed first, the wires laid into place, the cover then snapped on.   As with other raceways, channels may contain power wires, or telephone and similar wires, but not in the same channel.   As with the other types, fittings such as elbows and so on are available.

Fig. 29-18. "Pancake" type of surface metal raceway, for installation on floors. (*The Wiremold Co.*)

All the basic materials described are probably used more often for receptacles than for any other purposes.   Especially in stores, schools, laboratories, and similar locations, there is often a need for very many receptacles spaced quite close together.   Much installation labor is involved if they are installed using ordinary wiring methods.   To reduce the cost of labor, use a special raceway that has been developed, shown in Fig. 29-19.   The covers are prepunched for receptacles so that spe-

Fig. 29-19. Raceway with receptacle outlets at regular intervals. If many receptacles are needed, this saves much time over other installation methods. (*The Wiremold Co.*)

cial fittings for the receptacles are not needed.   This type of raceway (called multioutlet assembly in the Code) is available prewired with the receptacles in place, or may be had empty, with the receptacles prewired on long lengths of wire.   Several spacings are available, with receptacles from 6 to 60 in. apart.   A typical installation is shown in Fig. 29-20.

The Code covers the subject of multioutlet assemblies in Art. 353. It may be installed like other types of surface metal raceway, but if it runs through a wall or partition, no receptacle may be within the wall or partition.   Moreover, it must be so installed that the cover can be re-

Fig. 29-20. A typical installation of the material shown in Fig. 29-19.

moved on all portions outside the wall or partition, without disturbing the portions within the wall or partition.

This kind of material lends itself well for use in the two special appliance circuits required by Code in kitchens and other rooms; these were discussed in Chap. 13. It is equally suitable for home workshops or other locations where numerous receptacles are desirable.

Wiring Troughs. Sometimes it is necessary to run heavy wires for short distances, with many taps in the wire. For an example, see Fig. 26-4 in Chap. 26. The service wires must run to each of six disconnect switches. Unless very special switches are used, it is not permissible to run the wires to the first switch, make splices within it and continue to the second, make more splices there and continue to the third, and so on. Instead, use a wiring trough of the general type shown in Fig. 29-21. The trough is first installed, the wires laid into it later, and taps then made to run to each switch. The troughs have hinged or removable covers, making for an easy installation. Many fittings such as elbows, crossovers, end sections, and so on are available. Such troughs are available in lengths of 1, 2, 3, 4, 5, and 10 ft, and in several cross sections such as 2½ by 2½ in. up to 8 by 8 in. Raintight types are available for outdoor use.

If the troughs are used as just outlined, they become part of the service entrance, and grounding bushings and bonding jumpers must be installed where the entrance wires enter the trough, and also at each switch fed from such a trough.

It should not be supposed that such troughs are used only for the exact purpose outlined above. Indeed they are frequently used for long runs, especially in the larger sizes, and especially when many wires must be installed. Their use can be more economical than conduit, and certainly many times more flexible.

The Code conditions for such troughs are found in Code Art. 362, under the heading of "Wireways." The wires installed may occupy no more than 20% of the interior cross-sectional area of the trough, except in locations of taps or splices, where the 20% becomes 75%. The area of any size and kind of wire can be found in Code Tables 5, 6, and 7.

If the derating factors for the ampacity of wires specified in Note 8 of

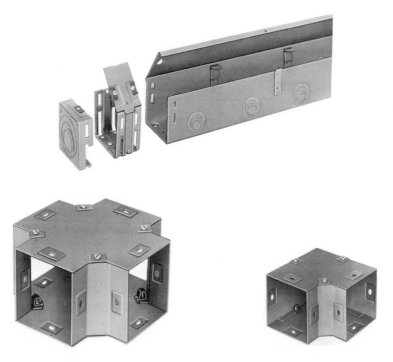

Fig. 29-21. Wiring troughs of this kind are very handy, especially for larger wires. (*Square D Co.*)

Tables 310-12 to 310-15 are applied to the wires in the trough, there is no limit to the number of wires that may be installed.   It is not necessary to derate if 30 or fewer wires are installed.   The 20% "fill" factor however must always be observed.

For the Tables 5, 6, and 7, and other conditions for wireways, see your copy of the Code.

30

Nonresidential Lighting

Chapter 14 discussed the fundamentals of lighting, and the kinds of incandescent and fluorescent lamps used mostly in residential lighting and in smaller nonresidential projects.   In this chapter we shall discuss other lamps and lighting equipment used mostly in nonresidential work, and factors involved in designing good lighting systems for such locations.

Extended-service Lamps.   In a home or in small offices it is a simple matter to replace a burned-out lamp.   In larger establishments, a maintenance man may have to drop what he is doing, come a block carrying a ladder, and then replace the lamp, a very expensive procedure.   Alternately, in a factory, the lamps may be installed 20 or 30 ft above floor level, requiring scaffolding or special ladders, again very expensive procedure.   In such locations it may be wise to use lamps designed for a voltage higher than the actual voltage, for example, 130-volt lamps on 115- or 120-volt circuits.   This will greatly extend their life, but will reduce the lumens per watt, thus increasing the cost of light.   But this higher cost may be fully justifiable because it is more than offset by the savings in maintenance costs.

But using 130-volt lamps on 115- or 120-volt circuits leads to a good deal of guesswork as to the actual watts consumed, the lumens of light produced, and the life.   It is more logical to use what are called "extended-service" lamps.   These are designed for 2,500-hr life when used at their rated voltage, and will consume their rated watts.   They produce about 15% fewer lumens per watt than ordinary lamps.

Group Relamping.   In buildings other than homes it often becomes an expensive procedure to replace lamps one at a time, as already explained.   If a hundred were replaced at one time, the cost per lamp would be very much less.   For that reason, many large establishments make it a practice to replace all lamps in a building at one time, whether burned out or not.   The extra cost of the new lamps is less than the cost of replacing one at a time.

"R" and "PAR" Lamps.   These lamps have a silver or aluminum reflector deposited directly on the inside of the glass bulb, as a permanent part of the lamp.   Being silver or aluminum, they are excellent reflectors.   Being part of the lamp itself, they are permanently bright and untarnished.   They cannot get dusty or dirty, thus eliminating maintenance costs, and preventing the loss in efficiency that comes with dirty separate reflectors.   They reduce initial cost because there is no need for separate reflectors or elaborate fixtures.   They are designed for much longer life than ordinary lamps of corresponding size.

Since such lamps have integral reflectors, it follows that they are not suitable for ordinary lighting in residential work.   They are however very widely used in commercial work, for example in lighting merchandise displays in stores, from flush ceiling fixtures.   They are equally suitable for lighting in industrial installations, especially buildings with very high ceilings.   Having built-in reflectors, they do away with the need for separate reflectors, which must be cleaned regularly for reasonable efficiency.   There are two types: R and PAR, both shown in Fig. 30-1.

Type R lamps are available in sizes from 30 to 1,000 watts, and have a substantially standard bulb of the required shape.   In the spotlight type the light is concentrated into a more or less circular beam from 35 to 60° in diameter.   In the floodlight type it is from 80 to 120° in diameter.   Most Type R lamps have bulbs made of ordinary or "soft" glass and can be used only indoors, because cold rain or snow falling on the hot glass would crack it. A few of them have "hard" glass which withstands rain or snow. Type R lamps are widely used in lighting merchandise displays, lighting special areas such as walls, pictures, or other points deserving special lighting. In homes they are sometimes used in bathrooms, over game tables, and in similar locations.

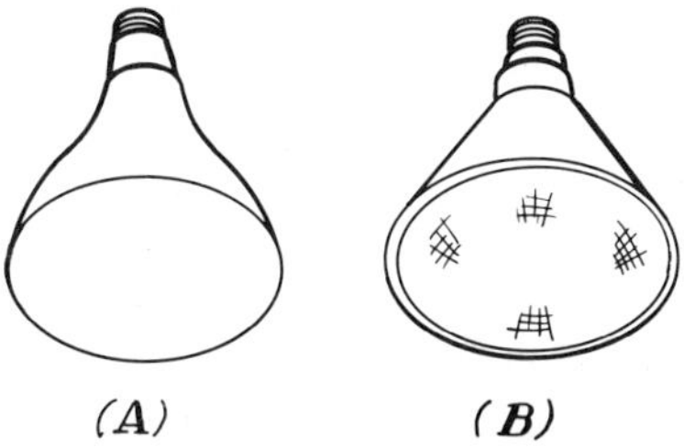

Fig. 30-1. Type R and PAR lamps have internal reflectors, permanently clean.

(A)   *(B)*

Type PAR lamps are available in sizes from 75 to 500 watts (or up to 1,000 watts in the "Quartzline" construction which will be described later). They are constructed using special two-part glass bulbs. The two-piece construction permits more accurate positioning of parts, with the result that beams can be controlled to a greater degree than in Type R lamps, in turn permitting better concentration of light, and narrower beams. In the floodlight type most of the light is concentrated into a beam about 60° in diameter, but some sizes are available with wider or narrower beams. In the spotlight type the light falls in a beam about 30° in diameter, while in the larger sizes the beam is rather oval in shape, ranging from 13 by 20° to 30 by 60°. Some sizes are available in a choice of beams. They are designed for a 2,000-hr life.

All PAR lamps are made of a special hard glass that is not affected by cold rain or snow, consequently may be used indoors or outdoors. They are widely used in lighting buildings or landscapes, construction projects, signs, sports lighting, or similar undertakings. On farms they are widely used as yard lights. While PAR lamps cost about twice as much as Type R lamps, and several times as much as ordinary lamps, that higher cost is more than offset by the elimination of expensive reflectors and their maintenance.

"Tungsten Iodide" Lamps. As you have yourself observed, an ordinary lamp blackens during its life; near the end of its life the inside of the bulb is quite black. As the lamp is used, the tungsten of which the filament is made gradually evaporates; the evaporated tungsten deposits on the inside of the bulb, blackening it.   That in turn reduces the transparency of the bulb, responsible in great part for the fact that the light output of an ordinary lamp is very considerably reduced toward the end of its life.

It has been found that a bit of iodine introduced into the bulb will, through a rather complicated bit of chemical action, prevent the blackening.   The tungsten that evaporates during its use is redeposited on the filament, and the bulb remains clear.   Such lamps are called "tungsten-iodide" type.   Depending on the manufacturer, they are called by trade names such as "Quartzline"®[1], "IQ", or "Tungsten-Halogen."   Such lamps operate at a very high temperature which would soften any kind of glass; so the bulbs are made of quartz, which withstands exceedingly high temperatures without softening.

Such lamps are usually designed to produce about the same number of lumens per watt as ordinary lamps, but with a life of two or three times that of ordinary lamps.   Their lumen output is maintained at almost their initial value; ordinary lamps toward the end of their life usually produce only about 85% of their initial output.

Tungsten-iodide lamps are not used for general lighting in homes or offices, but rather for special purposes as, for example, lighting of large areas in factories, floodlighting, sports lighting, and similar applications.

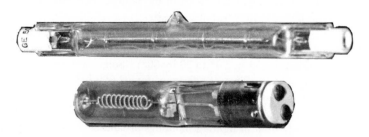

Fig. 30-2. Tungsten-iodide lamps are long and slim, but are sometimes enclosed in larger glass bulbs, sometimes of the PAR shape. (*General Electric Co.*)

In the general-purpose type, tungsten-iodide lamps are available in sizes from 150 to 2,000 watts.   The bulb is tubular, less than $\frac{1}{2}$ in. in diameter, and from $2\frac{13}{16}$ to $10\frac{1}{4}$ in. in length.   Bases vary a great deal with the size and type; two common types are shown in Fig. 30-2.

[1] Registered trademark, General Electric Co.

One of these has a single contact at each end. They are also available in the PAR shape, flood or spot, in which the quartz tube is enclosed in a larger glass bulb. Special-purpose lamps are available in other wattages, different-sized bulbs and different kinds of bases, including some types that have merely a wire lead at each end.

The actual area occupied by the light-producing filament is very, very small, making for an exceedingly bright and small source of light. This in turn makes it possible to concentrate the light into very narrow, intense beams. With the proper reflectors the beam may be as narrow as 6° in one direction, but quite wide in the other, making possible a much higher degree of illumination in a given area than is possible using ordinary lamps of equal wattage. This makes the lamp especially suitable for lighting athletic fields, lighting the field proper very well, yet not producing glare for the spectators sitting on the side of the field opposite the lamps. But because of the intense concentration of light, such lamps must be used only in fixtures or floodlights, or locations where it is impossible to look directly at the lamps from a short distance; doing so would damage the eyes. Their high operating temperature too makes the use of fixtures or floodlights designed for these lamps, a necessity.

Size of Light Source. For general seeing, we need lots of light, but any old source of light is not necessarily suitable for a particular purpose. In Chap. 14 we discussed the subject of surface brightness. In general, for routine lighting in homes, offices, and similar locations, we need plenty of light, but from a source where the light does not come from a small area, for then we would get the effect of reading in direct sunlight; reading in the shade of a tree is much more comfortable than in direct sunlight, even if the degree of illumination is much lower. In other words, we need light sources of low surface brightness.

But there are times when we need much light in one area, not for general seeing, but to emphasize for example a piece of sculpture, a particular piece of merchandise in a store, the outside of a building, or a landscape. Then we use a reflector to concentrate the light, to direct all the light into a small area. Using the best of reflectors, it is impossible to concentrate the light from an ordinary lamp into a really small area. The reason for this is that the source of the light is not small enough. The smaller the source of light (not the size of the bulb, but the area of the filament or arc tube that produces the light in the bulb), the greater the degree of concentration that the reflector can accomplish, the narrower the beam.

Therefore to concentrate the maximum of the available light into a very narrow beam, we must use a lamp in which the filament is concentrated in as narrow or small an area as is possible.   One way of accomplishing this is discussed in the following paragraphs.   (It may be interesting for you to know that the filament in an ordinary 60-watt lamp is over a yard long.   It is first wound into coil form which greatly reduces the over-all length.   Then the coil is coiled so that the over-all dimensions are small enough to fit into the bulb of the 60-watt lamp.)

12-volt Lighting.   If you need light concentrated into a very small area, use special 12-volt lamps, operated from a transformer that steps the usual 115-volt circuit voltage down to 12 volts.   These are available in the PAR shape with internal reflector, as already discussed, in sizes from 25 to 240 watts.   The available beam shapes vary from flood to very narrow spot.   The narrowest available type concentrates 50% of the available light in a beam as narrow as 4 by 6°.

This concentration is made possible by the fact that the light source, the filament, in a 12-volt lamp is very short.   The 115-volt lamp has a very long, thin filament.   The 12-volt lamp of the same wattage has a very much *shorter* filament, but very much *thicker*.   The *area* occupied by the filament is very much smaller indeed in the 12-volt lamp, and makes possible the very narrow beam already described.

Other Incandescent Lamps.   Since a large manufacturer of lamps may make well over 10,000 kinds, obviously only the more commonly used kinds can be described here.   To show the range, you can buy a "grain-of-wheat" lamp used in surgical instruments and consuming a fraction of a watt, or a large lamp consuming 10,000 watts.   You can buy a lamp for deep-sea diving, not damaged by 300 lb of water pressure per square inch on the bulb.   You can buy "black-light" lamps that produce no visible light but when their "black light" falls on properly painted surfaces, they glow in spectacular colors.   You won't find these kinds of lamps on dime-store counters.

More Information on Fluorescent Lamps.   Chapter 14 discussed only the bare fundamentals of ordinary fluorescent lamps.   This chapter will discuss many other details concerning such lamps: kinds of bases, various starting methods, efficiencies, and other characteristics.

Bases on Fluorescent Lamps.   Figure 30-3 shows the various kinds of bases used on fluorescent lamps, depending on their size and starting method.   Later paragraphs will define the particular kind used on each kind of lamp.   The original variety is the bi-pin shown at *A* of Fig. 30-

3; there are three diameters (miniature, medium, and mogul) depending on the diameter of the lamp.   At *B* is shown the recessed double-contact type (made in two diameters), and at *C* the single-contact type. The 4-pin type shown at *D* is used only on circular lamps.

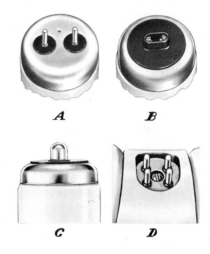

Fig. 30-3. Common types of bases used on fluorescent lamps. (*General Electric Co.*)

Types of Fluorescent Starting.   There are three types of starting in common use: preheat, instant-start, and rapid-start.   Lamps designed for one type of starting will not (with a few exceptions) fit sockets designed for a different kind.   In every case a ballast must be used, and the ballast must be carefully matched to the kind of starting and to the particular lamp under discussion, as well as the circuit voltage.

The *preheat* type of starting was the kind used when fluorescent lighting was first introduced, and is still widely used.   It was described in Chap. 14.   Lamps for preheat starting have a filament or cathode at each end of the tube.   When the lamp is first turned on, the current heats these filaments or cathodes.   Then the starter opens, current flows as an arc through the tube, limited to proper value by the ballast; this flow of current keeps the cathodes hot as long as the current flows, as long as the lamp is lighted.   Fluorescent lamps designed for preheat starting have bi-pin bases, and are available in sizes from 4 to 100 watts.

Lamps of the *instant-start* type also have a filament or cathode at each end, but each one is short-circuited inside the lamp.   Obviously then a current can't be made to flow through the filament or cathode before the lamp lights. No starter is needed, but the ballast is of a variety such that the open-circuit voltage before the lamp is turned on is from 450 to 600 volts; when the lamp is turned on, that voltage is high enough to start the lamp instantly. The amperage consumed by the lamp is such that the ballast under load drops the voltage to a figure that is proper for the lamp. After lighting, the cathode is kept hot by the arc current in the tube. Instant-start lamps are mostly of the "Slimline" type using a single-contact base. A few of the earlier type used a medium or mogul bi-pin. They are made in sizes from 40 to 75 watts.

More recent lamps are of the *rapid-start* type. (The difference in time between the instant-start and the rapid-start is very small; the former is truly instant; in the latter there is a lag of a fraction of a second.) Rapid-start lamps require no starter; the ballast must be of the proper type. The lamp has a filament or cathode at each end, not short-circuited as in the instant-start type; the cathode is kept heated while the lamp is in operation, by special 3½-volt windings in the ballast, plus the heat of the arc current in the tube. This reduces the voltage necessary to start the lamp. Starting is based on much the same principle as in the instant-start. The ballast delivers an open-circuit voltage of about 250 to 400 volts at the instant of starting and then controls the voltage to the proper value for normal operation. Rapid-start lamps do not have a uniform kind of base. Larger sizes have the recessed-double-contact type; smaller sizes the bi-pin. Rapid-start lamps are available in sizes from 30 to 220 watts, and include the so-called High-output and "Power Groove" ® 2 types.

It should be noted that there is now available a 40-watt 48-in. lamp of the "Preheat/Rapid-Start" type. It may be used in either a fixture with a ballast (and starter) designed for preheat lamps, or in a fixture with a ballast (no starter) designed for rapid-start lamps.

The first fluorescent lamps made were of the preheat variety; then the trend was to the instant-start as that type was developed. Now the trend is to the rapid-start in new installations.

A subvariety of the instant-start is the cold-cathode type. In this kind there is no filament in the lamp; instead, at each end there is a cathode which is a metal "thimble" containing electron-emitting materials, as is found on the filaments or cathodes of the usual fluorescent varieties, all of which can be called the hot-cathode type. The construction at the ends is substantially that found in the tubes of neon

2 Registered Trademark, General Electric Co.

signs.   Starting is based on the same principle as in ordinary instant-start lamps.   As the lamp is turned on, the ballast delivers a high open-circuit voltage to start the lamp, but the voltage is very high, on the order of 800 to 1,000 volts.   For that reason ordinary fixture wire, which is limited to usage at not over 300 volts, cannot be used in wiring for such lamps.   Wire designed for a higher voltage must be used between the ballast and the lamps.

Efficiency of Fluorescent Lamps.   The lumens per watt delivered by a fluorescent lamp depends on many factors and particularly on the color of the light produced by the lamp.   From Chap. 14 you will remember that there are many kinds of "white."   Tables showing the lumens produced are generally based on the cool-white color.   The nominal wattage ratings of fluorescent lamps are based on the power consumed by the lamp itself, not including the power consumed by the ballast.   The power consumed by the ballast, as compared with that consumed by the lamp, varies a great deal.   If the fixture and ballast are designed for a single lamp, the power consumed by the ballast is a much higher percentage of the total than it is in fixtures designed for two (or four) lamps.   The proportion for the smaller lamps is much bigger than for the larger lamps.   In the smallest sizes and in single-lamp fixtures, the ballast may consume a third of the total watts of the circuit.   In the largest sizes and with two lamps per ballast, it may be as little as 3 or 4%.   For the most commonly used lamps it is probably safe to use a figure of 15%.   Actually then the total watts in a circuit serving fluorescent lamps will be in the range of 12 to 17% higher than the watts indicated by the ratings of the lamps.

The light output of a brand-new fluorescent lamp drops rapidly, then stabilizes.   Efficiency tables are based on the output after 100 hr of operation; after 100 hr, the output drops slowly but continuously so that as the lamp approaches the end of its life, the output may be 20 to 30% below the 100-hr figure.   During its life the lamp delivers an average of perhaps 85% of its 100-hr rating.

The efficiency of the lamps also varies with the type and size; in general the larger sizes produce more lumens per watt than the smaller ones.   The actual efficiency varies from 45 to 80 lumens per watt; note these ratings are based on the power consumed by the lamp itself, not including the power consumed by the ballast.   As pointed out in Chap. 14, the *de luxe* cool white and warm white produce about 30% fewer lumens per watt than the other types.

Lumens per Foot of Length.   Entirely aside from the lumens *per watt*, the lumens *per foot of length* of a fluorescent lamp are another important consideration.   Most of the ordinary "garden varieties" produce from 600 to 750 lumens per foot.   Thus a fixture with four of the 40-watt 48-in. lamps will have altogether 16 ft of total lamp length, and will produce about 12,000 lumens.   To produce the high footcandle levels of lighting demanded today for efficient work in offices and other areas, many lighting fixtures must be installed; the ceiling can be literally covered with fixtures.

Why not produce lamps that will produce more lumens per foot of length?   Such lamps are available.   Lamps called "high-output" produce from 1,000 to 1,050 lumens per foot; others called "extra-high-output" or "Power Groove" produce from 1,700 to 1,900 lumens per foot.   (Note that "high output" does not refer to lumens *per watt* but rather to lumens *per foot of length.*)   Such lamps are available in 48-, 60-, 72-, and 96-in. lengths.   Using such lamps will not result in an installation consuming fewer watts for any given footcandle level of lighting, but will permit a smaller number of fixtures and lamps to be used. This leads to a definite saving in installation cost, less maintenance, and higher footcandle levels of lighting.

The output per foot of length is controlled by the amperage flowing through the lamp.   The current in the ordinary 40-watt lamp is about 0.430 amp.   In the high-output lamp it is about 0.800 amp, while in the very-high-output type it is about 1.500 amp.   The current is controlled by the proper selection of ballasts.

Fig. 30-4.   The U-shaped fluorescent lamp has advantages in some types of lighting.   (*General Electric Co.*)

U-shaped Fluorescent Lamps.   These lamps, shown in Fig. 30-4, can be used in place of the usual tubular type to provide attractive ceiling patterns and lighting effects.   The length of the tube is about 48 in., but it is bent into the U shape so that the space occupied by the lamp is only about 24 in.   Two or three of these can be installed in a single 24-

by 24-in. ceiling module. The output in lumens per watt is somewhat less than that of the ordinary 40-watt 48-in. straight lamp.

Fluorescent Type Designations. The numbering scheme used for fluorescent lamps can be quite confusing to the average user. Two basic schemes are used, one for lamps with bi-pin bases, and another for lamps with single-pin or *recessed* double-contact bases.

A typical designation for an ordinary bi-pin lamp is "F30T8." The "F" means "fluorescent"; "30" means "30 watts"; "T" means "tubular"; "8" means the tube is eight eights or 1 in. in diameter. (In the combination "Preheat/Rapid-Start" type, which is 12 eights or $1\frac{1}{2}$ in. in diameter, you would expect the designation "T12" but it does not appear, the lamp designation being merely "F40.") The type designation is often followed by additional letters indicating color or other special construction.

In lamps with single-pin or recessed double-contact bases, the number in the designation, instead of standing for the wattage of the lamp, indicates the length in inches; the wattage designation does not appear at all. Thus a "F48T12" lamp is a fluorescent 48 in. long, in a tubular bulb twelve-eights or $1\frac{1}{2}$ in. in diameter.

Effect of Voltage on Fluorescent Lamps. Ordinary incandescent lamps must be selected for the circuit voltage on which they are to be operated. For maximum efficiency a 115-volt lamp should not be operated on a 110- or 120-volt circuit. Chapter 14 outlined the very considerable effect that off-voltage has on the life and efficiency of such lamps.

Fluorescent lamps on the other hand are not rated in volts; the same lamp is used on a 115-volt circuit as on a 230-volt circuit. However, the ballast used must be carefully selected to match the voltage of the circuit on which the ballast and the lamp are to be used. Suppose the proper ballast has been selected and is used with a lamp on a 120-volt circuit. What is the effect on the lamp if the circuit voltage changes? No precise data can be given because the effect is dependent on the size of the lamp, the kind of lamp, and the particular ballast used.

On a 40-watt fluorescent lamp, however, a 10% overvoltage will increase the light output about 10% and reduce the efficiency about 5%, but affect the life of the lamp very little. However the ballast will tend to overheat, tending toward reducing its life. A 10% undervoltage will have the opposite effect, but the lamp may not start properly. In other words, fluorescent lamps are not so radically affected by incorrect voltages as are incandescent lamps.   But for the reasons cited, it is very important to match the rated voltage of the ballast with the line voltage.

Color of Light from Fluorescent Lamps.   As already mentioned in Chap. 14, there are many kinds of "white" fluorescent lamps.   The "de luxe" varieties of white are about 25% less efficient than the other kinds of white.   The cool-white lamp represents about 75% of all the fluorescent lamps sold and is quite satisfactory where color discrimination is not overly important.

In general the cool white and de luxe cool white produce light that simulates natural light; the warm white and de luxe warm white produce light that is more nearly like the light produced by incandescent lamps and emphasizes the red, orange, and brown colors.

For lighting merchandise displays in stores, the de luxe cool white probably gives the best over-all effect, more or less simulating natural daylight.   But if the merchandise on display tends toward the red, orange, and brown colors, the de luxe warm white is preferred by many.

For decorative purposes and specialized lighting, some fluorescent lamps are available in colors such as blue, green, gold, pink, and red. Ordinary incandescent lamps in color are most inefficient, for they are ordinary lamps with color on the glass that absorbs most of the light, allowing only the desired color to pass through.   Fluorescent lamps on the other hand have the proper phosphor on the inside of the tube, which creates primarily the color desired.   For that reason they are many times as efficient as colored incandescent lamps.

Tabulation of Characteristics of Fluorescent Lamps.   The table on page 531 shows the characteristics of the more common types of lamps. Necessarily this must be an abbreviated table, and those who need more information about a particular lamp or about lamps not shown can obtain it from the manufacturers of lamps.

Electric-discharge Lamps.   In any ordinary incandescent lamp, the current flows through a tungsten filament, heating it to a high temperature, often above 4000°F.   At that temperature light is produced.

In an electric-discharge lamp the current flows in an arc inside a glass or quartz tube from which the air has been removed, and a gas or mixture of gases introduced.   The most ordinary example is the fluorescent lamp already discussed.   Here the current is relatively low, from a small fraction of an ampere to a maximum of $1\frac{1}{2}$ amp.   It flows through a relatively long length, up to 96 in.   The gas pressure in the tube is relatively low, almost a vacuum.

Variety	Watts	Bulb type	Diameter, inches	Length, inches	Type of starting	Lumens ° Total	Lumens ° Per watt	Type of base
Ordinary	15	T-8	1	18	Preheat	850	57	Medium bi-pin
	30	T-8	1	36	Preheat	2,100	70	Medium bi-pin
	15	T-12	1½	18	Preheat	770	52	Medium bi-pin
	20	T-12	1½	24	Preheat	1,220	61	Medium bi-pin
	40	T-12	1½	48	Preheat/ Rapid	3,120	78	Medium bi-pin
Slimline	38	T-12	1½	48	Instant	2,900	73	Single-pin
	55	T-12	1½	72	Instant	4,400	80	Single-pin
	75	T-12	1½	96	Instant	6,200	83	Single-pin
High output	60	T-12	1½	48	Rapid	4,000	67	Recessed double contact
	85	T-12	1½	72	Rapid	6,450	76	Recessed double contact
	110	T-12	1½	96	Rapid	9,000	82	Recessed double contact
Extra-high output	110	PG-17	2⅛	48	Rapid	6,900	63	Recessed double contact
	165	PG-17	2⅛	72	Rapid	10,900	66	Recessed double contact
	220	PG-17	2⅛	96	Rapid	15,000	68	Recessed double contact

° Total lumens and lumens per watt are for cool-white lamps after 100 hr of use.   The average during the useful life of the lamps will be about 15% less.   The figures are based on watts consumed by the lamps, not including power consumed by the ballasts.

There are other kinds of electric-discharge lamps in which the current flow is as high as 10 amp, but it flows through a very short arc tube, just a few inches long.   The gas pressure in the tube is very much higher than in a fluorescent lamp.   Such lamps are called High Intensity Discharge (HID) lamps.   There are three different kinds in common use, and they will be described separately in later paragraphs.

In all HID lamps, the arc tube operates at a very high temperature; indeed the high temperature is necessary to the proper operation of the lamp.   Air currents must not be allowed to affect the temperature; so the arc tube, which is quite small, is enclosed in a much larger glass bulb which determines the over-all dimensions of the lamp.   None of these lamps can be connected directly to an electrical circuit, but they must be provided with a ballast to match the type and size of lamp involved, and the voltage of the circuit.   When first turned on, the lamp does not light instantly; it may require up to 10 min for the lamp to reach full brilliancy.   If turned off, it must be allowed to cool off for some minutes before it can be relighted.

This fact must be considered if HID lamps are planned for locations where power failures are frequent, or where violent voltage fluctuations occur, for considerable time is lost in relighting the lamps, once they have gone out.   For this reason, often a few incandescent lamps are installed as "insurance" in areas that are otherwise lighted only by HID lamps.   Even if a power interruption is very short, the HID lamps go out.   When the power is restored a few seconds or few minutes later, the incandescent lamps provide some degree of illumination instead of total darkness, during the time it takes for the HID lamps to restart.   In this way the incandescent lamps tend to prevent accidents, or panic, during the time that total darkness would otherwise prevail.

While all HID lamps have long life, that life depends on the number of times the lamp is turned on.   Hours of life shown are based on burning the lamp at least 5 hr every time it is turned on.   If burned continuously, the life is much longer.   Note too that the number of lumens produced diminishes rather quickly at first, before leveling off.   The number of lumens shown for each lamp is the figure after burning 100 hr, as in the case of fluorescent lamps; the lumens per watt are based on the watts consumed by the lamps not including the ballasts.   In the larger sizes the ballast watts are very roughly 10% of the lamp watts.

Advantages of HID lamps include, among others, high efficiency (lumens per watt), very long life, high wattages (and high number of lumens) from single fixtures, thus reducing installation and maintenance costs.   They are not generally used for ordinary lighting in residential and office work, but rather in factories, service stations, gymnasiums, parking lots, street lighting, floodlighting, and, in general, locations where large areas are to be lighted.

HID lamps are made in many types, three of which are in common use: (1) mercury vapor, (2) metallic-additive type, and (3) high-pressure sodium type.   All operate on the principles already outlined, and the details of each will be discussed separately.

Mercury-vapor Lamps.   These lamps, usually called just mercury lamps, were introduced in more or less their present form in 1934. Since that time they have been vastly improved.   Their life is probably ten times that of the original lamps; output in lumens per watt has been greatly increased; their present cost is a fraction of their original price. Mercury lamps have a short arc tube with some argon gas in it, and some mercury which is a liquid.   When the lamp is turned on, an arc develops through the argon gas, in a short *part* of the tube.   This slowly heats the mercury so that it gradually vaporizes, and then the arc develops through the entire length of the arc tube.   When all the mercury is vaporized, the lamp operates at full brilliancy.   This starting procedure takes about 10 minutes.   Every lamp requires a ballast, carefully matched to the lamp being used, and the voltage of the circuit.

Mercury lamps are available in sizes from 50 to 3,000 watts; the 175- and 400-watt sizes are the most common.   Figure 30-5 at C shows the 400-watt size, which has an over-all length of about 11¼ in.   The size of the bulb varies a great deal with the wattage of the lamp.   Figure 30-6 shows several shapes, *not* in proportion to the actual sizes.   Many of the sizes are available in the "R" (reflector) type.

The life of mercury lamps is extremely long.   In sizes above 100 watts it is at least 24,000 hr (almost 3 years of continuous burning); for the smaller sizes it is about 10,000 hr.   Their efficiency in lumens per watt is much greater than that of incandescent lamps, very roughly about the same as that of larger fluorescent lamps.   However, the efficiency drops off faster than in other lamps.   In the 400-watt size, at the end of 6,000 hr the output is from 85 to 95% of the original value; after 10,000 hr it is 70 to 90%; after 18,000 hr it is 60 to 80%.   Therefore in planning an installation using mercury lamps, the number of lamps

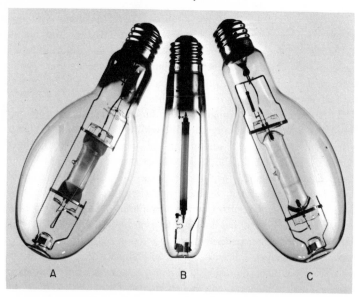

Fig. 30-5. Three types of HID lamps: metallic-additive type, high-pressure sodium halide, and mercury. (*General Electric Co.*)

planned should be based not on their initial output, but rather on the average during the time they are retained in the system, or on a minimum level of illumination.

The comparatively high light output coupled with relatively small size and especially their long life make them specially suitable where it

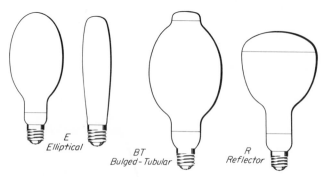

Fig. 30-6. Mercury lamps are made with many shapes of bulb. (*General Electric Co.*)

would be difficult and expensive to install fluorescent lighting providing an equivalent amount of light.   Since it is usually very expensive to replace a lamp, the long life of the mercury type suggests their use in such locations.

Not too many years ago, the only type of mercury lamp available produced light that had very little red and orange in it, the result being a greenish-bluish light that was acceptable for many purposes, but totally unsuitable where color was a factor.   People had a corpselike appearance under this light; blue and green colors were emphasized and strengthened, orange and red were distorted to appear brown or black. The only way to correct this shortcoming was to use a combination of mercury lamps (deficient in red) and incandescent lamps (strong in red).   This original type of lamp is still in use alone, where color is not too important, for example in street lighting.

Now you must remember that in a mercury lamp with a clear glass bulb, you see the peculiar color just described, but the arc also produces a good deal of *invisible* ultraviolet light, which is stopped by the outer glass bulb.   You will also remember from earlier discussions concerning fluorescent lamps, that invisible ultraviolet light striking phosphor powders on the inside of the fluorescent tube, makes the phosphor glow to produce visible light.   Mercury lamps are now made with a phosphor coating on the inside of the glass bulb.   This phosphor coating permits most of the greenish-bluish light to pass through it, but the phosphor also creates a good deal of orange-red light in addition, so that the mixture of the two kinds of light becomes more nearly the kind of light we are accustomed to.   There are two kinds of such mercury lamps.   One is called merely "color-corrected" or "color-improved" and is acceptable where color discrimination is not too essential.   The other is called de luxe white, and the color is very much like that of cool-white fluorescent lamps, quite acceptable for most purposes.   There is not much difference in lumens per watt, in the three types.

Metallic-additive Lamps.   This is a new type first introduced in 1964. They are known by various trade names (depending on the manufacturer), such as "Multi-vapor"®3, "Metalarc", or "BOC."   In appearance they are quite similar to ordinary mercury lamps; Fig. 30-5 at *A* shows the 400-watt size.   The principle of operation is substantially that of the mercury lamp, except that the arc tube, in addition to argon

3 Registered trademark, General Electric Co.

gas and mercury, also contains three other ingredients: sodium iodide, thallium iodide, and indium iodide. This leads to a very high efficiency of about 80 lumens per watt, and that without phosphor on the inside of the bulb. The color of the light is generally quite acceptable, approaching that of cool-white fluorescent lamps. Suitable ballasts are required, and the starting time is about 10 minutes.

Metallic-additive lamps are presently available only in two sizes: 400- and 1,000-watt. Their life is much less than that of mercury lamps: 7,500 hr for the 400-watt, and 6,000 hr for the 1,000 watt. At the end of about two-thirds of their life, the lumen output drops to about 60 to 70% of their initial output. Their cost is several times that of mercury lamps.

In spite of these apparent disadvantages as compared with the mercury type, they have two important advantages. Their output in lumens per watt is about 60% greater than that of mercury lamps, thus reducing the cost of the light, and the number of fixtures required. The other advantage applies primarily if the light is to be directed into a relatively small area with reflectors. The mercury type in the color-corrected or de luxe white style presents a large area of light source to the reflector, so that narrow beams are not possible. The metallic-additive type has a clear glass bulb, and the light source has an area only a small fraction as large as the area of the outer bulb of the mercury type, thus making relatively narrow beams quite practical.

High-pressure Sodium Halide Lamps. This is the newest of the HID lamps, having been introduced only in 1965. Again there are several trade names, depending on the manufacturer, such as "Lucalox" ® 4, or "Ceramalux." The arc tube operates at an exceedingly high temperature, around 1300°C; so it is made not of glass or quartz, but of a very special ceramic material that becomes quite translucent at the temperature involved. The arc tube, in addition to xenon gas and mercury, also contains sodium. While lamps containing sodium normally produce a very yellowish light, the combination of the particular ingredients in the arc tube of the high-pressure sodium halide lamps produces light that is quite rich in orange and red, much like that of warm white fluorescents, or the smaller sizes of ordinary incandescent lamps.

The principle of starting is quite different from that of other HID

[4] Registered trademark, General Electric Co.

lamps; special ballasts are required.   It lights to full brilliancy in less time than other HID lamps.   It is at present available only in two sizes: 275- and 400-watt.   The 400-watt size is shown at *B* in Fig. 30-5; it is 9¾ in. long.   Its light output is the highest of all known electric-light sources: over 100 lumens per watt, almost double that of fluorescent or mercury lamps, about five times that of ordinary 500-watt incandescent lamps.   The life of the lamp is about 8,000 hr.   Its output in lumens is maintained at a very high level, at about 90% after 6,000 hr.

While this is an expensive lamp (about four times as much as a mercury lamp of the same size) its very high output in lumens per watt makes the cost of the electric power for light very low.   The total lumens from a lamp are so high that a smaller number of fixtures or reflectors is often possible than with less efficient lamps.   The small size of the arc tube permits narrow beams to be projected, making the lamp very suitable for spotlighting or floodlighting.   It is often used in buildings such as factories with high ceilings, also in locations such as parking lots, athletic fields, outsides of buildings, and similar applications. As is usual with any new type of lamp, undoubtedly many improvements will take place over a period of years, making the lamp suitable for many more applications than now exist.

Comparison, Three HID Types: The characteristics of the three types of HID lamps are summed up in the table that follows.   All data are based on the 400-watt size of each type.

The metallic-additive and the high-pressure sodium halide types being such new lamps, it is reasonable to expect that improvements in

	Mercury	Metallic-additive	High-pressure sodium halide
Life, hr	24,000	7,500	8,000
Lumens per watt	50	80	105
Percent of initial lumens after ——hours	10,000 hr 70–90%	4,000 hr 60–70%	6,000 hr 90%
Color of light	Good °	Good, green and yellow emphasized	Good, orange and red emphasized
Cost of lamp	Lowest	Medium	Highest

° In color-corrected or de luxe white type.

all characteristics will appear, as experience in use and manufacture accumulates.   It seems likely too that the cost of the lamps will also gradually be reduced.

Luminaires.   Everybody knows what a lighting fixture is.   In engineering jargon and in many technical publications, fixtures are called *luminaires.*   In this book they will be called simply *fixtures.*

Good Lighting.   Chapter 14 outlined the fundamental fact that one lumen of light falling on one square foot of area always produces one footcandle of illumination.   From this statement it is easy to jump to some very wrong conclusions.   For example, if a room with 100 sq ft of floor area is lighted by a lamp producing 5,000 lumens, obviously there are 50 lumens for every square foot of floor area, but the illumination will be very much less than 50 footcandles.   That is because only a part of the light that is generated reaches the place where it is to be used.

Based on the preceding paragraph, you can see that theoretically all you need to do to determine the number of watts required per square foot, for any number of footcandles, is to use the simple formula

$$\frac{\text{Number of footcandles wanted}}{\text{Lumens per watt of lamps used}} = \text{watts per square foot needed}$$

But that is a very theoretical figure, and in practice you will have to provide from three to five times as many watts as the answer in the formula above indicates.   If you multiply the answer by four, you will probably be "in the ball park" and not hopelessly wrong.

The Code in Sec. 220-4(a) requires a minimum of anywhere from ¼ to 5 watts per sq ft for lighting in various occupancies.   It would be very useful to be able to translate "watts per square foot" directly into a predictable level of footcandles, but unfortunately it can't be done. Of all the light produced by a lamp, part is absorbed by the fixture, part by the room surfaces.   The part that reaches the surfaces to be lighted depends on fixture design, the size and shape of the room, and the reflectances of ceiling, wall, and floor areas.   There is no dependable simple method that will give you the level of illuminations in a room, knowing only the light sources.   Watts per square foot has been used as the basis for rule-of-thumb estimates in the past, but accurate predictions using this method depend on years of experience.

For these reasons, it is unfortunately necessary to go through somewhat lengthy but not overly difficult calculations to reach the answer, and the procedure will be outlined in following paragraphs.

Coefficient of Utilization. The ratio of the number of lumens of light reaching the area where it is wanted, to the total lumens generated, is called the "coefficient of utilization" or "utilization factor." It may be as little as 15% or even less, and rarely exceeds 85%. Important factors are the particular type of fixture (direct or indirect lighting), the color and cleanliness of walls and ceilings, the size and shape of the room. These will be discussed separately, not necessarily in the order of their importance, for what is the most important factor in one installation may not be so important in the next.

Room Ratio. In any room, we want the light on our desks or on our machinery. Black walls would be very dreary, so light-colored walls are used. Of the light that falls on the walls, some is reflected into the room, but some is absorbed and lost. Naturally we want to lose as little as possible, so let us consider the influence of walls. A room 10×10 ft has a floor area of 100 sq ft. If the ceiling height is also 10 ft, the wall area is 400 sq ft, or *four* square feet of wall per square foot of floor. If the room is 20×20 ft, the floor area is 400 sq ft, and if the ceiling height remains 10 ft, the total wall area is 800 sq ft, or *two* square feet of wall per square foot of floor. The ratio is twice as good as in the case of the smaller room. Stating it another way, the larger room has less light-absorbing *wall* area per square foot of *floor* area than the smaller room. Therefore, as the size of a room increases, more and more of the light that is produced is usefully used, and less and less of it is wasted by absorption on the walls. Likewise, the higher the mounting height of the lighting fixtures, the more the light produced will fall on the walls, and a greater portion of it will be lost. Therefore, in general, low mounting heights produce the greater efficiency if all other factors are properly considered.

The room ratio (or room index as it was once called) is numerically expressed by numbers ranging from approximately 5.0 to 0.50; the higher the number, the better the factor. As a matter of fact, the ratio is based on a simple formula $(W \times L) / [H \times (W + L)]$ in which W is the width of the room in feet, L is the length, and H is the height of the fixture above the floor (except when, in the case of the particular fixtures to be used, less than 40% of the light output goes downward, in which case H becomes the ceiling height).

An abbreviated table showing room ratios is given in Fig. 30-7.   You will note that the ratio improves with the size of the room, but deteriorates as the mounting height (or ceiling height) increases.   More complete information can be found in technical literature available from manufacturers of lighting equipment.

Reflectance of Ceilings.   If no light is reflected from the ceiling, in other words if good reflectors direct *all* the light downward, the reflectance of the ceiling is of no importance.   Whether it is a clean white or a dirty black makes little difference.

If, however, a totally indirect lighting system is used, in which all the light is directed to the ceiling from which it is reflected downward, the reflectance of the ceiling is all-important.   A dirty ceiling regardless of color will absorb a large part of the light thrown on it.   The ceiling must have good reflectance and so must be of an appropriate color (see Chap. 14), and it must be kept clean.

The color and cleanliness of the ceiling are important depending on how much of the light that falls on it is to be reflected downward. However, it is little affected by room ratio.

Reflectance of Walls.   What has been said about the ceiling also holds true for the walls, but to a lesser extent.   The higher the room ratio, the less important the reflectance of the walls (for the ratio of wall area to floor area is a large factor in determining the room ratio in the first place).

Type of Fixture Used.   Regardless of the type of fixture used, a goodly share of the light produced is lost right in the fixture, being absorbed by glassware or plastic substitutes, and by reflecting surfaces.   Losses range from 20 to 40%.   This is mentioned merely as a factor in helping you understand why most installations have what appear to be relatively low coefficients of utilization.

The type of fixture will be determined by the kind of lighting wanted. Direct-lighting fixtures direct 100% of the useful light downward; indirect-lighting fixtures direct 100% of the useful light upward.   In between are endless variations.   The type of fixtures selected will greatly affect the over-all efficiency or coefficient of utilization of the installation.   The type of room, its purpose (foundry or style-show) will usually determine the type of lighting that is suitable and proper.

Maintenance Factor.   A lighting installation using brand-new fixtures, new lamps, and freshly painted walls, and providing, for example, 50 footcandles when installed, will produce less and less light as time

DIMENSIONS OF ROOM		Luminaire Mounting Height for Systems with 40% or More of Fixture Output Downward										
		8	9	10	11	12	13	15	17	19	23	27
Width (feet)	Length (feet)	Ceiling Height for Systems with Less than 40% of Fixture Output Downward										
		10½	12	13½	15	16½	18	21	24	27	33	39
8	10	0.8	0.7	0.6	0.5	0.5						
	14	0.9	0.8	0.7	0.6	0.5	0.5					
	18	1.0	0.9	0.7	0.7	0.6	0.5					
	25	1.1	0.9	0.8	0.7	0.6	0.5	0.5				
	30	1.2	1.0	0.8	0.7	0.7	0.6	0.5				
	40	1.2	1.0	0.9	0.8	0.7	0.6	0.5	0.5			
10	10	0.9	0.8	0.7	0.6	0.5	0.5					
	14	1.1	0.9	0.8	0.7	0.6	0.6	0.5				
	18	1.2	1.0	0.9	0.8	0.7	0.6	0.5				
	25	1.3	1.1	1.0	0.8	0.8	0.7	0.6	0.5			
	30	1.4	1.2	1.0	0.9	0.8	0.7	0.6	0.5	0.5		
	40	1.5	1.2	1.1	0.9	0.8	0.8	0.6	0.6	0.5		
12	12	1.1	0.9	0.8	0.7	0.6	0.6	0.5				
	16	1.3	1.1	0.9	0.8	0.7	0.7	0.6				
	20	1.4	1.2	1.0	0.9	0.8	0.7	0.6	0.5	0.5		
	30	1.6	1.3	1.1	1.0	0.9	0.8	0.7	0.6	0.5		
	50	1.8	1.5	1.3	1.1	1.0	0.9	0.8	0.7	0.6	0.5	
14	14	1.3	1.1	0.9	0.8	0.7	0.7	0.6	0.5			
	20	1.5	1.3	1.1	1.0	0.9	0.8	0.7	0.6	0.5		
	30	1.7	1.5	1.3	1.1	1.0	0.9	0.8	0.7	0.6		
	40	1.9	1.6	1.4	1.2	1.1	1.0	0.8	0.7	0.6	0.5	
	60	2.1	1.8	1.5	1.3	1.2	1.1	0.9	0.8	0.7	0.6	0.5
	80	2.2	1.8	1.6	1.4	1.3	1.1	1.0	0.8	0.7	0.6	0.5
16	16	1.5	1.2	1.1	0.9	0.8	0.8	0.6	0.6	0.5		
	20	1.6	1.4	1.2	1.0	0.9	0.9	0.7	0.6	0.5		
	30	1.9	1.6	1.4	1.2	1.1	1.0	0.8	0.7	0.6	0.5	
	40	2.1	1.8	1.5	1.3	1.2	1.1	0.9	0.8	0.7	0.6	0.5
	60	2.3	1.9	1.7	1.5	1.3	1.2	1.0	0.9	0.8	0.6	0.5
	80	2.4	2.1	1.8	1.6	1.4	1.3	1.1	0.9	0.8	0.7	0.5
18	20	1.7	1.5	1.3	1.1	1.0	0.9	0.8	0.7	0.6		
	30	2.1	1.7	1.5	1.3	1.2	1.1	0.9	0.8	0.7	0.6	
	40	2.3	1.9	1.7	1.5	1.3	1.2	0.9	0.8	0.7	0.6	0.5
	60	2.5	2.1	1.9	1.6	1.5	1.3	1.1	1.0	0.8	0.7	0.6
	80	2.7	2.3	2.0	1.7	1.6	1.4	1.2	1.0	0.9	0.7	0.6
	100	2.8	2.4	2.0	1.8	1.6	1.5	1.2	1.1	0.9	0.7	0.6
20	20	1.8	1.5	1.3	1.2	1.1	1.0	0.8	0.7	0.6	0.5	
	30	2.2	1.8	1.6	1.4	1.3	1.1	1.0	0.8	0.7	0.6	0.5
	40	2.4	2.1	1.8	1.5	1.4	1.3	1.1	0.9	0.8	0.7	0.6
	60	2.7	2.3	2.0	1.8	1.6	1.4	1.2	1.0	0.9	0.7	0.6
	80	2.9	2.5	2.1	1.9	1.7	1.5	1.3	1.1	1.0	0.8	0.7
	100	3.0	2.6	2.2	2.0	1.8	1.6	1.3	1.2	1.0	0.8	0.7
	120	3.1	2.6	2.3	2.0	1.8	1.6	1.4	1.2	1.0	0.8	0.7
25	30	2.5	2.1	1.8	1.6	1.4	1.3	1.1	0.9	0.8	0.7	0.6
	40	2.8	2.4	2.1	1.8	1.6	1.5	1.2	1.1	0.9	0.8	0.6
	60	3.2	2.7	2.4	2.1	1.9	1.7	1.4	1.2	1.1	0.9	0.7
	80	3.5	2.9	2.5	2.2	2.0	1.8	1.5	1.3	1.2	0.9	0.8
	100	3.6	3.1	2.7	2.4	2.1	1.9	1.6	1.4	1.2	1.0	0.8
	120	3.8	3.2	2.8	2.4	2.2	2.0	1.7	1.4	1.3	1.0	0.8
30	30	2.7	2.3	2.0	1.8	1.6	1.4	1.2	1.0	0.9	0.7	0.6
	40	3.1	2.6	2.3	2.0	1.8	1.6	1.4	1.2	1.0	0.8	0.7
	60	3.6	3.1	2.7	2.4	2.1	1.9	1.6	1.4	1.2	1.0	0.8
	80	4.0	3.4	2.9	2.6	2.3	2.1	1.7	1.5	1.3	1.1	0.9
	100	4.2	3.6	3.1	2.7	2.4	2.2	1.9	1.6	1.4	1.1	0.9
	120	4.4	3.7	3.2	2.8	2.5	2.3	1.9	1.7	1.5	1.2	1.0
	140	4.5	3.8	3.3	2.9	2.6	2.4	2.0	1.7	1.5	1.2	1.0
35	40	3.4	2.9	2.5	2.2	2.0	1.8	1.5	1.3	1.1	0.9	0.8
	60	4.0	3.4	3.0	2.6	2.3	2.1	1.8	1.5	1.3	1.1	0.9
	80	4.4	3.8	3.3	2.9	2.6	2.3	1.9	1.7	1.5	1.2	1.0
	100	4.7	4.0	3.5	3.1	2.7	2.5	2.1	1.8	1.6	1.3	1.1
	120	4.9	4.2	3.6	3.2	2.9	2.6	2.2	1.9	1.6	1.3	1.1
	140	5+	4.3	3.7	3.3	3.0	2.7	2.2	2.0	1.7	1.4	1.1
40	40	3.6	3.1	2.7	2.4	2.1	1.9	1.6	1.4	1.2	1.0	0.8
	60	4.4	3.7	3.2	2.8	2.5	2.3	1.9	1.7	1.5	1.2	1.0
	80	4.9	4.1	3.6	3.1	2.8	2.5	2.1	1.8	1.6	1.3	1.1
	100	5+	4.4	3.8	3.4	3.0	2.7	2.3	2.0	1.7	1.4	1.2
	120	5+	4.6	4.0	3.5	3.2	2.9	2.4	2.1	1.8	1.5	1.2
	140	5+	4.8	4.1	3.7	3.3	3.0	2.5	2.1	1.9	1.5	1.3
50	50	4.5	3.9	3.3	2.9	2.6	2.4	2.0	1.7	1.5	1.2	1.0
	70	5+	4.5	3.9	3.4	3.1	2.8	2.3	2.0	1.8	1.4	1.2
	100		5+	4.5	3.9	3.5	3.2	2.7	2.3	2.0	1.6	1.4
	140		5+	4.9	4.3	3.9	3.5	2.9	2.5	2.2	1.8	1.5
	170			5+	4.5	4.1	3.7	3.1	2.7	2.3	1.9	1.6
	200			5+	4.7	4.2	3.8	3.2	2.8	2.4	2.0	1.6
60	60	5+	4.6	4.0	3.5	3.2	2.9	2.4	2.1	1.8	1.5	1.2
	80		5+	4.6	4.0	3.6	3.3	2.7	2.4	2.1	1.7	1.4
	100		5+	5.0	4.4	4.0	3.6	3.0	2.6	2.3	1.8	1.5
	140			5+	4.9	4.4	4.0	3.4	2.9	2.5	2.1	1.7
	170				5+	4.7	4.2	3.6	3.1	2.7	2.2	1.8
	200				5+	4.9	4.4	3.7	3.2	2.8	2.3	1.9

Fig. 30-7. A partial table of room ratios. (*General Electric Co.*)

goes on.   This is due to quite a variety of causes.   First of all, all lamps produce fewer lumens per watt after a period of use, as compared with the lumens per watt produced when new.   Fixtures gather dust which absorbs light.   Walls and ceilings darken, which means that they absorb more light.   For these reasons, every lighting job must be calculated for a higher level of illumination than is expected during average use.   A maintenance factor of 70% is probably average; in other words, an installation which will provide 50 footcandles in a brand-new building will later provide more nearly 35 footcandles on the average.   Therefore if 50 footcandles are wanted during normal usage, figure the job for about 70 footcandles.

Coefficient of Utilization.   Complete and elaborate tables can be obtained from manufacturers of lighting equipment, showing the coefficient of utilization that may be expected in any given room, using any one of dozens of types of fixtures.   A very abbreviated table is shown in Figs. 30-8 and 30-9, showing what coefficient to expect from six different types of equipment.   The column headed "Typical Distribution" shows the percentage of light falling on the ceiling, and going downward, for each type of fixture.   The percentages at the heads of the various columns show the reflectances of floor, ceiling, and walls.

You will note that the coefficients for the particular fixtures shown vary from 17 to 88%.   What is the significance of this coefficient? When you know this factor, you can reasonably accurately predict how many watts will be needed to produce a given level of illumination, using fixtures of the type you used in arriving at the coefficient.

Assume that you have determined that, for a particular room, the coefficient will be 45%.   Also assume that the maintenance factor is 70%.   Since 70% of 45% is 31½%, use the nearest even number of 30%.   Now be arbitrary and say that 300-watt incandescent lamps will be used for lighting.   The table on page 225 shows that this lamp produces 6,240 lumens.   But you can utilize only 30% of that total, and 30% of 6,240 is 1,872 lumens.   If you want 50 footcandles, 1,872/50 gives you an answer of about 37; therefore you must provide one 300-watt lamp for every 37 sq ft of floor area, or about 9 watts per sq ft. Such an installation will provide more nearly 70 footcandles when new, but only 50 footcandles after a period of time.

An ordinary 40-watt fluorescent lamp produces about 3,100 lumens; 30% of 3,100 gives us 930 lumens.   Since 930/50 gives us about 18, we must provide one 40-watt fluorescent lamp for every 18 sq ft of floor area

10% REFLECTANCE

Typical Distribution	Room Ratio	Ceiling 70%		Ceiling 50%		Ceiling 30%	
		Walls 50%	30%	50%	30%	30%	10%
1 General service filament lamp in deep-bowl porcelain enameled reflector with ventilated husk for good maintenance. Maintenance Factors: Good — 0.80, Med. — 0.70, Poor — 0.60 Maximum Fixture Spacing = Mounting Height × 1.1 0↑ / 100↓	0.6	.36	.29	.35	.28	.28	.24
	0.8	.44	.37	.43	.37	.37	.32
	1.0	.51	.44	.50	.44	.43	.39
	1.25	.58	.51	.56	.50	.50	.46
	1.5	.62	.56	.61	.55	.55	.51
	2.0	.68	.63	.67	.62	.61	.58
	2.5	.72	.67	.70	.66	.65	.62
	3.0	.75	.71	.73	.70	.68	.66
	4.0	.79	.76	.77	.74	.73	.71
	5.0	.82	.79	.80	.77	.76	.74
11 Recessed downlight for general service filament lamp. Lens and aluminum reflector provide medium-spread distribution. Maintenance Factors: Good — 0.75, Med. — 0.65, Poor — 0.55 Maximum Fixture Spacing = Mounting Height × 0.8 0↑ / 100↓	0.6	.40	.35	.39	.35	.39	.35
	0.8	.46	.42	.46	.42	.45	.41
	1.0	.50	.46	.50	.46	.49	.46
	1.25	.54	.51	.54	.50	.53	.50
	1.5	.57	.54	.56	.53	.55	.53
	2.0	.60	.58	.60	.57	.59	.56
	2.5	.62	.60	.62	.60	.60	.59
	3.0	.64	.62	.63	.61	.62	.60
	4.0	.65	.63	.65	.63	.63	.62
	5.0	.66	.65	.66	.64	.64	.63
15 R-40 flood lamp in downlight with spun aluminum brightness-controlling reflector. Maintenance Factors: Good — 0.80, Med. — 0.75, Poor — 0.70 Maximum Fixture Spacing = Mounting Height × 0.6 0↑ / 100↓	0.6	.64	.59	.63	.59	.63	.59
	0.8	.70	.66	.70	.66	.69	.66
	1.0	.74	.70	.74	.70	.73	.70
	1.25	.78	.74	.77	.74	.76	.74
	1.5	.80	.77	.79	.76	.78	.75
	2.0	.83	.80	.82	.80	.81	.79
	2.5	.85	.82	.84	.82	.83	.81
	3.0	.86	.84	.85	.83	.84	.82
	4.0	.87	.86	.87	.85	.85	.84
	5.0	.88	.87	.88	.86	.86	.85

Fig. 30-8. A partial table of coefficients of utilization. (*General Electric Co.*)

Coefficient of Utilization — 10% REFLECTANCE (Floor)

Fixture	Typical Distribution	Description	Ceiling → 80%	80%	70%	70%	50%	50%
		Walls → / Room Ratio	50%	30%	50%	30%	50%	30%
9 Two-lamp louvered industrial type fixture for Power Groove lamps. Reflector sides and louvers of white enameled steel provide 30° crosswise, 35° lengthwise shielding. Open top relieves brightness contrasts and improves maintenance. MF's for similar slimline and high-output units will be about 5 points higher. Maintenance Factors: Good — 0.70, Med. — 0.63, Poor — 0.55. Maximum Fixture Spacing= Mounting Height × 1.0	$\frac{30\uparrow}{70\downarrow}$	0.6	.29	.24	.28	.24	.23	.20
		0.8	.37	.32	.36	.31	.30	.29
		1.0	.44	.39	.41	.38	.36	.33
		1.25	.50	.45	.47	.43	.41	.39
		1.5	.54	.50	.51	.47	.45	.42
		2.0	.61	.56	.57	.52	.50	.48
		2.5	.64	.60	.60	.56	.53	.51
		3.0	.67	.63	.63	.59	.55	.54
		4.0	.70	.67	.65	.63	.59	.57
		5.0	.73	.70	.68	.65	.61	.60
24 Typical two-lamp, direct-indirect unit for 40-watt, slimline, or high-output lamps. Metal louvers and metal or dense plastic sides provide approximately 30° x 30° shielding. Maintenance Factors: Good — 0.75, Med. — 0.70, Poor — 0.65. Maximum Fixture Spacing= Mounting Height × 1.2	$\frac{45\uparrow}{55\downarrow}$	0.6	.24	.18	.23	.18	.21	.17
		0.8	.31	.26	.30	.25	.28	.24
		1.0	.37	.32	.36	.31	.33	.29
		1.25	.43	.37	.42	.36	.37	.33
		1.5	.48	.42	.45	.40	.41	.36
		2.0	.55	.49	.52	.46	.45	.42
		2.5	.59	.54	.55	.51	.49	.45
		3.0	.62	.57	.59	.54	.51	.48
		4.0	.66	.63	.62	.59	.53	.51
		5.0	.69	.65	.64	.61	.55	.53
28 Wide, shallow surface-mounted fixture for four 40-watt or slimline lamps with translucent side panels and metal louvers providing 30° crosswise, 45° lengthwise shielding. Maintenance Factors: Good — 0.75, Med. — 0.70, Poor — 0.65. Maximum Fixture Spacing= Mounting Height × 1.1	$\frac{10\uparrow}{90\downarrow}$	0.6	.26	.22	.26	.21	.25	.21
		0.8	.33	.28	.32	.27	.31	.27
		1.0	.38	.33	.37	.31	.35	.31
		1.25	.42	.38	.42	.37	.40	.36
		1.5	.46	.41	.45	.41	.43	.39
		2.0	.50	.46	.49	.46	.47	.44
		2.5	.53	.49	.52	.48	.49	.47
		3.0	.55	.52	.54	.51	.51	.49
		4.0	.58	.55	.56	.54	.54	.52
		5.0	.60	.57	.58	.56	.56	.54

Fig. 30-9. Another partial table of coefficients of utilization. (*General Electric Co.*)

in this particular room, in order to provide a maintained 50-footcandle level of illumination.

Note well that if you use this formula for any specific footcandle level, and the installation is new (all fixtures clean and new, all lamps new, walls and ceilings new or newly painted) you will obtain far *more* footcandles than the number you used in the formula.   But after the lamps become older their output in lumens diminishes, reflectors and walls and ceilings become dusty or dirty, and you will then obtain the footcandle level that was your original goal.

Formula for Determining Total Lumens Required.   From the known factors it is fairly easy to calculate the lumens required for any particular lighting problem.   The formula is

$$\text{Lumens} = \frac{\text{footcandles} \times \text{square feet}}{\text{coefficient of utilization} \times \text{maintenance factor}}$$

Assume that an office 20 ft square, floor area 400 sq ft, is to be lighted to 50 footcandles.

Assume that fluorescent fixtures each with two ordinary 40-watt lamps are to be used.   The ceiling will probably be relatively low, so the fixtures will be mounted directly on the ceiling.   The walls and ceiling will probably be a light color with good reflectance.   The combination of all these things suggests a coefficient utilization of 50%.   A maintenance factor of 70% may be assumed.   The formula above then becomes

$$\text{Lumens} = \frac{50 \times 400}{0.50 \times 0.70} = \frac{20,000}{0.35} = 57,000 \text{ (approx)}$$

A 40-watt fluorescent lamp produces about 3,100 lumens, so you will need 57,000/3,100, or 18 lamps, making a total of 9 two-lamp fixtures.

Using the same formula, it is quite simple to estimate the total lumens required for any given area, with any type of lighting fixture, for any number of footcandles desired.

Zonal Cavity Method.   The method just outlined for determining total lumens required for a lighting installation should perhaps be called the "old" method.   A newer method called the "zonal cavity method" is coming into use, and produces more precise answers than the old method.   However, since nothing is ever free, the more accurate answers are to be had only at the cost of more effort and lengthier calculations.   In using the zonal cavity method, instead of considering only the ceiling height, you must consider separately: (1) height of the ceiling cavity, or distance from fixture to ceiling; (2) height of floor cavity, or distance from floor to work plane, which in an office would be the top of the desks; and (3) the room cavity, or distance from the work plane to the fixtures. The sum of these three of course is the ceiling height. Reflectances of ceiling and floor cavities must be considered separately.

While the new method provides greater accuracy, its use is more complicated than the old, and a full explanation would require far more space than here available. Be aware that there is a newer method, but use the old knowing it will produce reasonably accurate results—answers that have been considered quite acceptable in the past.

Comparative Lumen Output of Various Kinds of Lamps. The output of each kind of lamp in lumens per watt has already been outlined in other parts of this chapter in considerable detail. Remember that for fluorescent and HID lamps the figures given were for lamps after 100 hr of use and were based on the watts of the lamps only, not including the power consumed by ballasts or other auxiliary equipment. A brief comparison of the various types follows:

Incandescent (filament type)	14	to	23 lumens
Fluorescent	58	to	75 lumens
Mercury	50	to	57 lumens
Metallic-additive	75	to	80 lumens
High-pressure sodium halide	100	to	110 lumens

Future Lighting. When you enter a room lighted to 100 footcandles, it will appear well lighted; if you move from there into another room lighted to 500 footcandles, that room will appear very brightly lighted. But if you move from outdoor sunlight into a room lighted to 500 footcandles, this room will *not* appear very bright. As a matter of fact few people (even those who are in the lighting business) moving from sunlight into such a 500-footcandle lighted area will estimate the lighting at more than 200 footcandles.

Since people in general are quite accustomed to outdoor light which ranges from 500 to 10,000 footcandles, it would seem only normal that the trend toward higher and higher footcandle levels of lighting will continue. Who knows what level will be demanded 10 or 20 years from now? But as the demand for a higher level increases, how is the additional light to be provided? There is a limit to the number of lighting fixtures that can be installed in a given area. The answer can only be new light-producing sources. Lamp manufacturers are, of course, working on the problem and have many ideas under development, which will bear fruit over a period of years.

Selection of Lighting System. In choosing the type of lighting fixtures for any job, you will meet an endless variety of fixtures for the purpose. At one extreme is the direct-lighting variety in which all the light is directed downward, with no intention of letting any of it fall on the ceiling. At the other extreme is the indirect-lighting variety in which all the light is directed upward against the ceiling, to be reflected downward. In between are all possible combinations of the two. Which is most suitable for a particular installation? That question has no direct answer; only general principles can be outlined here.

Indirect lighting is possible when ceilings and walls have good and permanent reflecting ability. Sometimes it is used for special lighting effects. Usually the coefficient of utilization is low with indirect lighting, making the operating cost higher than with other methods.

Direct lighting becomes the only choice in any location where ceilings and walls do not have permanently good reflecting power. This includes heavy manufacturing areas, foundries, and similar locations. Direct lighting usually provides the best coefficient of utilization.

Except for locations where only direct lighting may be used, most installations are a combination of direct and indirect. Of the useful light produced, in one installation it may be 90% direct, 10% indirect; in another it may be 10% direct, 90% indirect. A combination is usually considered a logical system. Especially in offices or any other location where close attention to detail is required, a 100% direct-lighting system will prove quite unsatisfactory, for it will lead to uneven lighting including glare, shadows, and a generally cramped feeling. This has led to a trend toward fixtures so designed that, even if the fixture is mounted directly on the ceiling, a generous portion of the light falls on the ceiling, to be reflected downward. This produces a pleasing, comfortable, diffused light, which is essential to efficiency.

Location of Reflectors. Especially in direct-lighting installations, the spacing and location of reflectors become very important.

Usually it is sound practice to mount them as far apart as their height above the floor. This results in a reasonably even level of illumination at the work level of tables or benches, as shown in Fig. 30-10.   If such reflectors are mounted too far apart, dark areas will result, as shown in Fig. 30-11.

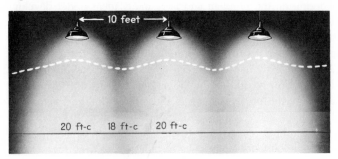

Fig. 30-10. Reflectors properly spaced provide even lighting. (*General Electric Co.*)

Point-of-work Lighting.   It is quite customary to provide additional localized lighting, beyond the general lighting, where work of a very exacting nature is performed.   Such locations are common in manufacturing establishments: assembly of very small parts, inspection stations,

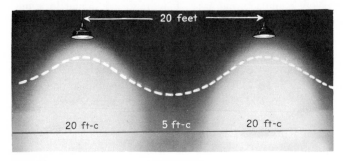

Fig. 30-11. Reflectors wrongly spaced result in dark areas.   (*General Electric Co.*)

toolmakers' stations, and many others.   In work such as is shown in Fig. 30-12, a level of 500 footcandles is by no means too high; investment in such lighting pays dividends.   Special reflectors or reflector-type lamps described earlier in this chapter are often used.

The illumination in the concentrated area, however, should not be too great as compared with the immediately surrounding area, for

Fig. 30-12. In areas where exacting and detail work is performed, provide 300 to 500 footcandles of light. (*General Electric Co.*)

otherwise the contrast will be extreme, glare will become a factor, and much of the advantage of the additional lighting will be offset.

Lighting of Athletic Fields. The lighting of an athletic field is a much bigger project than is usually realized. For example, consider what is required for a baseball field for major-league games. Not many years ago, 150 footcandles for the infield and 100 footcandles for the outfield were considered acceptable. Now 500 and 200 footcandles respectively are being recommended. For good color-television broadcasting, a minimum of 300 footcandles is recommended.

To produce such levels of illumination requires an extraordinary amount of power. Half a dozen fields in the United States now have lighting installations consuming over 2,000 kw of power. Most of these installations consist of a large number of incandescent lamps installed on towers. It is quite usual to burn lamps at an overvoltage, for example 110-volt lamps on 120-volt circuits. From previous discussions you will remember that this greatly decreases the life of the lamp, but also increases the total light, and the lumens per watt. Since the lamps are used relatively few hours per year, the short life is not a major drawback.

But why not use lamps producing more lumens per watt, for example the HID lamps discussed in this chapter?   That of course is the trend, as new installations are made, or as old installations are replaced.   For example, at Cleveland Municipal Stadium the original installation consisted of 1,318 incandescent lamps each consuming 1,500 watts, a total of roughly 2,000 kw.   The installation was revamped, retaining 398 of the 1,500-watt lamps, but replacing 920 of them with 1,000-watt metallic-additive lamps.   The total power consumption dropped from about 2,000 kw to about 1,500 kw.   In spite of the reduction in power, the lighting on the infield increased from about 180 footcandles to almost 300 footcandles, and in the outfield from 140 to almost 200 footcandles.

From this example you can see that by using HID lamps, or a combination of incandescent and HID, it is possible to greatly increase the footcandles of lighting without increasing the total kilowatts of power consumed, and thus avoid the extremely expensive procedure of installing large wires, conduits, transformers, and so on, that would be required if additional incandescent lighting were added.

Even fields for lesser events than major-league baseball require more power for lighting than most people would think necessary.   For football, even a high school game should have at least 75 kw, while college football requires 150 to 500 kw.   Even a tennis court should have 10 kw for ordinary recreational playing and 20 kw or more for more serious games.   Really meaningful figures cannot be given, for so much depends on the kinds of lamps used: incandescent or HID.

From this it should be apparent that the lighting of athletic fields in general involves problems much greater than might be offhand expected.   The design of such installations is not for the beginner.

31

Wiring for Motors

Chapter 15 covered the wiring of ordinary types of motors as used in residential and farm applications. This chapter will cover the wiring of commercial and industrial motors in considerable detail, so that the motor when installed will have proper operating characteristics, proper protection, and proper control.

Sections 430-1 to 430-153 of the Code cover all phases of this work. This portion of the Code is decidedly complicated and involved but can be broken down to be relatively simple, although the wiring of motors from a fraction of a horsepower to hundreds of horsepower can never be condensed to a few simple rules. Only installations at 600 volts or less will be discussed.

In studying this chapter, you will make it easier for yourself if you will think in terms of a particular motor as you read the chapter; as you read the chapter again, think of a much smaller or much larger motor, or a different kind.

This chapter does not pretend to cover every last detail covered by the Code in the 37 pages of its Art. 430.   Rather it is intended to pertain to the more ordinary installations, which should include a large proportion of motors being installed.   For the more difficult and intricate installations, consider this chapter a sort of preview of the kinds of problems that will be met; you will not be called upon to design such installations until you have had considerable experience in the more ordinary jobs.

Sealed (Hermetic-type) Refrigeration Motors.   Motors of this kind must be treated differently from ordinary motors.   They will be discussed in a separate section toward the end of this chapter.

Switches.   Various types of switches will be mentioned here as in the Code.   A good understanding of these various types is essential; study the following definitions, which are quoted from the Code:

> **General-use Switch.**   A switch intended for use as a switch in general distribution and branch circuits.   It is *rated in amperes* and is capable of interrupting its rated current at its rated voltage.
>
> **Motor-circuit Switch.**   A switch, *rated in horsepower,* capable of interrupting the maximum operating overload current of a motor of the same horsepower rating as the switch at the rated voltage.
>
> **Isolating Switch.**   A switch intended for isolating a circuit from the source of power.   It has no interrupting rating and is intended to be operated only after the circuit has been opened by some other means.
>
> **Circuit Breaker.**   A device designed to open and close a circuit by nonautomatic means, and to open a circuit automatically on a predetermined overload of current, without injury to itself when properly applied within its rating.

Motor Branch Circuit.   The elements that make up a motor branch circuit are as follows:

A. Motor-branch-circuit conductors.   The wires from the panelboard to the motor.

B. Motor-branch-circuit overcurrent protection.   To protect the wires, the controls, and the motor against overloads *due to short circuits and grounds only.*

C. Disconnecting means.   Totally to isolate the motor and its controls when necessary to work on the motor or its controls or the driven machinery.

D. Motor-running overcurrent protection.   To protect the motor, the overcurrent device itself, and the wires against damage caused by overloads *other than short circuits or grounds.* Such devices are *not* capable of opening short-circuit currents.

E. Controller. To start and stop the motor, or reverse it, possibly control its speed, etc.

Figure 31-1 shows all the elements involved. In practice several of these elements often are combined into a single device; these cases will be considered one at a time. *The first part of this chapter will cover these elements under the simplest possible conditions: one motor, and nothing except that motor, on a branch circuit.*

Motor Current. The amperage consumed by a motor varies a great deal with the circumstances. In most motors it is very high when the motor is first turned on and gradually drops as it speeds up, then drops to normal value when the motor reaches full speed and carries its rated horsepower load, but increases if the motor is overloaded. That introduces many problems in overcurrent protection. It will be wise to review the subject.

Full-load Current. This is the amperage consumed by the motor while it is delivering its rated horsepower. The Code also calls it "motor-running" current. It is also the amperage stamped on the name-plate of the motor. However, when a motor installation is being planned, the motor itself usually has not been obtained. Therefore consult Code Tables 430-147 to 430-150, which specify the amperage to be used in calculations for different kinds of motors at various voltages. Tables 430-148 and 430-150 cover the more ordinary kinds of motors, and will be found in the Appendix.

If you are using 208-volt motors, increase the amperage shown in the tables for 230-volt motors by 10%, as required by the footnotes under the tables.

Overload Current. As already discussed in Chap. 15, no motor should be installed with the expectation that it will deliver more than its rated horsepower *continuously.* But when called upon to do so, the motor will deliver more than its rated horsepower, but will consume an increased amperage. If that higher amperage is permitted to flow indefinitely, in other words if the motor is overloaded continuously, it will overheat, leading to short life. If the overload is large enough, the motor will burn out quickly. The purpose of what the Code calls "motor-running overcurrent protection" is to disconnect the motor from

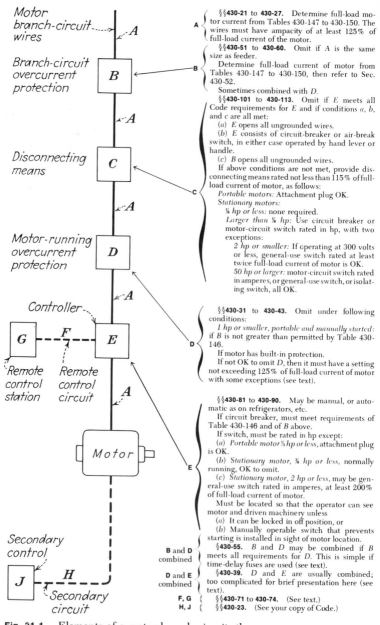

Motor branch-circuit wires — *A*

§§**430-21** to **430-27.** Determine full-load motor current from Tables 430-147 to 430-150. The wires must have ampacity of at least 125% of full-load current of the motor.

Branch-circuit overcurrent protection — *B*

§§**430-51** to **430-60.** Omit if *A* is the same size as feeder.

Determine full-load current of motor from Tables 430-147 to 430-150, then refer to Sec. 430-52.

Sometimes combined with *D*.

Disconnecting means — *C*

§§**430-101** to **430-113.** Omit if *E* meets all Code requirements for *E* and if conditions *a*, *b*, and *c* are all met:

(*a*) *E* opens all ungrounded wires.

(*b*) *E* consists of circuit-breaker or air-break switch, in either case operated by hand lever or handle.

(*c*) *B* opens all ungrounded wires.

If above conditions are not met, provide disconnecting means rated not less than 115% of full-load current of motor, as follows:

Portable motors: Attachment plug OK.

Stationary motors:

⅛ *hp or less:* none required.

Larger than ⅛ *hp:* Use circuit breaker or motor-circuit switch rated in hp, with two exceptions:

2 *hp or smaller:* If operating at 300 volts or less, general-use switch rated at least twice full-load current of motor is OK.

50 *hp or larger:* motor-circuit switch rated in amperes, or general-use switch, or isolating switch, all OK.

Motor-running overcurrent protection — *D*

§§**430-31** to **430-43.** Omit under following conditions:

1 *hp or smaller, portable and manually started:* if *B* is not greater than permitted by Table 430-146.

If motor has built-in protection.

If not OK to omit *D*, then it must have a setting not exceeding 125% of full-load current of motor with some exceptions (see text).

Controller — *E*

G — *F* — *E*

Remote control station   *Remote control circuit*

§§**430-81** to **430-90.** May be manual, or automatic as on refrigerators, etc.

If circuit breaker, must meet requirements of Table 430-146 and of *B* above.

If switch, must be rated in hp except:

(*a*) *Portable motor* ⅓ *hp or less,* attachment plug is OK.

(*b*) *Stationary motor,* ⅛ *hp or less,* normally running, OK to omit.

(*c*) *Stationary motor,* 2 *hp or less,* may be general-use switch rated in amperes, at least 200% of full-load current of motor.

Must be located so that the operator can see motor and driven machinery unless

(*a*) It can be locked in off position, or

(*b*) Manually operable switch that prevents starting is installed in sight of motor location.

Motor

Secondary control

J — *H*

Secondary circuit

B and D combined   §**430-55.** *B* and *D* may be combined if *B* meets all requirements for *D*. This is simple if time-delay fuses are used (see text).

D and E combined   §**430-39.** *D* and *E* are usually combined; too complicated for brief presentation here (see text).

F, G { §§**430-71** to **430-74.** (See text.)

H, J { §§**430-23.** (See your copy of Code.)

Fig. 31-1. Elements of a motor branch circuit; the details are shown in abbreviated form.

the line if the overload is heavy enough to damage the motor, if it were allowed to continue indefinitely. It will also disconnect the motor if it fails to start.

Starting Current. When a motor is first connected to the line, the amperage consumed is very much higher than after the motor comes up to speed. The ratio of starting current (also called "locked-rotor" or "stalled-rotor" current) to running current, varies greatly with the size and type of motor and may be as high as 600%. An ordinary single-phase 1-hp 230-volt motor consuming only 8 amp while running at full speed and full load, will require as much as 48 amp for a second or two when first started. For larger motors the ratio is usually lower, but starting current is nevertheless always higher than running current.

The ratio of starting current to running current for any given motor also depends on the kind of machinery the motor is driving. If the load is hard to start, the starting amperes will be higher than if the load is easy to start. Moreover, the higher starting amperage will persist for a longer time, for the motor will not come up to full speed as quickly as with an easy-starting machine.

Any AC motor of ½ hp or more manufactured since 1940 carries on its name-plate a "Code Letter" that indicates the approximate kilovolt-amperes consumed *per horsepower* with the motor in a locked-rotor condition. Locked-rotor means that the motor is locked so that the rotor cannot turn. Of course, a motor is never operated that way, but at the moment when a motor is first connected to the power line, it is not turning, so that locked-rotor condition does exist until the motor starts to turn. The amperage consumed by the motor until it starts to turn is very high. Then the amperage drops off until the motor reaches full speed, at which time normal amperage is established. From the Code letter you can determine the maximum amperage required by a motor while starting, which is useful information in establishing the size of the various components in a motor circuit. The table in Code Sec. 430-7(b) covering the Code letters is shown on the next page.

Generally speaking, the larger the motor, the lower the locked-rotor current per horsepower. For example, in general-purpose 60-cycle 1,800 rpm 3-phase motors, the 1-hp size in most brands has a Code letter L; the 2-hp has K; the 5-hp has H; the 7½- and 10-hp have G; 15- to 100-hp have F. Single-phase capacitor motors range from L in the case of ½-hp size to E, F, or G on the 5-hp size. Single-phase repulsion-start induction-run motors range from H on the ½- and 1-hp sizes to E on the 5-hp size.

Code letter	Kilovolt-amperes per horsepower with locked rotor	Code letter	Kilovolt-amperes per horsepower with locked rotor
A	Under 3.14	L	9.00 to 9.99
B	3.15 to 3.54	M	10.00 to 11.19
C	3.55 to 3.99	N	11.20 to 12.49
D	4.00 to 4.49	P	12.50 to 13.99
E	4.50 to 4.99	R	14.00 to 15.99
F	5.00 to 5.59	S	16.00 to 17.99
G	5.60 to 6.29	T	18.00 to 19.99
H	6.30 to 7.09	U	20.00 to 22.39
J	7.10 to 7.99	V	22.40 and up
K	8.00 to 8.99		

To establish the approximate maximum amperes required by a specific motor while starting, first determine the actual Code letter on the name-plate of the motor; assume that it is J and that the motor is 3 hp. Refer to the table and you will find that J shows from 7.10 to 7.99 kva per horsepower. Call the average 7.55 kva; or 7,550 volt-amp. For 3 hp the total is 7,550 × 3 or 22,650 volt-amp. If the motor is single-phase, divide by the voltage. This (22,650/230) produces a result of about 98 amp. In other words as the motor is first thrown on the line, there is a momentary inrush of about 98 amp, diminishing gradually to about 17 amp (Table 430-148) when the motor is running at full speed and delivering its rated 3 hp.

If the motor is a 3-phase motor, first multiply the volts by 1.73 (see Chap. 3); in the case of a 230-volt motor, the result is 397.9. Use the number 400, which is easy to remember. Then divide 22,650 by 400, giving about 57 amp, which is the inrush when the motor is first turned on, diminishing to about 8 amp when the motor reaches full speed.

MOTOR-CIRCUIT WIRES
(Code Secs. 430-21 to 430-27)

Must Carry 125% of Running Current. The Code in Sec. 430-22 specifies that these wires (whether feeder or branch-circuit) must have an ampacity of at least 125% of the full-load current of the motor, as set forth in Code Tables 430-147 to 430-150. There are a number of exceptions when the wire must be larger and a few where it may be smaller. The exceptions pertain to types of motors that are not very common and for that reason will not be discussed here, especially since Code Sec. 430-22 quite clearly defines the exceptions.

Voltage Drop. The Code specifies only the *minimum* size wires that may be used in a motor circuit. It does not take into consideration the voltage drop that always occurs. Measure or figure the distance from the motor back to the panelboard, measured along the wire; then calculate the voltage drop for the minimum size wire (how to do this was covered in Chap. 7), and if it exceeds 2½% of the voltage at which the motor operates, use larger wire. Remember, if the voltage drop as calculated is 5%, in order to reduce it to 2½%, wire with double the cross-sectional area (twice the circular mils) must be used. It is not ordinarily considered good practice to permit voltage drop greater than 2½%, for this simply means that a substantial percentage of the power paid for is wasted, which may mean a considerable sum during the life of the motor. Too great a voltage drop also means the motor will not deliver full power, start heavy loads, or accelerate so rapidly as under full voltage. Remember that there is additional voltage drop ahead of the panelboard.

Don't overlook the starting current of the motor. If you have figured your wire size for 2½% drop during normal running, but the motor consumes four times normal current while starting, the voltage drop during that starting period will be 10%, not 2½%. That may be of little importance if the motor starts without load, the load being applied after the motor comes up to speed. But if the motor must start against a heavy load, this factor can become quite important.

MOTOR-CIRCUIT OVERCURRENT PROTECTION
(Code Secs. 430-51 to 430-63)

This branch-circuit overcurrent protection is necessary to protect the wires of the circuit against overloads greater than the starting amperage, in other words, against short circuits and grounds. At the same time, it protects the motor controller, which ordinarily is designed to handle only the amperage consumed by the motor and which would be damaged by a short circuit or ground if branch-circuit protection were not separately provided. Grounds are often practically equivalent to short circuits.

As in other branch circuits, the protection in a motor branch circuit may take the form of either fuses or circuit breakers. The maximum amperage rating permitted by Code for the overcurrent device in a *motor* branch circuit is sometimes higher when fuses are used than when circuit breakers are used.

Maximum Rating of Overcurrent Protection. This subject is discussed by Code Sec. 430-52. A circuit serving a single motor of the types commonly used, may be protected by fuses rated not over 300% of the full-load current of the motor, or by a circuit breaker not over 250%. To determine the maximum rating for less commonly used motors, first refer to Tables 430-147 to 430-150 to determine the full-load current of the motor. Then refer to Tables 430-152 and 430-153 (see Appendix) to determine the maximum rating. Note that if the calculation leads to a nonstandard rating of a fuse or breaker, the next larger standard rating may be used. If the maximum rating so determined will not carry the starting current of the particular motor, a rating not exceeding 400% of the full-load current of the motor may be used.

Since Sec. 430-22 requires that the motor-branch-circuit wires must have an ampacity of *at least* 125% *of the motor* full-load current, and since Sec. 430-52 permits overcurrent protection in some cases up to 400% of the motor full-load current, it is evident that the branch-circuit overcurrent protection may be as much as 400/125, or 320% of the ampacity of the wire. This is contrary to general practice; hence it should be well understood that this is permitted only in the case of wires serving motors.

All the foregoing has reference to the maximum setting permitted for the overcurrent device. In practice it should be set as low as possible and still carry the maximum current required by the motor while starting or running. However, Sec. 430-57 requires that when fuses are used, the fuseholder must be capable of holding the *largest* fuse permitted. Occasionally this will necessitate using an adapter to permit, for example, 60-amp fuses to be used in fuseholders designed for the larger 70- to 100-amp fuses. However, if time-delay fuses are used, this requirement is waived. It makes sense to use only time-delay fuses.

Motor-branch-circuit Overcurrent Device Omitted. If the branch-circuit wires are the same size as the feeder, all the way up to the motor-running overcurrent device, the branch-circuit overcurrent device may be omitted entirely. Obviously the branch-circuit wires are protected by the feeder overcurrent protection. It would be correct to say that there is no feeder and that what was considered a feeder becomes the motor branch circuit (assuming of course that the motor is the one and only load on the circuit).

Motor-feeder Overcurrent Protection.   Before discussing this topic, which is covered by Secs. 430-61 to 430-63 of the Code, it is important to review just what a feeder is.   The Code defines feeders as "circuit conductors between the service equipment, or the generator switchboard of an isolated plant, and the branch-circuit overcurrent device." The overcurrent device used to protect the motor for *running protection*, which will be discussed later in this chapter, is not the overcurrent device referred to; the feeder ends at the *branch-circuit* overcurrent device.

In large installations the requirements for power are such that each individual branch circuit cannot possibly be run back to a common point at the service entrance.   Instead, feeders run from the service to panelboards at various locations, and the individual circuits start from these panelboards.

The feeders are subject to the requirements of Code Secs. 430-61 to 430-63.   The requirements are simple.   A feeder that supplies one motor and nothing else must be provided with overcurrent protection not greater than that calculated as just outlined for the motor-branch-circuit protection.

Feeder Taps in Inaccessible Locations.   At times it is necessary to tap the motor-branch-circuit wires to a feeder at a point not readily accessible.   To locate the branch-circuit overcurrent protection at such a point would render it equally inaccessible, which would be completely impractical.   The simplest way of getting around this difficulty is to make the tap from the feeder the same size as the feeder proper, running it to a readily accessible point, where the motor branch circuit then begins, and where the overcurrent protection is then installed.   The wire in the branch circuit from that point need only be large enough as required for the motor, just as if the motor branch circuit ran all the way back to the panelboard.

However, the Code in Sec. 430-59 permits an important exception. If the tap between the feeder and the branch-circuit overcurrent protection is not over 25 ft long, if it is protected against physical damage, it may be much smaller.   Its ampacity must be at least one-third that of the feeder, and of course large enough for the motor involved.

Moreover, per Sec. 240-15 Exception 5, if the tap is not over 10 ft long and is protected by conduit, it need be only large enough to handle the load it will have to carry, *provided* it does not extend beyond the "Switchboard, panelboard or control device which it supplies." The "control device" can be the disconnecting means for the motor.

DISCONNECTING MEANS, RUNNING OVERCURRENT PROTECTION, AND CONTROLLER

These three components in a motor circuit may be three separate devices, but more often two are combined into a single device, and sometimes all three are combined into a single device. In some cases one or more of the components may be omitted. We shall first discuss the requirements of each component separately, as if each were a separate device; this will automatically lead to a discussion of how some of the components can be combined with others. Later we shall discuss other situations that permit such combining.

"Out of Sight. . . ." The Code in many cases requires that the motor may not be out of sight of the controller and the controller in turn may not be out of sight of the disconnecting means. These statements are self-explanatory, but do note that per Sec. 430-4 a distance of 50 ft or more means "out of sight" regardless of circumstances.

DISCONNECTING MEANS
(Code Secs. 430-101 to 430-113)

The purpose of the disconnecting means is totally to isolate the motor and its controller, when, for example, it is necessary to work on the motor, the controller, or the machinery driven by the motor.

Requirements for Disconnecting Means. The Code requires that the disconnecting means must have a capacity of at least 115% of the running current of the motor. It must plainly indicate whether it is in the open or closed position. It must be located within sight of the controller unless it can be locked in the off position. It must open all the ungrounded wires simultaneously (it may also open a grounded wire provided it opens all the ungrounded wires at the same time). It must disconnect the motor and the controller; it may be in the same case with the controller.

The disconnecting means must be a manually operated switch *rated in horsepower* or a circuit breaker, with the following exceptions:

1. For stationary motors of ⅛ hp or less, the branch-circuit overcurrent device is sufficient, no separate disconnecting means being required.

2. For stationary motors of 2 hp or less and 300 volts or less, a general-use switch, *rated only in amperes,* may be used if it has an amperage rating at least *twice* the full-load current rating of the motor. Additionally, a "general-use AC-only" switch described in Chap. 4 may be used, but only if the full-load running current of the motor does not exceed 80% of the switch rating. Naturally the switch must be rated for the voltage involved.

3. For stationary motors of more than 50 hp, a motor-circuit switch rated also in amperes, a general-use switch, or an isolating switch may be used.

4. For portable motors, the attachment plug and receptacle are sufficient.

Per Code Sec. 430-102, the disconnecting means must be located within sight of the controller.

Types of Disconnecting Means. Any kind of circuit breaker may be used. If a switch is used, it may be the general type shown in Fig. 31-2; similar but larger switches are suitable for larger motors. These

Fig. 31-2. This switch may be used as the disconnecting means for small motors. (*Square D Co.*)

switches do not need to be fused. However, if properly fused, they may also serve as the running overcurrent protection under conditions that will be outlined later. Sometimes the switch will also serve as a controller as will also be discussed later. The disconnecting switch must open all ungrounded wires to the motor.

Service Switch as Disconnecting Means. In the rare instances when an installation consists of a single motor and nothing else, the service switch, per Sec. 430-106, serves as the disconnecting means, provided it is within sight of the controller. If it is not within sight, it must be capable of being locked in the open position.

MOTOR-RUNNING OVERCURRENT PROTECTION
(Code Secs. 430-31 to 430-43)

Need for Running Protection. A motor that requires, for example, 10 amp while delivering its rated horsepower may require 40 amp or more while starting. Once the motor has come up to speed, there will probably be times when the machine which the motor drives will require more power than the rated horsepower of the motor; the motor may be entirely capable of delivering more horsepower for a nominal period of time, consuming a correspondingly greater amperage while doing so. Under overload this motor usually drawing 10 amp may require 15 amp, or more. If permitted to draw 15 amp continuously, the motor will probably be damaged.

So far in this discussion only the motor-branch-circuit overcurrent device has been under consideration, and that may under certain conditions be rated as high as 400% of the full-load current, or 40 amp in this case. Obviously a 40-amp overcurrent device will protect the 10-amp motor against short circuits and grounds, but it will in no way protect the motor against an overload which makes the motor consume 15 instead of 10 amp.

Therefore the Code requires (in addition to the motor-branch-circuit overcurrent protection) a separate overcurrent device that protects the motor against nominal increase in amperage, which if continued long enough will damage the motor.

Types of Running Overcurrent Devices. The motor-running overcurrent device may consist of fuses. The fuses must carry the starting current for a short period, must carry the normal running current indefinitely, but must blow if an overload current occurs and is maintained long enough to damage the motor. Ordinary fuses will seldom serve the purpose, but time-delay fuses are suitable for most purposes.

A circuit breaker may be used in place of fuses and serves the purpose in approximately the same way as time-delay fuses.

Quite often the motor-running overcurrent protection is combined with the controller in a single device which most people simply call a motor starter. The overcurrent portion of the starter can be called an overcurrent *device;* the device contains one or more overcurrent *units.* For example, a 3-phase starter may have two or three contacts to close the circuits in as many wires when the motor is started. It will also contain two or more overcurrent *units,* each in a separate wire, although the device collectively is only one overcurrent *device.* However, no matter if only one overcurrent *unit* operates, due to an overcurrent, it will by virtue of the mechanical design open all the contacts of the starter, stopping the motor. The operating principle will be discussed later in this chapter.

The Code requirement as to the number of overcurrent *units* and their location in the several wires to the motor can be found in Sec. 430-37 of the Code. For single-phase motors, only one is required, in any ungrounded wire. For 3-phase motors, two are required, in any ungrounded wires. However, if the motor is located in an isolated, inaccessible, or unattended location (as, for example, in the case of irrigation-pump motors), then three units are required, one in each wire. Actually, while only two are required for most locations, the trend is toward using three units in all cases.

"Continuous Duty." The Code in Sec. 430-33 defines this as follows: "Any motor application is considered to be for continuous duty unless the nature of the apparatus which it drives is such that the motor cannot operate continuously *with load* under any condition of use." That seems to leave some room for debate as to when a motor is *not* continuous-duty. One might argue that a motor on a home water system with an automatic controller that stops the motor when the pressure in the water tank reaches its proper value, is not a continuous-duty motor. However, if a sprinkler is accidentally left on for a day, the motor will probably run continuously. If in doubt, protect the motor as specified for a continuous-duty type.

Inherent Overcurrent Protection. Some motors have built-in overcurrent protection, making the need for separate overcurrent devices unnecessary. This built-in device is a component of the motor which operates not only by the heat created by the current flowing through it, but also by heat conducted to it from the windings of the motor. If the motor is already hot from operating for a long time at full load, then the device will disconnect the motor more quickly when an overload arises, than if the motor started cold and immediately became overloaded to the same degree. In this respect such built-in devices as "Thermotrons" or "Thermoguards" provide better protection than separate devices. When such devices operate, they must be manually reset after the motor cools off.

Continuous-duty Motors. The motor-running overcurrent device required is defined in Code Secs. 430-31 to 430-43 and depends on many factors, as follows:

More than 1 hp. Whether manually started, or automatically as on refrigerators, water pumps, etc., the maximum rating of the overcurrent device is defined in the Code as a percentage of the full-load current of the motor, and depends on the type of motor, as follows:

Motors marked with a Service Factor (defined in Chap. 15) of 1.15 or higher 125%
Motors marked with a Service Factor lower than 1.15 . . . . . 115%
Motors marked with a temperature rise of 40°C or less . . . 125%
Motors marked with a temperature rise greater than 40°C. . . 115%
Hermetic motors: see discussion toward end of this chapter.
All other motors 115%

If applying the percentages outlined above leads to an amperage for which there is no standard overcurrent device available, the 125% figure may be increased to not over 140%, and the 115% figure to not over 130%.

One horsepower or less, not permanently installed, manually started. If the motor is out of sight of the starter location, it must be protected as will be explained below for *automatically* started motors. If, however, it is in sight from the starter location, the branch-circuit overcurrent protection is considered adequate if it does not exceed the value already discussed. However, it is permissible to operate any such motor at 125 volts or less on any circuit protected with 20-amp overcurrent protection.

One horsepower or less, not portable, *manually started.* Protect such motors in the way described in the next paragraph for automatically started motors.

One horsepower or less, automatically started. For protecting such motors, a choice of several methods is available, as follows:

1. Separate overcurrent devices, rated as discussed in an earlier paragraph for motors over 1 hp.

2. Inherent overcurrent protection, as already discussed.

3. The motor shall be considered as properly protected when it is

part of an approved assembly which does not normally subject the motor to overloads and which is also equipped with other safety controls (such as the safety combustion controls of a domestic oil burner) which protect the motor against damage due to stalled-rotor current. Where such protective equipment is used, it shall be indicated on the nameplate of the assembly where it will be visible after installation.

4. Motors of the general type of clock motors are not harmed if they do not start and are considered protected by the branch-circuit overcurrent device.

MOTOR CONTROLLERS
(Code Secs. 430-81 to 430-90)

A motor controller according to Code definition is "any switch or device normally used to start and stop the motor." The controller may be a manually operable device, or it may be an automatic device as found on refrigerators, oil burners, and similar appliances.

Requirements for Controllers. In general the controller "shall be capable of starting and stopping the motor which it controls, and for an alternating-current motor shall be capable of interrupting the stalled-rotor current of the motor." These basic requirements are of chief interest to manufacturers. Users and contractors will find that approved controllers furnished by manufacturers will automatically meet these requirements if the proper selection is made by the user. Nevertheless it will be well to be familiar with them.

The Code requires that the controller have a horsepower rating not less than the horsepower of the motor, with some exceptions. The most important ones are:

1. Stationary Motor of 1/8 hp or Less. For a stationary motor rated at 1/8 hp or less that is normally left running and is so constructed that it cannot be damaged by overload or failure to start, such as clock motors and the like, the branch-circuit overcurrent device may serve as the controller.

2. Portable Motor of 1/3 hp or Less. For a portable motor rated at 1/3 hp or less, the controller may be an attachment plug and receptacle.

3. Stationary Motor of 2 hp or Less. For a stationary motor rated at 2 hp or less, *and 300 volts or less,* the controller *may* be a general-use switch having an ampere rating at least twice the full-load current rating of the motor.   A "general-use AC-only" may also be used, provided the full-load current of the motor does not exceed 80% of the rating of the switch; the switch must have a voltage rating at least as high as that of the motor.

4. Circuit Breaker as Controller.   A branch-circuit circuit breaker, rated in amperes only, may be used as a controller.   When this circuit breaker is also used for running overcurrent protection, it must also meet the general requirements for running overcurrent protection.

The controller need open only enough conductors to start and stop the motor.   If, however, it serves also as the disconnecting means as will be discussed shortly, then it must meet certain other requirements. In this connection note that it is permissible that the grounded wire be opened by the controller provided only that the device is so constructed that all the ungrounded wires are opened simultaneously with the grounded wire.

In general the controller must be located so that the motor and its driven machinery shall be within sight of the point from which the motor is controlled unless *one* of the following points is complied with:

1. The controller or its disconnecting means is capable of being locked in the open position.

2. A manually operable switch, which shall prevent the starting of the motor, is placed within sight of the motor location.   Obviously this provision is included to make it safe to work on the motor or its driven machinery without the danger of having the motor started from some remote point by some person who does not know that the motor or its machinery is being worked on.   This switch may be directly in the motor circuit or it may be a very simple toggle switch (such as is used for lighting purposes) in the remote-control circuit of the motor, in a wire to the push-button starting station controlling the motor.

3. Special permission is given by the inspector.

How Motor Starters Operate.   While there are literally dozens of types of starters, they can be classed into one of two types: manual and magnetic.   Figure 31-3 shows one of the manual type; the start and stop buttons are built into the cover.   When you push the start or stop button, a mechanical linkage to the buttons closes or opens the main contacts.   The overload relays for running-current overcurrent protection are part of the starter, and will be explained in the next paragraphs.

If the starter trips, stopping the motor because of overload, reset the starter by pushing the "stop/reset" button.

Fig. 31-3. Starters of this type not only control a motor, but protect it against overload. (*Square D Co.*)

A magnetic starter is shown in Fig. 31-4, together with a separate push-button station that can be located at a distance from the motor. Inside the starter is a magnetic coil that closes the starter contacts to

Fig. 31-4. The push-button station at left, may be placed at a distance from the starter. (*Square D Co.*)

start the motor, when the start button is pushed. When the circuit to the coil is interrupted, as when pushing the stop button, the coil loses its energy, and the starter contacts open, stopping the motor. The motor can be restarted by pushing a separate "reset" button on the starter.

The principle of operation can be better understood by studying the diagram of Fig. 31-5, showing how such a magnetic starter serves both as a controller and as a motor-running overcurrent device.   The diagram is for a 3-phase motor.   The wires shown as heavy lines carry the

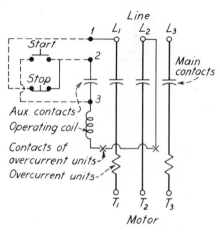

Fig. 31-5.   Circuit diagram for the starter and push-button station shown in Fig. 31-4.

full motor current; the light lines represent small wires carrying only a very small current flowing through the start-stop buttons.   Note the difference between the two buttons: The start button is a normally open type, and pushing the button *closes* a circuit; the stop button is of the normally closed type, and pushing the button *opens* a circuit.

The starter is an electrically operated switch.   When the operating coil is energized by pushing the start button, it closes the three main contacts and starts the motor.   It also closes a small auxiliary contact at the same time.   In the diagram the operating coil is not energized and the motor is not running.   Study the circuit and you will see that pushing the start button lets current flow through the operating coil, which energizes it; it then closes all four contacts.   That starts the motor, and as you remove your finger from the start button, the circuit opens but the coil remains energized because the circuit to the coil, at first energized by current flowing through the start button, now remains energized by current flowing through the auxiliary contacts.   Do note that the current that energizes the coil at all times flows through the stop button, which is normally closed.

When the time comes to stop the motor, push the stop button; this interrupts the current flowing through the coil, deenergizing it, and all contacts open, stopping the motor. When the motor has stopped, remove your finger from the stop button; the motor is then ready for the next start.

Should the voltage fail or drop to a very low value while the motor is running, the coil will not have enough power to keep the contacts closed. The contactor will "drop out," stopping the motor, just as if the stop button had been pushed. This is a safety feature, for a motor running at greatly reduced voltage will probably burn out. Moreover, should the power fail completely, the motor of course would stop, but if the contactor did *not* drop out, the motor would restart automatically when the power is restored—a dangerous situation when motor-driven machinery starts unexpectedly.

Now consider the motor-running overcurrent protection built into the starter. The starter incorporates several overcurrent units already discussed; the number required varies, but in this particular case two are used, as shown in the diagram. These overcurrent units carry the full amperage of the motor and are basically heating elements. Near each of these overcurrent units is a small, normally closed contact sensitive to heat. These contacts are in series with the coil. While the motor is consuming only its normal amperage, the heat developed in the overcurrent units is not enough to trip these contacts, but when the motor for any reason consumes more than its normal amperage, the heat in the overcurrent unit makes the contact near it open. The contact is in series with the coil, so when the contact opens, it is the same as pushing the stop button, and the motor stops.

The starter may have two or three overcurrent units, but all of them are in series with the coil. No matter which one opens, it stops the motor. Note that when the motor stops because of overload, it cannot be started again until the contacts are manually closed by pushing a "reset" button in the cover of the starter.

The horsepower rating of a starter defines the largest motor that it will control; it may of course also be used with a smaller motor. But the overcurrent units (called heater coils or thermal overload units) in the starter are rated not in horsepower but in amperes, and must be selected to match the running current of the motor. A typical heater coil is shown in Fig. 31-6.   For a given physical size of starter, dozens of different ampere ratings of units are available, all the same physical size. It is a simple matter to install the units in the starter.

Fig. 31-6.   A heater coil used in motor starters, for motor protection.   (*Square D Co.*)

COMBINING SEVERAL COMPONENTS
OF MOTOR BRANCH CIRCUIT

In order to combine several components or omit one or more components, specific Code requirements must be met.   If these cases have not already been discussed in the several sections concerning individual components, they will be discussed here.

Disconnecting Means Omitted.   All permissible cases have already been discussed.

Running Overcurrent Protection Omitted.   In addition to the cases already discussed, it is permissible, per Sec. 430-31(b) of the Code, to omit it "where it might introduce additional or increased hazards as in the case of fire pumps."

Controller Omitted.   All permissible cases have already been discussed.

Branch-circuit and Running Overcurrent Protection Combined.   As already discussed, the branch-circuit overcurrent protection *may* have an ampere rating several times as great as that of the running overcurrent protection.   It is, however, required that its ampere rating be as low as possible.   Some motors start so readily that their starting current is not greatly in excess of the running current.   In that case, per Sec. 430-55, if the branch-circuit overcurrent protection has an amperage *not higher than* that permitted for the running overcurrent protection, the running overcurrent protection may be omitted.   In practice this means that a circuit breaker or time-delay fuses must be used for the motor-branch-circuit overcurrent protection; ordinary fuses will not serve the purpose.

Running Overcurrent Protection and Controller Combined.   If the conventional type of motor starter shown in Fig. 31-3 is used, you will find that most of them combine the functions of controller and running overcurrent protection. If a fused switch of the general type shown in Fig. 31-2 is used as a controller, and if the fuses are of the time-delay type with an ampere rating not greater than permitted for running overcurrent protection, then per Sec. 430-90 the switch serves the purpose of controller and running overcurrent protection.

Disconnect, Running Overcurrent Protection, and Controller Combined. The switch discussed in the preceding paragraph will, per Sec. 430-111, serve the purpose of all three components provided it opens all ungrounded conductors. A circuit breaker of an amperage rating not exceeding that permitted for the running overcurrent protection and opening all ungrounded conductors will also serve the same purpose. In each case the switch or breaker must be of a type operated by applying the hand to a handle, and in each case the branch-circuit overcurrent protection must open all ungrounded conductors in the circuit. Fuses are acceptable for the purpose.

Branch-circuit Overcurrent Protection and Controller Combined. Per Sec. 430-83 (Exception No. 2) when a circuit breaker is used as the motor-branch-circuit overcurrent protection, it may also be used as the controller for the motor.

Branch-circuit Overcurrent Protection, Running Overcurrent Protection, and Controller Combined. If the circuit breaker just discussed has an ampere rating not greater than permitted for separate running overcurrent protection, it will serve the purpose of all three components.

Disconnect and Controller Combined. If the controller takes the form of a motor starter of the general type of Fig. 31-3 or 31-4 (which then usually includes the running overcurrent protection), the disconnecting means, which may be a circuit breaker or a switch, is often installed in the same cabinet with the controller. This, then, is not really a case of combining several components in the sense of having one serve the purpose of the other, but rather a case of locating several separate components in a single enclosure.

REMOTE-CONTROL CIRCUITS
(Code Secs. 430-71 to 430-74)

Often a motor must be started and stopped from a point some distance from the motor. You can locate the controller at that point, run the circuit wires first to the controller, then to the motor.   If the controller is a hand-operated switch, or a manual starter of the type shown in Fig. 31-3 with push buttons in the cover, there is no choice.

If the controller must be located at some distance from the motor, this leads to extra cost of running heavy wires farther than in a direct line to the motor.   Because of that fact the controller is often of the remote-control magnetic type, with start and stop buttons located at a distance. Then the wires carrying the full motor current run directly to the starter located near the motor, and then to the motor.   Only small wires run from the starter to the start-stop push-button stations, which may be located any distance from the starter.   These wires carry only a few amperes even in the case of a very large motor.   This method saves much material and labor, adds greatly to convenience, and is commonly used.

The wiring from the control station to the controller proper is known as "motor-control wiring" and is covered by Code Secs. 430-71 to 430-74.   Overcurrent protection for them is subject to the following conditions:

1. No overcurrent protection is necessary provided that the setting of the branch-circuit overcurrent device in the circuit serving the motor in question is not over 500% of the ampacity, per Table 310-12, of the wires connecting the push-button control station with the controller proper.

2. No overcurrent protection is necessary if the controller device and the point of control (start and stop buttons, pressure switch, thermostatic switch, etc.) are both located on the same machine and the control circuit does not extend beyond the machine.

3. No overcurrent protection is necessary if the opening of the control circuit would create a hazard, as, for example, the control circuit of a fire-pump motor.

4. In the absence of any of the above three points, protect the wires of the control circuit with overcurrent devices rated at not over 500% of their ampacity per Table 310-12.   The overcurrent device, however, must *not* be of the time-delay type (Sec. 240-5, Exception 5).

If damage to the motor-control circuit would constitute a hazard, the wires of the remote-control circuit must be installed in conduit or otherwise suitably protected against mechanical injury.

The control device must be so wired that it is disconnected automatically whenever the motor is disconnected by the disconnecting means as defined in the Code and as covered in a previous portion of this chapter.

MISCELLANEOUS

Protection of Live Parts. Most motors in common use and built today do not have exposed live parts. Some motors constructed years ago and still in use may have exposed live parts. If there are such parts, then the motor must be guarded against accidental contact by a guard-rail or some similar device.

Grounding. Per Sec. 430-142 of the Code, stationary motors must be grounded if any of four conditions exist: (a) if supplied by metal-enclosed wiring; (b) if located in wet places and not isolated or grounded; (c) if in a hazardous location; (d) if operated with any terminal at more than 150 volts *to ground.*

If the motor is permanently connected through any kind of metal conduit, or by armored cable, grounding is automatic and no further action need be taken. This assumes the motor is provided with a metal junction box mounted on the motor, and that the conduit or armor is physically anchored to it.

If nonmetallic-sheathed cable is used, the bare grounding wire of the cable must be grounded to the frame of the motor, which can be done by connecting it under one of the screws or bolts that hold the junction box to the motor.

If the motor is permanently installed but fed by a cord and plug, the cord of course must contain the extra grounding wire, and the receptacle must be the grounding type.

Portable motors must be grounded if they operate at a voltage above 150 volts *to ground.* The simplest procedure is to use a cord with the extra (green) grounding wire, and a grounding-type receptacle. But whether it is required or not, it is good practice to ground all portable motors regardless of voltage.

Controller cases must always be grounded unless they are part of an ungrounded portable motor.

SEALED (HERMETIC-TYPE) REFRIGERATION COMPRESSOR MOTORS

Motors of the ordinary type are cooled by radiating the heat that develops during operation, into the surrounding air of the area in which

they operate.   Motors which the Code calls "sealed (hermetic-type) refrigeration compressor motors" are closely coupled to the compressor
and sealed in a common case with the compressor (see Fig. 31-7): The
refrigerant in the case is in a gaseous state, and that motor runs in that

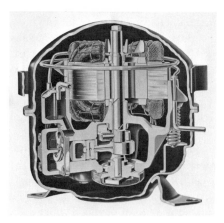

Fig. 31-7. A typical hermetic-type motor, in same
case with refrigeration compressor. (*Tecumseh
Products Co.*)

refrigerant.   As the refrigerant comes into the compressor, it is at a
temperature *below* room temperature and, of course, cools the motor.
    Because of that unusually efficient cooling, a relatively small motor
can safely consume more amperes (and deliver more horsepower) than
an ordinary motor of the same physical size, cooled in the usual way.
But when such a motor is started after a period of idleness, the refrigerant is at room temperature and provides relatively poor cooling; the
motor then heats up faster than an ordinary motor of the same physical
size.   For that reason, overcurrent protection on hermetic motors must
be somewhat different from that on ordinary motors if full protection is
to be expected.
    Hermetic motors are used specifically on compressors of refrigeration
units, including those used for air conditioning.
    On an ordinary motor, the name-plate shows the horsepower of the
motor, also the number of amperes it consumes while delivering that
horsepower.   On a hermetic-type motor the name-plate does not show
a horsepower rating at all.   It shows two amperage ratings: first, the
number of amperes consumed by the motor while driving its compressor

at its rated output and, second, the locked-rotor or starting current of that motor.

All the Code rules discussed so far regarding motor-running overcurrent protection, controllers, and disconnecting means correlate these components to the horsepower rating of the motor.   But hermetic-type motors have no horsepower ratings.   How then is an intelligent selection of these components to be made?   The Code covers the situation.

First of all, on ordinary motors, you do not need to see the name-plate of the motor as long as you know its general characteristics.   The Code Tables 430-147 to 430-153 provide the answer.   But if you are dealing with a hermetic-type motor, you must have the name-plate information of the particular motor to be installed: its full-load amperage and its locked-rotor amperage.

With these two figures before you, proceed as outlined below.   Assume that the motor is a 60-cycle 230-volt 3-phase motor with name-plate ratings of 25-amp full-load current and 125-amp locked-rotor current.

Motor-branch-circuit Wires.   Instead of using Code Tables 430-147 to 430-150, use the actual name-plate full-load amperes and add 25%.   For the 25-amp motor under discussion, that makes a total of 31.25 amp.   Use wire with an ampacity of 31.25 amp or more, per Code Table 310-12.   You may wish to use a larger wire to minimize voltage drop.

Motor-branch-circuit Overcurrent Protection.   Proceed as in the case of ordinary motors, keeping the rating of the fuse or breaker as low as possible, but still capable of carrying the starting current shown on the name-plate of the motor.   Per Code Table 430-153, the normal limit is 175% of the running current of the motor, but per Sec. 430-52(b) this may be increased to 225% if necessary to carry the starting current.

Disconnecting Means.   Follow Code Sec. 430-110(b).   For ordinary motors, as discussed earlier in this chapter, the starting point is the horsepower rating of the motor and the switch usually is rated in horsepower.   But the hermetic-type motor has no horsepower rating, so you must work backwards.   The motor under discussion has a full-load current of 25 amp.   Turn to Code Table 430-150; look in the "230-volt" column.   You won't find a motor with a 25-amp rating, but you will find that a 7½-hp motor of the ordinary type has a 22-amp rating and a 10-hp size has a 28-amp rating.   You will then have to call the 25-amp hermetic-type motor temporarily a "10-hp" motor, based on its 25-amp full-load rating.

Turn to Code Table 430-151, which is concerned with the locked-rotor amperage.   Go to the 3-phase 230-volt column and look for the 125-amp figure that you originally determined from the name-plate of the motor.   There won't be a 125-amp figure, so use the next larger one, 132 amp.   From there go left to the first column, and there you will find the figure 7½ hp.   So now you call the motor temporarily a "7½-hp."

When Tables 430-150 and 430-151 give two different answers, use the larger one of the two; in this case provide a disconnecting means which is suitable for an ordinary 10-hp motor, as discussed earlier in this chapter.

Controller.   When you have determined the answer for the disconnecting means, you also have the answer for the controller.   For this motor, provide the same controller that you would provide for an ordinary 10-hp motor.   In practically all cases, the controller will come with the air-conditioning equipment, so the problem is solved in advance.

Motor-running Overcurrent Protection.   Proceed as with ordinary motors, but base your conclusions on the actual full-load amperage found on the name-plate of the motor, not on the theoretical amperage derived from Code tables.   The maximum rating is 140% of the full-load amperage of the motor if the overcurrent device consists of overload relays, but only 125% if it is of a different type.   If the calculated rating comes to an amperage for which there is no standard device available, it may be increased to not over 140%, regardless of the type of device.

TWO OR MORE MOTORS ON ONE CIRCUIT

The Code differentiates between one or more motors connected to a *general-purpose* branch circuit (an ordinary circuit in a home) and two or more motors connected to a single *motor* branch circuit.

Several Motors on *General-purpose* Branch Circuit.   If each motor is 1 hp or smaller and has a full-load current of 6 amp or less, then per Secs. 430-42 and 430-53(a) two or more may be connected to a general-purpose circuit protected by 20-amp or smaller overcurrent protection, if the circuit operates at 125 volts or less, or if protected at 15 amp at higher voltages but not over 600 volts.

Each motor still requires individual overcurrent protection if it would be required, if the motor were a single motor on its own circuit.

If the motors are *not* continuous-duty (for example, home workshop motors) they do not require the individual overcurrent protection. But do remember the Code definition in Sec. 430-22: "Any motor application is considered to be for continuous duty unless the nature of the apparatus which it drives is such that the motor will not operate continuously *with load* under any condition of use."

Motors larger than discussed above may still be connected to general-purpose branch circuits, but each must have a controller and running overcurrent protection of a type "approved for group installation." This will be defined later in this chapter.

If the motor is connected to the circuit by means of a cord and plug and receptacle, *and* individual running overcurrent protection is omitted, the plug and receptacle may have a rating not over 15 amp at 125 volts or 10 amp at 250 volts. If the motor does have individual running overcurrent protection, it must be part of the motor or appliance, and the plug and receptacle then may have a rating permissible for the particular circuit involved, as outlined in Chap. 5.

Several Motors on One *Motor* Branch Circuit.   It is best to provide an individual circuit for each motor. When several motors are to be used in the same location, it is a simple matter to provide a feeder to that location to serve a number of individual branch circuits, one for each motor.

Nevertheless the Code permits several motors on one circuit, but certain factors change considerably. These will be discussed here in the same sequence in which these elements appeared in the first part of this chapter, where a single motor on a branch circuit was discussed.

"Approved for Group Installation."   Motor controllers and motor-running overcurrent protection devices are rated as already discussed, and are entirely capable of handling the normal starting and running current of a motor. But if a short circuit develops in the wires between the controller or motor-running overcurrent device and the motor, the amperages involved could be many times greater than even the starting current of the motor. Such short-circuit currents can literally explode the controller or the running-current overcurrent device. But if the motor is properly installed, the *branch-circuit* overcurrent device (which may not be rated at more than 400%, and usually is much less than 400%, of the motor-running current) will open the circuit, before the controller or motor-running protection can be damaged.

The preceding paragraph pertains to *one* motor on a branch circuit.

When several motors are on one circuit, the branch-circuit overcurrent protection must be adequate for the amperage required by the entire group of motors, and then will be more than 400% of the running current of the *smallest* motor in the group.   Under these circumstances a short circuit in the wires between the controller or motor-running overcurrent device, and the smallest motor, would probably damage the device, for the branch-circuit overcurrent protection would not open the circuit quickly enough.   Special controllers and running-circuit overcurrent devices (including thermal cutouts and overload relays) have been developed that will safely handle many times the current that could be handled by an ordinary similar device; such devices are designated and are marked "Approved for group installation."   They are also marked with the maximum rating of the fuse or breaker that may be used as the branch-circuit protection in the circuit in which they are installed.

Motor-circuit Wires.   When more than one motor is involved, first determine the minimum required ampacity of the wires (whether branch circuit or feeder) if only the *largest* motor of the group were involved. Then add the full-load current (Tables 430-147 to 430-150) for each of the other motors in the group.   Finally, add the amperage required for the lighting or appliance load, if any.   The total is the minimum ampacity, from which it is easy, with the aid of Table 310-12, to determine the size of wire required as a minimum.

For example, consider one each 3-, 1½-, and ½-hp single-phase 230-volt motors.   In Table 430-148 the full-load currents for these three motors are, respectively, 17, 10, and 4.9 amp.   If only the 3-hp motor were on the circuit, the minimum ampacity of the wires would be 125% of 17 amp, or 21.2 amp.   Add 10 and 4.9 amp, making a total of 36.1 amp.   The minimum ampacity of the wires must be 36.1 amp.

If two of the motors are the same size, consider one of them as the "largest" and the others as the smaller motors.

If the circuitry is so arranged that not all the motors can run at the same time, the wire size need be only large enough for the largest motor, or the group consuming the highest amperage, at any one moment.

Motor-feeder Overcurrent Protection.   The Code in Sec. 430-62 specifies that the feeder for a group of motors is to be protected by an overcurrent device of an amperage rating no greater than the *branch-circuit* overcurrent device for the largest motor served by the feeder, plus the full-load current of each of the other motors in the group.   If two motors are of the same size, consider one of them the "largest," the other as one of the smaller ones. There may be times when several motors must be *started* at the same time. In that case both feeder wires and overcurrent protection may have to be increased; consult your local inspector.

If a lighting or appliance load is also served by the same feeder (which will rarely be the case), simply add the amperage required for that purpose to the amperage for the motors.

Motor-branch-circuit Overcurrent Protection. Motors that may be connected to general-purpose circuits, as already discussed, may also be connected to a circuit serving only motors, under the same conditions. But several types of motors that may *not* be connected to a *general-purpose* circuit, may nevertheless be connected to a *motor* circuit. The Code discusses this in Sec. 430-53.

Each motor must have a controller and motor-running overcurrent protection, both approved for group installation. The branch-circuit overcurrent protection must be rated at an amperage not more than permitted for the largest motor in the group if it were the only motor on the circuit, plus the running current of each of the other motors in the group. Moreover it must not exceed the amperage stamped on the approved-for-group-installation overcurrent device for the *smallest* motor in the group.

If a room air-conditioning unit is involved, even if it has several motors, it is to be considered as a single motor, provided it is cord-connected, rated at not more than 40 amp at 230 volts single phase, with the full-load current shown on the over-all name-plate, instead of the separate name-plates of the individual motors; the rating of the branch-circuit overcurrent device may not be more than the ampacity of the circuit wires, or the rating of the receptacle, whichever is less.

Problems will arise as to the size of the wires and protection for the wires from the point where they branch off from the branch-circuit wires proper (see Fig. 31-8). The wire from A to B to C is the branch-circuit wire, the size of which is determined as already outlined. A is the branch-circuit overcurrent device, also determined as already outlined. From B to D, from C to E, and from C to F are the taps supplying the individual motors. If these taps are the same size as the branch-circuit wires, no overcurrent protection is required at points B and C. If the tap is the minimum size, as outlined in the first part of this chapter covering a single motor, if it has an ampacity at least one-third that of the branch-circuit wire *A-B-C*, and if it is not over 25 ft long and is protected against mechanical injury, then no overcurrent protection is required at points *B* and *C*.   Under all other conditions overcurrent protection is required at these points; determine the size just as if you were considering a single motor on a branch circuit, as outlined in the first part of this chapter.

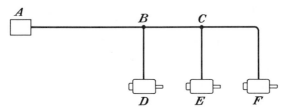

Fig. 31-8.   Problem of several motors on one circuit.

As a matter of fact, installing overcurrent protection at the points *B* and *C* completely changes the complexion of the picture, for then the wire *A-B-C* is no longer a branch-circuit wire but becomes a feeder; *A* becomes the feeder overcurrent protection; the wires *B* to *D* and *C* to *E*, also *C* to *F*, become branch-circuit wires.   The overcurrent protectors at points *B* and *C* become branch-circuit overcurrent devices, and, instead of a group of motors on a single branch circuit, we now have individual motors on individual branch circuits, fed by a common feeder *A-B-C*.

Disconnecting Means.   Each motor may have its own disconnecting means, but the Code permits a group to be handled by a single disconnecting means under one of three conditions:

1. If several motors drive different parts of a single machine, as, for example, metal- or woodworking machines, cranes, hoists, etc.

2. If several motors are in a single room within sight of the disconnecting means.

3. If several motors are protected by a single branch-circuit overcurrent device.

The rating of such a common disconnecting means must not be smaller than would be required for a single motor of a horsepower equal to the sum of the horsepowers of all the individual motors (or a full-load amperage equal to the sum of the full-load amperages of all the individual motors).

Running Overcurrent Protection. Select a running current overcurrent protective device for each motor, as you would if that motor were the only motor on the circuit, except that it must be the type approved for group installation. This is required by Code Sec. 430-53(c) (1).

Controllers. Each controller must be the type approved for group installation. Several motors may be handled by a single controller under the same conditions as outlined for disconnecting means.

32

Wiring Schools and Churches

Statistics show that 9% of all students in the elementary grades of school have defective vision. In high schools the figure is 24%, and in colleges it is 31%. It is a well-established fact that, if good lighting is provided from the first grade onward, the percentages of students with defective vision are greatly reduced. Moreover it is equally well established that good lighting contributes materially toward raising the average grades of the students. Many of those who have been lagging behind respond when proper illumination is provided; the number of failures drops. Good lighting is therefore a tremendous asset for the students themselves and likewise has been found to pay good dividends in the form of reduced over-all cost of providing education on a "per student per year" basis.

Footcandles Required. The number of footcandles required varies greatly depending on the nature of the work being done in a particular area. Classrooms require at least 50 footcandles for reading printed matter in a reasonably large type size, to at least 100 footcandles for smaller type, or handwritten or duplicated material. Blackboards should have over 100 footcandles.

Domestic-science rooms need only 50 footcandles except for sewing, for which 150 footcandles is not too much. Shops and drafting rooms need 100 footcandles or more. For halls and corridors 20 footcandles is probably adequate. In auditoriums 15 footcandles will suffice except of course for the stage. Cafeterias need at least 30 footcandles, gymnasiums 30 to 50 footcandles.

It would be very wise to provide higher levels of illumination than recommended in auditoriums and cafeterias because often these areas, sooner or later, will be pressed into use for study halls, lecture rooms, for giving examinations and similar purposes; this might be temporary use, or more or less permanent, as the number of students increases beyond the original estimate.

It is safe to say that only a small percentage of all schools are lighted to a degree even remotely approaching these recommended levels of lighting. Figure 32-1 shows a well-lighted schoolroom. Students in well-lighted schools will make more progress than those in poorly lighted schools; good lighting is a good investment.

Fig. 32-1. A well-lighted school, using fluorescent lamps. Note the bright ceiling. (*General Electric Co.*)

Types of Fixtures Recommended. A modern installation will automatically be of the fluorescent type, furnishing an abundance of light with minimum surface brightness and glare. If the ceiling height is sufficient, a type of fixture which lets a good deal of the light fall on the ceiling will probably furnish most acceptable lighting. Special attention, of course, must be paid to the ceilings, walls, and all reflecting surfaces. Ceilings and walls must be clean, of a light color to reflect light well, but with a flat finish to avoid glare and to provide well-diffused light.

Switch Control. In a fairly large classroom, the level of illumination from natural light will vary greatly in different parts of the room. See Fig. 32-2, which shows how the footcandle level drops as the distance

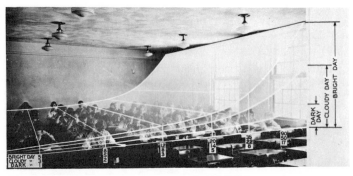

Fig. 32-2. The footcandle level of illumination drops very rapidly, as the distance from windows increases. (*General Electric Co.*)

from the windows increases. The row of desks next to the windows may have 100 footcandles of light; the row on the opposite side of the room may have only 5 footcandles.

Considering this fact, it may be well to consider having switches that control each row of lights separately, so that the row farthest from the window can be turned on independently of the rest. The more usual custom, however, is to have all the lights on most of the time.

Code Requirements for Schools. For lighting purposes the Code requires 3 watts per sq ft of area, except in assembly halls and auditoriums where only 1 watt is required, and corridors where the minimum is $\frac{1}{2}$ watt. The demand factor is always 100%. If you provide only what the Code requires, the result may be lighting that is inadequate, depending to a large degree on the type of fixtures selected. Ignore the Code minimum; install enough circuits of the ampacity required to provide the footcandles that are recommended, and you will more than meet Code requirements.

In addition to the lighting circuits, you must provide circuits for all the other loads, and in a modern school that constitutes a very significant total. Take into consideration not only the usual loads such as heating, ventilation, and possibly air conditioning, but also such other loads as motion-picture or slide-projection equipment, appliances in domestic-science departments, special loads in laboratories and shops, and similar equipment.

Don't overlook plenty of receptacle outlets, including those in halls and corridors, where they are essential for the maintenance crews.

Churches. The Code requires 1 watt per sq ft for lighting, with the usual 100% demand factor regardless of size. That will not provide the footcandles that are now considered minimum.

Proper attention must be given to what is considered good lighting in a church. Illumination below 10 to 15 footcandles in the main worship area would be considered totally inadequate. To provide this level may require 2 to 4 watts per sq ft, depending on many factors such as ceiling height, reflectance of ceilings and walls, and especially the type of lighting fixtures selected.

In older churches, one frequently sees lighting fixtures that seem to have been designed primarily to conform to the architectural scheme of the church, with less thought given to the matter of good lighting. Sometimes such fixtures are designed to use a number of small lamps, providing much less light than a single lamp consuming the same total wattage.

Today in newer churches, the major part of the lighting is usually provided by simple, unobtrusive fixtures at the ceiling. They are installed pointing forward so that they are not at all visible to the people in the congregation, unless they happen to look toward the rear of the church. By using lamps of the PAR type, good light distribution is obtained without using large clumsy reflectors. But because the lamps are not easy to replace it may be well to burn them at less than their rated voltage, thus prolonging their life, and certainly lamps should be used with longer than usual rated life, for example the Quartzline type.

Additional general lighting is often provided by suspended fixtures or lanterns that fit into the architectural scheme of the particular church.

The nave or auditorium of the church having been properly lighted, the fact that in most churches the sanctuary and the altar constitute the focal point of the worshipers must not be overlooked.   This portion of the church should be lighted to a level considerably above that of the rest of the church; levels of 50 to 100 footcandles or higher are the rule. This lighting is usually provided with floodlights concealed from the congregation.   This may require a minimum of 1,500 watts, upward to a much higher figure, depending entirely on the area to be lighted, and the intensity desired.

See Fig. 32-3, which shows the First Presbyterian Church in Oklahoma City; it is a splendid example of a well-designed church lighting

Fig.   32-3. An   unusually   well-lighted   church. Note   the   combination   of   down-lighting,   and   fixtures   for   general   illumination.   (*General   Electric Co.*)

project.   But each church will present an individual problem, and the design of the installation should be left to one well versed in the art of church lighting.

If the church has a choir loft, an absolute minimum of 50 footcandles should be provided.   An organ motor will probably require a special circuit.   The heating plant will require its own circuit.   Throughout the church, miscellaneous loads also exist.   Conveniently located receptacle outlets should not be overlooked, to operate fans, pulpit lights, public-address systems, and similar equipment; they are required too for maintenance work.   Outdoor receptacles are most desirable, for example to permit lighted displays during festival seasons.

33

Wiring Offices

An office may be a small room with a desk and chair and one lighting fixture, or a modern well-appointed office with really adequate lighting, an assortment of electrical equipment including dictating machines, calculators, duplicators, water coolers, and similar equipment, and often with a room-size air conditioner in addition. It is difficult to cover the subject adequately in a single chapter, and suggestions made here must of necessity be quite general.

Code Requirements. For lighting purposes, the Code requires 5 watts per sq ft, with 100% demand factor for the entire area. Providing 5 watts per sq ft will probably not provide the footcandles considered acceptable in a modern office. Assuming that fluorescent lighting is installed, 5 watts may provide from 50 to 80 footcandles, depending on the many factors outlined in Chap. 30.

The absolute minimum should be 75 footcandles, and the alert office manager knows that the extra cost of equipment required to provide a much higher level is soon offset by the increased efficiency of the personnel. For regular office work a level of 100 footcandles is now considered more or less standard. For accounting work or similar work where prolonged attention to detailed figures is necessary, the recommendation is 150 footcandles, and for drafting, it is 200 footcandles.

The specific load which is planned, rather than the Code 5-watts-per-sq-ft minimum, will in all probability determine the total wattage. Once the wattage and the amperage have been determined, feeder sizes and similar data can easily be figured as in other cases that have already been discussed.

In planning an office installation of some significant size, do consider the advantages of 277-volt lighting as discussed in another chapter.

Receptacle Outlets. The Code has no specific requirement as to the number of such outlets in offices; most offices have entirely too few. Every individual office space should have a minimum of two; one every 10 linear ft of wall space is not too many. The need for such receptacle outlets is greater than ever before, considering the wide use of equipment such as dictating machines, electric typewriters, calculators, and similar equipment.

If the building is not totally air-conditioned, extra circuits for room air conditioners should be provided, for surely the occupant will sooner or later want to install such equipment.

Don't overlook receptacles in spaces that are not individual offices, for equipment such as water coolers, duplicating equipment, maintenance equipment, and so on.

Types of Fixtures. Office work is usually of a continuous and exacting nature. Besides a sufficient quantity of light there is needed the right kind of lighting—freedom from glare, low surface brightness, avoidance of shadows. The factors that make good lighting as outlined in Chap. 30 must be considered.

Flexibility. If a sizable floor area is involved, which will accommodate a considerable number of people, it is not likely that original arrangement of the individual office space will long remain unchanged. Large office space is often subdivided into many individual offices by movable partitions about 7 ft high. These partitions are shifted as the need arises, and, as the individual office spaces change, the lighting equipment must be moved. This is frequently accomplished by originally providing flush ceiling outlet boxes on which the original lighting equipment is installed. When it becomes necessary to move this equipment, surface metal raceway of the type described in Chap. 29 is used in the way there described. Similar material is used to provide receptacle outlets for dictaphones and other equipment at the exact point needed.

Signaling Equipment. In larger offices it will be necessary to provide a raceway, usually concealed, for low-voltage wires for telephones, buzzer systems, call systems, and similar purposes. This raceway, conduit or surface raceway, must never contain wires that are part of the regular electrical system operating at the usual voltages. A review of this subject in Chap. 26 would be advisable. To bring such low-voltage wires to the individual desks in an office, surface metal raceway of the floor type which was shown in Fig. 29-18 in Chap. 29 is very commonly used.

Control Equipment. In relatively small offices, a single panelboard or fuse cabinet with switches in the individual circuits will probably serve the purpose. In larger installations it is often necessary to install more elaborate equipment so that an entire group of fixtures can be controlled by a single switch. Many fixtures such as those located in individual offices, of course, must be provided with individual switches.

34

Wiring Stores

Wiring in stores may vary from a simple installation in a small store to a complicated and highly specialized installation in a large department store. The latter type of job is entirely beyond the scope of this book, which will cover only the wiring of small stores.

Footcandle Recommendations. The minimum illumination in the average store in a neighborhood location must be at least 80 footcandles in areas open to the shopping public if a good selling display is to be obtained. In larger stores and in better locations the minimum is higher, never less than 100 footcandles. If self-service counters are involved, 150 to 200 footcandles will be necessary. For showcases, wall cases, and feature displays, double the figures.

If these levels of illumination at first seem high, remember that mediocre lighting may be sufficient to sell the customer the things that he has decided to buy before he comes into the store. No merchant, however, will be a great success if he depends only on that type of business, if he sells only the specific things his customers ask for. Good lighting draws the customer's attention to the merchandise on display, emphasizes the points which lead to a wish for ownership and an urge to buy. Additional sales of customers' "wants" rather than "needs" will result, and good lighting will prove to be a good investment rather than expense.   See Fig. 34-1, which shows a modern, well-lighted store.

Fig. 34-1.   An example of a well-lighted store.   Note the combination of general lighting from fluorescents, and down-lighting from flush ceiling fixtures.   (*General Electric Co.*)

Note the combination of fixtures used, each for its own specialized purpose.   General lighting is provided by recessed fluorescent fixtures. Show cases are down-lighted from flush ceiling fixtures using PAR incandescent lamps.   Lighting for the wall cases is also from flush incandescent fixtures.

In show windows 100 footcandles is usually considered a bare minimum, and in larger cities 200 to 500 footcandles is not unusual.   On a dark street a window lighted to 100 footcandles will appear very bright and well lighted, but if the street is itself well lighted, a window with 200 footcandles may appear very ordinary indeed, especially if another window next door is provided with a still higher level of illumination.

In order to light a show window to the point where it will attract attention in the daytime, it is necessary to provide sufficient light inside the window to minimize the reflections that arise from the natural outdoor light. This requires an unusual level of illumination, for 300 to 1,000 footcandles must be provided.

Code Requirements. The Code in Sec. 220-2(a) requires 5 watts per sq ft for general lighting, for the store proper, the area open to shoppers. But if you plan your installation based on the Code minimum, it is not likely that you will achieve the number of footcandles needed from the standpoint of good merchandising, as already discussed. Ignore the Code minimum, determine the number and wattage of the lighting fixtures you will need, and the circuit capacity for them.

Don't overlook the Code requirement discussed in an earlier chapter: If the load is to be in use for three or more hours at a time, it is considered "continuous" and the circuit may be loaded only to 80% of rating of the branch-circuit overcurrent devices, and the ampacity of the wires in the circuit.

As in the case of other occupancies, specific loads such as motors, appliances, and similar equipment must be added. The demand for electric power in the average store will certainly grow rather than decrease; so it is well to leave a very generous margin for future expansion, when figuring the service entrance, feeders, spare circuits on panelboards, and all similar factors.

Show-window Lighting. Provide enough capacity for the number and size of lamps you plan to use. The requirements of Code Sec. 220-2(c) Exception 2 are met if you provide 200 watts for every linear foot of show window, measured horizontally at the base. For better stores this is not enough. Small reflectors with Type R or PAR lamps are very effective. The installation must be made so that the light source can't be seen from the street, if an effective display of merchandise is to be obtained. Figure 34-2 shows a well-lighted window.

The show-window lights can be controlled manually, but it is better to provide a time switch of the general type shown in Fig. 34-3, which will automatically turn the lights on and off at a preset time. The particular switch shown has a "Sunday and Holiday Cutoff" so that any particular days of the week can be automatically skipped. It also has an "astronomical dial," which means the switch can be set to turn the lights on and off at a particular interval before and after sundown; once set, it automatically compensates for longer or shorter days as the seasons change.

Fig. 34-2.   A well-lighted window.

Type of Lighting Equipment.   For a most modern store located in perhaps the most competitive area of a large city, there is little choice except to design the lighting system specifically to suit the exact size, structure, and layout of the building; the type of merchandise sold; the effects desired; and many other factors.   Only the most general rules can be given in a book of this kind, and final selection of the equipment must be left to one competent to analyze the dozens of factors involved.

No general rules can be given concerning the types of fixtures to be used.   They might be fluorescent or incandescent or a combination thereof.   They might be suspended from the ceiling, mounted on the ceiling, or flush in the ceiling.   With a low ceiling and fixtures mounted directly on the ceiling, select fixtures that direct most of the light downward.   With higher ceilings and fixtures suspended some distance below the ceiling, select fixtures that let a considerable portion of light go upward to light the ceiling.   In any event let some light fall on the ceiling in order to avoid a somewhat cavernous appearance that would be the result of dark ceilings.

Fig.   34-3. Automatic   time switches are most convenient in controlling show-window lighting.   (*Paragon Electric Co.*)

Often it is desirable to direct extra light upon some particular feature item on display.   This cannot be accomplished by using fluorescent lamps.   Often flush ceiling fixtures with incandescent lamps are used for the purpose.

In general, store lighting should be installed based on the recommendations of a lighting specialist well versed in the subject.

Types of Fluorescent Lamps.   For general lighting and for the most light per watt of power consumed, use cool white lamps.   On the other hand, the kind of merchandise on display will also be a major factor in the proper selection of lamps.   For light that is most nearly like natural light, use de luxe cool white; if the merchandise happens to have mostly "warm" colors such as red, brown, orange, or tan, the de luxe warm white will enhance its appearance.   In food stores the de luxe cool white emphasizes the crisp appearance of green vegetables and gives good appearance to meats, both lean and fat.   Prepared foods have acceptable appearance under either de luxe cool white or de luxe warm white.

35

Wiring Miscellaneous Occupancies

In a book of this kind, it is not possible to provide complete information regarding the wiring of the kinds of structures discussed in this chapter. Only a very basic and incomplete outline can be given, to acquaint you with the nature of the problems involved. The design of wiring installations in such structures should be undertaken only by those who have the necessary experience and qualifications for the kind of work involved. Nevertheless, this chapter should provide you with a preview of the kinds of problems involved.

In determining the service entrance, switchboards and panelboards, feeders, and circuits for any kind of occupancy, it is necessary to estimate the lighting load, to which must be added the appliance load, if any, and the motors. The Code in Sec. 220-2(a) (see Appendix) specifies the watts per square foot that must be allowed for the lighting load. The calculations based on the minimum are then relatively simple, but are not likely to provide the footcandles of light now considered acceptable.

Wet Locations. In locations that are more or less permanently wet, conduit and other metal parts of the wiring system rust and deteriorate far more rapidly than in normal dry locations.   Study Code Sec. 300-5, for it is important.   Section 300-5(c) makes special reference to dairies, laundries, and similar locations, including locations where walls are frequently washed, or where there are surfaces of absorbent material: The entire wiring system including conduit, cable, boxes, fittings, and so on must be installed so that there is at least $\frac{1}{4}$ in. of space between it and the supporting surface.

Factories.   It is impossible to specify the power that will be required for lighting purposes without knowing exactly what kind of project is involved.   A factory may be anything from a foundry where 50 footcandles might be considered good lighting to a watchmaking establishment where 300 footcandles would be considered too little.   Furthermore, the complete operation must be broken down into smaller parts.   The lighting in the boiler room need not be so good as that in the office.   In very few cases, however, will less than 30 footcandles be considered sufficient.

The motor load can be determined from the specification of the building.   The demand factor that is to be used with the motors will depend entirely upon the type of installation involved, and good common sense will be far more valuable than any number of printed pages in a book.

Theaters and Assembly Halls.   The Code in Art. 520 outlines the requirements for electrical installations in theaters and assembly halls, which are defined as "buildings, or part of a building, designed, intended, or used for dramatic, operatic, motion-picture, or other shows, and night clubs, dance halls, armories, sporting arenas, bowling alleys, public auditoriums, television studios, and like buildings used for public assembly."

Code requirements are rigid and severe, but simply covered in Art. 520, which should be studied in detail by those who intend to design or install an electrical installation in a theater.   Only the most essential of the requirements can be covered here.

If the area intended for public assembly has a capacity of 200 persons or more, the wiring must be in conduit, ALS cable, or MI cable. In smaller establishments, armored cable or nonmetallic-sheathed cable may be used.

Projectors of the professional type (those using 35-mm film or wider) must be installed in accordance with Code Art. 540.   Among the requirements are installation in an approved booth; operation by a professional operator; in some cases Type AA wire with 200°C or 392°F limits is required.   The wiring for sound is covered by Art. 640.

Receptacles for permanently installed arc lamps must have ampacity of at least 50 amp and must be wired with No. 6 or larger wire.   Receptacles for incandescent lamps may be rated as little as 20 amp and must be wired with No. 12 or heavier wire.

In dressing rooms all lights *and receptacles* must be controlled by switches.   All lights within 8 ft of the floor must be provided with open-end guards, riveted or otherwise permanently sealed or locked in place.   Each switch controlling a receptacle must have a pilot light, showing that the receptacle is turned on.

The requirements outlined by no means constitute all requirements for theaters.   Study Art. 520 well for all the requirements.

Signs.   This topic is covered by Code Art. 600.   The switch that controls a sign must open all ungrounded wires to the sign, and must be located within sight of the sign, unless it is of the type that can be locked in the "off" position.   Obviously this provision is intended to protect persons working on the sign.   All controls, including flashers, must be enclosed in metal cases.

All signs must be grounded, except that they may be completely insulated from the ground if "inaccessible to unauthorized persons."   This does not apply to small portable signs supplied from a circuit operating at a voltage of less than 150 volts to ground.   The maximum current permitted is 20 amp per circuit.

The wiring in signs may be open wiring on insulators, in conduit, or in troughs formed by the construction of the sign, or armored cable.   Wires must be of the lead-covered type, unless the wire is in conduit, so arranged as to be raintight and to drain in case moisture penetrates.   The wiring may be no lighter than No. 14 and, in case of those signs operating in excess of 600 volts, which includes all neon signs, must be selected with an insulation heavy enough for the voltage involved.   These details constitute but a small part of the Code requirements with respect to signs.   If you intend to do sign work, study Art. 600.

Remote Control and Signal Circuits.   The Code covers this subject in Art. 725.   Included are circuits such as the wires between a start-stop station for a motor and its controller; remote-control wiring as discussed in Chap. 20; thermostat wiring for heating systems; fire alarms; doorbells, chimes, and similar signaling devices; and many others.   Telephone, telegraph, messenger or fire and burglar-alarm systems *if* *connected to a central station* are not included, but are covered by a separate Art. 800.

The Code divides such circuits into two classes: Class 1 and Class 2. It will be easier to understand the subject if we discuss Class 2 systems first.

Class 2 systems are those supplied by transformers so designed that even under short-circuit conditions, the total power output is so small that there is no danger of fire. At times they may be protected by overcurrent protection as will be discussed. If a circuit does not qualify as Class 2, it automatically becomes Class 1.

Class 2 circuits are subdivided into five subgroups, as follows:

725-31. Limits of Class 2 Systems. Class 2 remote-control and signal systems, depending on the voltage shall have the current limited as follows:

(a) Maximum 15 Volts: 5 Amperes. Circuits in which the open-circuit voltage does not exceed 15 volts and having overcurrent protection of not more than 5-amperes rating. Where the current supply is from a transformer or other device having energy-limiting characteristics and approved for the purpose, or from primary batteries, the overcurrent protection may be omitted.

(b) 15 to 30 Volts: 3.2 Amperes. Circuits in which the open-circuit voltage exceeds 15 volts but does not exceed 30 volts and having overcurrent protection of not more than 3.2 amperes rating. Where the current supply is from a transformer or other device having energy-limiting characteristics and approved for the purpose, or from primary batteries, the overcurrent protection may be omitted.

(c) 30 to 60 Volts: 1.6 Amperes. Circuits in which open-circuit voltage exceeds 30 volts but does not exceed 60 volts and having overcurrent protection of not more than 1.6 amperes rating. Where the current supply is from a transformer or other device having energy-limiting characteristics and approved for the purpose, the overcurrent protection may be omitted.

(d) 60 to 150 Volts: 1 Ampere. Circuits in which the open-circuit voltage exceeds 60 volts but does not exceed 150 volts, and having overcurrent protection of not more than 1-ampere rating, provided that such circuits are equipped with current-limiting means other than overcurrent protection which will limit the current as a result of a fault to not exceeding 1 ampere.

(e) Maximum 150 Volts: 5 Milliamperes. Circuits in which the open-circuit voltage does not exceed 150 volts provided that such circuits are equipped with current limiting means, other than overcurrent protection, which are approved for the purpose and which will limit the current as a result of a fault to not exceeding 5 milliamperes.

The usual transformers used in these circuits may not exceed 100 volt-amp in rating.   Their primaries may be connected directly to any ordinary circuit with not over 20-amp overcurrent protection.   The low-voltage wires from the transformer may not run in the same raceway with power wires, unless separated by metal partition or divider.

If the system operates at 30 volts or less, there is no restriction on the kind and size of wire to be used.   If it operates at over 30 volts, the individual wires may be as small as No. 18 provided they have insulation at least equal to that found on Type TF fixture wire, and if No. 14 or heavier must be Type T or equivalent.   If, however, cables instead of individual wires are used, the requirements become more complicated and are covered by Sec. 725-42(c)(3); generally speaking, cables made for the purpose automatically meet the requirements.

Class 1 systems include specifically the wires between a start-stop station of a motor and its controller, and have already been discussed in Chap. 31.   Other examples are *local* (not central station) fire-alarm systems, special electric clock circuits, call systems, and all other systems where the supply system is *not* limited as outlined for Class 2 systems. The wires of the system may run in the same raceway as the wires of the regular wiring system, provided all wires are insulated for the same voltage; for example, if the signal wires are to carry only 30 volts but are installed in the same raceway with wires carrying 115 or 230 volts, all must be insulated for the higher voltage.   Wires as small as No. 18 may be used if installed in a raceway or in the form of a suitable cable.   Insulation on the wires must be the same as on Class 2 systems over 30 volts.   Overcurrent protection must be provided if the system operates at over 30 volts; the rating of the overcurrent device is the ampacity of the wire used.   Transformers if used to supply the power for such circuits must be limited to not over 1,000 volt-amp and not over 30 volts, and the secondary must be provided with overcurrent protection rated at not more than 250% of the ampere rating of the secondary.

Other requirements will be found in your copy of the Code.

Outdoor Lighting.   At times it will be necessary to install outdoor lighting, temporary or permanent.   The Code covers the subject in Art.

730.   The lighting might be for decorative purposes, or for lighting athletic fields, automobile parking lots, drive-ins, and similar locations. Overhead wiring must be a minimum of No. 10 for spans up to 50 ft, and not smaller than No. 8 for longer spans.   The minimum clearances above ground are the same as for service drops, as discussed in Chap. 17.

If all the lighting is on a pole, a common neutral may be used for all the circuits on the pole, provided there are not more than eight ungrounded wires.   The ampacity of the neutral must be: (a) if all the circuits are fed by a 2-wire 115-volt feeder, equal to the sum of the ampacities of all the hot wires; (b) if the circuits are fed by a 3-wire 115/230-volt feeder, equal to the sum of the ampacities of all the hot wires connected to one leg of the feeder; if unequal, use the larger of the two; (c) if fed by a 3-phase feeder, equal to the sum of the ampacities of all the wires connected to any one of the three hot wires; if unequal, naturally use the largest of the three.

At times you will want to install "festoon" lighting, which is defined in Code Sec. 730-6(b) as "a string of outdoor lights suspended between two points more than 15 ft apart."   The minimum size of wire that may be used is No. 12, unless supported by a messenger cable, usually a steel wire for strength; if the span is over 40 ft a messenger wire must be used regardless of the size of the electrical wire.   The wires must be supported by insulators at both ends.

The sockets are usually of the general class of "weatherproof sockets" shown in Fig. 35-1.   The individual leads of the sockets must be soldered to the wires, and the points of attachment must be staggered so that the soldered joint in one wire does not come directly opposite that in the other wire.   A pin-point socket of the type shown in Fig. 35-2 is most handy and saves much time, because soldering is not required. Lay the wires in the grooves of the socket and screw on the cover; the pin points in the socket puncture the insulation of the wire and make contact with the copper conductor of the wire.   Only *stranded* wire may be used with pin-point sockets.   Note the hook on each socket, for hanging on messenger cable.

Hazardous Locations.   This topic is covered by Code Arts. 500 to 517.   A location may be considered hazardous for a great variety of reasons.   If explosive vapors, such as those formed by gasoline, lacquers, or similar volatile products are present, obviously an explosion hazard exists.   The same is true where explosive dusts such as coal or grain dusts are present, as also is the case where there is a considerable amount of easily ignitable fibers, such as in a cotton mill.   Such materials can easily cause an explosion if ignited by electric sparks caused by switches, motors, pulling a plug out of a receptacle, a broken lamp, excessive heating, and many other causes.

Fig. 35-1. A weather-proof socket for outdoor use. (*General Electric Co.*)

Fig. 35-2. No soldering is necessary when using sockets of this type. Lay the wires in the grooves, screw on the cover. The sharp points puncture the insulation and make contact with the conductor. (*Pass & Seymour, Inc.*)

Wiring in Hazardous Locations.   Nowhere is properly installed wiring more important than in hazardous locations.   Poorly designed or carelessly installed wiring constitutes an invitation to explosions and fires that can be most disastrous.   The Code devotes 45 pages to the subject; so it should not be supposed that this book can do more than give a general outline of some of the most important points.   The student should not undertake wiring in such locations, but will want to know a little something about the subject.   The actual design and wiring must be undertaken only by the most experienced people, with lots of help from the electrical inspector.

In most hazardous locations, only conduit or MI cable may be used for wiring. Ordinary outlet boxes may not be used; explosion-proof fittings are used instead, to enclose all wires, switches, and other components. Such fittings must have threaded hubs which will engage at least five full threads of conduit. Covers may be either the screw type with five full threads, or alternately, fitting and cover must have mating flat machined surfaces accurate to within 0.0015 in. and at least ⅜ in. wide.

Installed in an explosive atmosphere, gases do seep into the interior of such fittings, and they do explode inside the fittings. The fittings are sturdy enough to withstand damage from such explosions, but it should not be supposed that they confine the explosion entirely to the inside of the fitting. The exploding gases do escape, but experience has shown that by the time they have passed through the crevices of five threads, or through the wide but very thin space between fitting and cover, they have cooled off sufficiently so that they will not set off an explosion on the outside.

Many types of explosion-proof fittings are available. Some are similar to ordinary conduit fittings shown in Figs. 29-1 and 29-2 but of special explosion-proof construction. A larger one is shown in Fig. 35-3,

Fig. 35-3. A typical explosion-proof fitting. (*Killark Electric Mfg. Co.*)

and others are available up to huge sizes to contain very large circuit breakers and similar equipment.

Where portable equipment must be used, special types of explosion-proof receptacles and plugs must be used. Both are shown in Fig. 35-4, and the combination is so designed that with the first pull of the plug, it can be withdrawn only part of the way. That first pull breaks the connection, and an arc, if it forms, will explode the small amount of vapor present in the interior.   This takes place during the brief period of time required to give the plug the twist that is necessary before it can be completely withdrawn.

Fig. 35-4. For portable equipment in hazardous areas, use explosion-proof plug and receptacle. (*Killark Electric Mfg. Co.*)

Fixtures of the explosion-proof type are designed so that the lamp is enclosed in a sturdy glass globe.   A typical unit of this type is shown in Fig. 35-5.

Fig. 35-5. An explosion-proof lighting fixture. (*Killark Electric Mfg. Co.*)

When a number of fittings or enclosures are installed using conduit, explosive gases or other materials can pass from one fitting or enclosure to another, through the conduit.   The conduit itself can also contain a substantial amount of such explosive material.   To minimize the quantity of explosive material that can accumulate in any one place, the Code in Sec. 501-5 requires that seals of the general type shown in Fig. 35-6 be installed, under many conditions there outlined.   The seal is installed in the run of conduit, and after the wires are pulled into place, is filled with a sealing compound that effectively prevents explosive materials from passing from one portion of the electrical installation, to another, through the conduit.

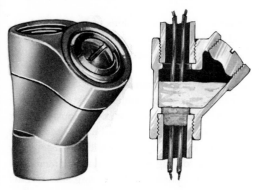

Fig. 35-6. By pouring sealing compound into a fitting of this kind, one run of conduit is sealed off from another, minimizing danger of explosion. (*Killark Electric Mfg. Co.*)

Code Classifications. In its treatment of hazardous locations, the Code in Art. 500 at first appears to be very complicated, but careful analysis reduces it to fairly simple proportions. The entire subject, however, is in no sense one which can be understood by casual reading, and careful study is necessary.

The Code divides different kinds of explosive *atmospheres* into seven different *groups*, depending on the nature of the explosive element.

The Code divides different kinds of *locations* into three *classes*, each subdivided into two *divisions*.

Types of Hazardous Atmospheres. For exact definition, see Code Sec. 500-2. The following is an *abbreviated* recap:

Group A: Atmospheres containing acetylene.

Group B: Atmospheres containing hydrogen, or gases or vapors of equivalent hazard such as manufactured gas.

Group C: Atmospheres containing ethyl ether vapors, ethylene, or cyclopropane.

Group D: Atmospheres containing gasoline, hexane, naphtha, benzine, butane, propane, alcohol, acetone, benzol, lacquer solvent vapors, or natural gas.

Group E: Atmospheres containing metal dust, including aluminum, magnesium, and their commercial alloys, and other metals of similarly hazardous characteristics.

Group F: Atmospheres containing carbon black, coal, or coke dust.

Group G: Atmospheres containing flour, starch, or grain dusts.

Types of Hazardous Locations.   The various kinds of hazardous locations are extensively defined in the three pages of Code Secs. 500-4 to 500-6.   The following is a *very abbreviated* recap:

Class 1, Div. 1.   Locations in which flammable gases or vapors in quantities sufficient to produce explosive or ignitable mixtures may be regularly expected (periodically, intermittently, or continuously) under *normal operating conditions,* or may exist frequently because of leakage or breakdown.   Examples: spray booths, locations with open tanks or vats of volatile flammable liquids; drying rooms or compartments for evaporation of flammable solvents; portions of cleaning and dyeing rooms where flammable liquids are used; many similar locations.

Class 1, Div. 2.   Locations similar to Class 1, Div. 1, except that the explosive conditions are not present under normal operating conditions, but only under *abnormal* conditions such as exist, for example, during a period of breakdown.   Details such as the amount of hazardous material that might escape in case of accident, the efficiency of the ventilating system, and similar factors will determine whether the location in question falls into Div. 1 or Div. 2.   The local inspector is the judge.

Class 2, Div. 1.   Locations where combustible and explosive **dust** may be *regularly* expected (periodically, intermittently, or continuously) under *normal operating conditions.*   Examples: grain-handling and -storage plants, flour mills, starch plants.

Class 2, Div. 2.   Locations similar to Class 2, Div. 1, except that the explosive conditions are not present under normal operating conditions, but only under *abnormal* conditions such as exist during breakdown.

Class 3, Div. 1.   Locations where **ignitible fibers or combustible flyings** are handled, manufactured, or used.   Examples: rayon, cotton, and similar textile mills; clothing-manufacturing establishments; woodworking plants.

Class 3, Div. 2.   Locations where such materials are stored or handled.

Residential Garages.   If the garage is designed to hold no more than three cars, and if it is at or above ground level, it is not considered a hazardous area, and any wiring method acceptable for the house is also acceptable for the garage.

If, however, the floor of such a garage is below ground level, it is considered a hazardous area as far as the first 18 in. above floor level are concerned, and any wiring within 18 in. of the floor must then be handled as in any other Class 1, Div. 2, location. But if you keep all wiring above that 18-in. line, wire the garage using any method acceptable for the house.

Commercial Garages. Any locations where cars and similar vehicles are serviced and repaired, or where more than three such vehicles may be stored, are considered commercial garages. Only some of the more important details of the wiring of such buildings can be covered here, and careful study of Art. 511 is necessary before proceeding to work on such premises.

All floors at or above ground level are considered Class 1, Div. 2, locations as far as the first 18 in. above each floor are concerned. But for any floor below ground level, the entire floor is considered Class 1, Div. 2, up to a point which is 18 in. *above* the bottom of *outside* doors or similar openings. Pits and depressions below floor level are also considered Class 1, Div. 2, locations, except that a single depression may be called Class 1, Div. 1, by the inspector.

The wiring within the Class 1, Div. 2, areas must, of course, be as prescribed by the Code for such locations, but all wiring in the remainder of the garage must be in a metallic raceway, or ALS or MI cable. Adjacent areas such as storerooms, offices, and so on are not considered hazardous areas provided their floors are at least 18 in. above the level of the garage floor.

Equipment located less than 12 ft above floor level, if of the type that can produce arcs, sparks, or particles of hot metal (cutouts, switches, charging panels, generators, motors, or other equipment having make-and-break or sliding contacts) must be of the totally enclosed type, or constructed to prevent escape of sparks or hot metal particles.

Permanently installed lighting fixtures located over lanes through which vehicles are commonly driven, or which otherwise might be subject to physical damage, must be located at least 12 ft above floor level. At lower levels, they must be the totally enclosed type, or constructed to prevent escape of sparks or hot metal particles.

Receptacles must be the grounding type. If located within an area defined as Class 1, they must be of the explosion-proof type, similar to that shown in Fig. 35-4.

Filling Stations.   This area is discussed in Code Art. 514.   The wiring to the island on which gasoline-dispensing pumps are found must be in rigid conduit, and all fittings within 4 ft of the ground level must be of the explosion-proof type.   A sealing fitting must be installed where the conduit enters the pump or other equipment on the island.   If the insulation on the wires is rubber or some other type which would be affected by gasoline spilled on it, the wires must be in lead-sheathed cable. Some brands of Type T or TW are also available in a special construction suitable for use in contact with gasoline.   The label on the coil will read "Gasoline- and Oil-resistant."   Such wire may be used without a lead sheath.   Do note, however, that ordinary Type T or TW is not suitable for this purpose; it is suitable only if it is the special type with the special label.   All circuits feeding dispensing pumps on pump islands must be controlled by switches that disconnect every wire, including the neutral (which, however, does not change the requirement that there must be no fuse in the neutral wire); if a circuit breaker is used, it must break all wires simultaneously.

Motors in Hazardous Locations.   All motors for use in hazardous locations must be of the explosion-proof type.   Obviously such motors can have no opening into the motor but must be totally enclosed and also constructed with extremely close tolerances on the mechanical fit of all parts.

Paint-spray Booths.   No electrical devices of any kind are permitted within the booth.   Here there is danger of an explosion not only from sparks but also by vapors present, which may explode upon contact with the relatively high temperatures found on lamps and similar devices under ordinary operating conditions.

Bibliography

HARRISON, WARD, AND K. A. STALEY: "Fundamentals of Illumination," General Electric Co., Nela Park, Cleveland.

HARRISON, WARD, AND C. E. WETT: "Illumination Design Data," General Electric Co., Nela Park, Cleveland.

SEGALL, B. Z.: "Electrical Code Diagrams," Peerless Publishing Co., New Orleans, La.

STETKA, FRANK, AND MERWIN BRANDON: "NFPA Handbook of the National Electrical Code," McGraw-Hill Book Company, New York.

"Farmstead Wiring Handbook," Industry Committee on Interior Wiring Design, New York.

"Handbook of Interior Wiring Design," Industry Committee on Interior Wiring Design, New York.

"IES Lighting Handbook," Illuminating Engineering Society, New York.

"National Electrical Code," National Fire Protection Association, Boston, Mass.

"Steel Electrical Raceways Design Manual," American Iron and Steel Institute, New York.

Appendix

This Appendix contains tables quoted from the National Electrical Code.[1] Some of the tables shown in the Code are not repeated here because they refer to subjects beyond the scope of this book. Others are omitted because they are rarely used and can be found in your copy of the Code. Portions of other tables are omitted for the same reason.

Certain sections of the Code are most important and specific, for which reason they are included here in full.

The Code contains some specific "examples" illustrating the application of Code requirements. They are included in this Appendix.

[1] Every student is urged to study the National Electrical Code.

Excerpts from the 1968 National Electrical Code are published with the permission of the National Fire Protection Association. Any further reproduction of this material is not authorized except with the permission of that association. Copies of the full text of the 1968 National Electrical Code are available from the National Fire Protection Association, 60 Batterymarch St., Boston. Mass. 02110, in a prepared bound edition at $2.00 per copy.

TABLE 220-2(a)   General Lighting Loads by Occupancies

Type of Occupancy	Unit Load per Square Foot, Watts
Armories and auditoriums	1
Banks.	2
Barbershops and beauty parlors	3
Churches	1
Clubs	2
Courtrooms.	2
Dwellings (other than hotels) °	3
Garages—commercial (storage)	½
Hospitals	2
Hotels and motels, including apartment houses without provisions for cooking by tenants °	2
Industrial commercial (loft) buildings	2
Lodge rooms.	1½
Office buildings	5
Restaurants.	2
Schools	3
Stores	3
Warehouses (storage)	¼
In any of the above occupancies except single-family dwellings and individual apartments of multifamily dwellings:	
Assembly halls and auditoriums	1
Halls, corridors, closets	½
Storage spaces	¼

° All receptacle outlets of 15-amp or less rating in single-family and multi-family dwellings and in guest rooms of hotels [except those connected to the receptacle circuits specified in Paragraph 220-3(b)] may be considered as outlets for general illumination, and no additional load need be included for such outlets.

TABLE 220-4(a)   Calculation of Feeder Loads by Occupancies

Type of occupancy	Portion of lighting load to which demand factor applies, wattage	Feeder demand factor, %
Dwellings—other than hotels	First 3,000 or less at Next 3,001 to 120,000 at Remainder over 120,000 at	100 35 25
Hospitals °.	First 50,000 or less at Remainder over 50,000 at	40 20
Hotels and motels including apartment houses without provision for cooking by tenants °	First 20,000 or less at Next 20,001 to 100,000 at Remainder over 100,000 at	50 40 30
Warehouses (storage)	First 12,500 or less at Remainder over 12,500 at	100 50
All others	Total wattage	100

° The demand factors of this table shall not apply to the computed load of subfeeders to areas in hospitals, hotels, and motels where entire lighting is likely to be used at one time, as in operating rooms, ballrooms, or dining rooms.

TABLE 220-5 Demand Loads for Household Electric Ranges, Wall-mounted Ovens, Counter-mounted Cooking Units, and Other Household Cooking Appliances over 1¾-kw Rating

(Column A to be used in all cases except as otherwise permitted in Note 4 on next page)

Number of appliances	Maximum demand, kilowatts (see notes) Column A (not over 12-kw rating)	Demand factors, % (see Note 4) Column B (less than 3½-kw rating)	Column C (3½- to 8¾-kw rating)
1	8	80	80
2	11	75	65
3	14	70	55
4	17	66	50
5	20	62	45
6	21	59	43
7	22	56	40
8	23	53	36
9	24	51	35
10	25	49	34
11	26	47	32
12	27	45	32
13	28	43	32
14	29	41	32
15	30	40	32
16	31	39	28
17	32	38	28
18	33	37	28
19	34	36	28
20	35	35	28
21	36	34	26
22	37	33	26
23	38	32	26
24	39	31	26
25	40	30	26
26–30	15 plus 1 for each range	30	24
31–40		30	22
41–50	25 plus ¾ for each range	30	20
51–60		30	18
61 & over		30	16

Notes to Table 220-5

Note 1.   Over 12-kw to 27-kw ranges all of same rating.   For ranges, in- dividually rated more than 12 kw but not more than 27 kw, the maximum demand in Column A shall be increased 5% for each additional kilowatt of rating or major fraction thereof by which the rating of individual ranges ex- ceeds 12 kw.

Note 2.   Over 12-kw to 27-kw ranges *of unequal ratings.*   For ranges indi- vidually rated more than 12 kw and of different ratings but none exceeding 27 kw an average value of rating shall be calculated by adding together the ratings of all ranges to obtain the total connected load (using 12 kw for any range rated less than 12 kw) and dividing by the total number of ranges; and then the maximum demand in Column A shall be increased 5% for each kilo- watt or major fraction thereof by which this average value exceeds 12 kw.

Note 3.   This table does not apply to commercial ranges.   See Table 220- 6(a) for demand factors for commercial cooking equipment.

Note 4.   Over 1¾-kw to 8¾-kw.   In lieu of the method provided in Col- umn A, loads rated more than 1¾ kw but not more than 8¾ kw may be con- sidered as the sum of the name-plate ratings of all the loads, multiplied by the demand factors specified in Columns B or C for the given number of loads.

Note 5.   Branch-circuit load.   Branch-circuit load for one range may be computed in accordance with Table 220-5.   The branch-circuit load for one wall-mounted oven or one counter-mounted cooking unit shall be the name- plate rating of the appliance.   The branch-circuit load for a counter- mounted cooking unit and not more than two wall-mounted ovens, all sup- plied from a single branch circuit and located in the same room shall be com- puted by adding the name-plate ratings of the individual appliances and treating this total as equivalent to one range.

TABLE 220-6(a) **Feeder Demand Factors for Commercial Electric Cooking Equipment; including Dishwasher Booster Heaters, Water Heaters, and Other Kitchen Equipment**

Number of units of equipment	Demand factor, %
1	100
2	100
3	90
4	80
5	70
6 and over	65

TABLE 220-6(b) **Demand Factors for Household Electric Clothes Dryers**

Number of dryers	Demand factor, %
1	100
2	100
3	100
4	100
5	80
6	70
7	65
8	60
9	55
10	50
11–13	45
14–19	40
20–24	35
25–29	32.5
30–34	30
35–39	27.5
40 and over	25

TABLE 310-2(a)   Conductor Application

Trade name	Type letter	Max operating temp	Application provisions
Rubber-covered fixture wire, solid or 7-strand	RF-1 °	60°C 140°F	Fixture wiring. Limited to 300 volts
	RF-2 °	60°C 140°F	Fixture wiring, and as permitted in Sec. 310-8
Rubber-covered fixture wire, flexible stranding	FF-1 °	60°C 140°F	Fixture wiring. Limited to 300 volts
	FF-2 °	60°C 140°F	Fixture wiring, and as permitted in Sec. 310-8
Heat-resistant rubber-covered fixture wire, solid or 7-strand	RFH-1 °	75°C 167°F	Fixture wiring. Limited to 300 volts
	RFH-2 °	75°C 167°F	Fixture wiring, and as permitted in Sec. 310-8
Heat-resistant rubber-covered fixture wire, flexible stranding	FFH-1 °	75°C 167°F	Fixture wiring. Limited to 300 volts
	FFH-2 °	75°C 167°F	Fixture wiring, and as permitted in Sec. 310-8
Thermoplastic-covered fixture wire, solid or stranded	TF °	60°C 140°F	Fixture wiring, and as permitted in Sec. 310-8, and for circuits as permitted in Art. 725
Thermoplastic-covered fixture wire, flexible stranding	TFF °	60°C 140°F	Fixture wiring, and as permitted in Sec. 310-8, and for circuits as permitted in Art. 725
Heat-resistant, thermoplastic-covered fixture wire, solid or stranded	TFN °	90°C	Fixture wiring, and as permitted in Sec. 310-8
Heat-resistant thermoplastic-covered fixture wire, flexible stranding	TFFN °	90°C	Fixture wiring, and as permitted in Sec. 310-8

° Fixture wires are not intended for installation as branch-circuit conductors except as permitted in Art. 725.

TABLE 310-2(a) **Conductor Application** (*Continued*)

Trade name	Type letter	Max operating temp	Application provisions
Cotton-covered heat-resistant fixture wire	CF °	90°C 194°F	Fixture wiring. Limited to 300 volts
Asbestos-covered heat-resistant fixture wire	AF °	150°C 302°F	Fixture wiring. Limited to 300 volts and indoor dry location
Fluorinated ethylene propylene fixture wire, solid or 7-strand	PF ° PGF °	150°C 302°F	Fixture wiring and as permitted in Sec. 310-8
Fluorinated ethylene propylene fixture wire	PFF ° PGFF °	150°C 302°F	Fixture wiring and as permitted in Sec. 310-8
Silicone rubber insulated fixture wire, solid or 7-strand	SF-1 °	200°C 392°F	Fixture wiring. Limited to 300 volts
	SF-2 °	200°C 392°F	Fixture wiring and as permitted in Sec. 310-8
Silicone rubber insulated fixture wire, Flexible stranding	SFF-1 °	150°C 302°F	Fixture wiring. Limited to 300 volts
	SFF-2 °	150°C 302°F	Fixture wiring and as permitted in Sec. 310-8
Heat-resistant rubber	RH	75°C 167°F	Dry locations
Heat-resistant rubber	RHH	90°C 194°F	Dry locations
Moisture and heat-resistant rubber	RHW	75°C 167°F	Dry and wet locations. For over 2,000 volts, insulation shall be ozone-resistant
Heat-resistant latex rubber	RUH	75°C	Dry locations
Moisture-resistant latex rubber	RUW	60°C 140°F	Dry and wet locations
Thermoplastic	T	60°C 140°F	Dry locations

° See footnote on page 617.

TABLE 310-2(a)   Conductor Application (Continued)

Trade name	Type letter	Max operating temp	Application provisions
Moisture-resistant thermoplastic	TW	60°C 140°F	Dry and wet locations
Heat-resistant thermoplastic	THHN	90°C 194°F	Dry locations
Moisture and heat-resistant thermoplastic	THW	75°C 167°F	Dry and wet locations
Moisture and heat-resistant thermoplastic	THWN	75°C 167°F	Dry and wet locations
Moisture and heat-resistant cross-linked thermosetting polyethylene	XHHW	90°C 194°F 75°C 167°F	Dry locations Wet locations
Moisture-, heat- and oil-resistant thermoplastic	MTW	60°C 140°F 90°C 194°F	Wet locations, machine-tool wiring. (see Art. 670 and NFPA Standard No. 79) Dry locations. Machine-tool wiring. (see Art. 670 and NFPA Standard No. 79)
Moisture-, heat-, and oil-resistant thermoplastic	THW-MTW	75°C 167°F 90°C 194°F	Dry and wet locations Special applications within electric discharge lighting equipment. Limited to 1,000 open-circuit volts or less. (size 14-8 only)
Thermoplastic and asbestos	TA	90°C 194°F	Switchboard wiring only
Thermoplastic and fibrous outer braid	TBS	90°C 194°F	Switchboard wiring only
Synthetic heat-resistant	SIS	90°C 194°F	Switchboard wiring only

Table 310-2(a)   Conductor Application (*Continued*)

Trade name	Type letter	Max operating temp	Application provisions
Mineral insulation (metal-sheathed)	MI	85° C 185° F	Dry and wet locations with Type O termination fittings
		250° C 482° F	For special application
Silicone-asbestos	SA	90° C 194° F	Dry locations
		125° C 257° F	For special application
Fluorinated ethylene propylene	FEP or FEPB	90° C 194° F	Dry locations
		200° C 392° F	Dry locations—special applications
Varnished cambric	V	85° C 185° F	Dry locations only. Smaller than No. 6 by special permission
Asbestos and varnished cambric	AVA	110° C 230° F	Dry locations only
Asbestos and varnished cambric	AVL	110° C 230° F	Dry and wet locations
Asbestos and varnished cambric	AVB	90° C 194° F	Dry locations only
Asbestos	A	200° C 392° F	Dry locations only. Only for leads within apparatus or within raceways connected to apparatus. Limited to 300 volts
Asbestos	AA	200° C 392° F	Dry locations only. Only for leads within apparatus or within raceways connected to apparatus or as open wiring. Limited to 300 volts
Asbestos	AI	125° C 257° F	Dry locations only. Only for leads within apparatus or within raceways connected to apparatus. Limited to 300 volts
Asbestos	AIA	125° C 257° F	Dry locations only. Only for leads within apparatus or within raceways connected to apparatus or as open wiring
Paper		85° C 185° F	For underground service conductors, or by special permission

TABLE 310-12   Allowable Ampacities of Insulated Copper Conductors
Not More than Three Conductors in Raceway or Cable or Direct Burial (Based on Ambient Temperature of 30° C 86° F)

Size	Temperature rating of conductor. See Table 310-2(a)						
AWG MCM	60°C (140°F)	75°C (167°F)	85°C (185°F)	90°C (194°F)	110°C (230°F)	125°C (257°F)	200°C (392°F)
	Types RUW (14-2), T, TW	Types RH, RHW, RUH (14-2), THW, THWN, XHHW, THW-MTW	Types V, MI	Types TA, TBS, SA, AVB, SIS, FEP, FEPB, RHH, THHN, XHHW†	Types AVA, AVL	Types AI (14-8), AIA	Types A (14-8), AA, FEP,° FEPB °
14	15	15	25	25 ‡	30	30	30
12	20	20	30	30 ‡	35	40	40
10	30	30	40	40 ‡	45	50	55
8	40	45	50	50	60	65	70
6	55	65	70	70	80	85	95
4	70	85	90	90	105	115	120
3	80	100	105	105	120	130	145
2	95	115	120	120	135	145	165
1	110	130	140	140	160	170	190
0	125	150	155	155	190	200	225
00	145	175	185	185	215	230	250
000	165	200	210	210	245	265	285
0000	195	230	235	235	275	310	340
250	215	255	270	270	315	335	
300	240	285	300	300	345	380	
350	260	310	325	325	390	420	
400	280	335	360	360	420	450	
500	320	380	405	405	470	500	
600	355	420	455	455	525	545	
700	385	460	490	490	560	600	
750	400	475	500	500	580	620	
800	410	490	515	515	600	640	
900	435	520	555	555			
1000	455	545	585	585	680	730	
1250	495	590	645	645			
1500	520	625	700	700	785		
1750	545	650	735	735			
2000	560	665	775	775	840		

° Special use only.   See Table 310-2(a).

† For dry locations only.   See Table 310-2(a).

These ampacities relate only to conductors described in Table 310-2(a).

‡ The ampacities for Types FEP, FEPB, RHH, THHN, and XHHW conductors for sizes AWG 14, 12, and 10 shall be the same as designated for 75°C conductors in this table.

For ambient temperatures over 30°C, see Correction Factors, Note 15.

Be sure to read Notes on pages 625 and 626.

TABLE 310-13 **Allowable Ampacities of Insulated Copper Conductors**
Single Conductor in Free Air
(Based on Ambient Temperature of 30° C 86° F)

Size	Temperature rating of conductor. See Table 310-2(a)							
AWG MCM	60°C (140°F)	75°C (167°F)	85°C (185°F)	90°C (194°F)	110°C (230°F)	125°C (257°F)	200°C (392°F)	
	Types RUW (14-2), T, TW	Types RH, RHW, RUH (14-2), THW, THWN, XHHW	Types V, MI	Types TA, TBS, SA, AVB, SIS, FEP, FEPB, RHH, THHN, XHHW †	Types AVA, AVL	Types AI (14-8), AIA	Types A (14-8), AA, FEP,° FEPB °	Bare and covered conductors
14	20	20	30	30 †	40	40	45	30
12	25	25	40	40 †	50	50	55	40
10	40	40	55	55 †	65	70	75	55
8	55	65	70	70	85	90	100	70
6	80	95	100	100	120	125	135	100
4	105	125	135	135	160	170	180	130
3	120	145	155	155	180	195	210	150
2	140	170	180	180	210	225	240	175
1	165	195	210	210	245	265	280	205
0	195	230	245	245	285	305	325	235
00	225	265	285	285	330	355	370	275
000	260	310	330	330	385	410	430	320
0000	300	360	385	385	445	475	510	370
250	340	405	425	425	495	530	....	410
300	375	445	480	480	555	590	....	460
350	420	505	530	530	610	655	....	510
400	455	545	575	575	665	710	....	555
500	515	620	660	660	765	815	....	630
600	575	690	740	740	855	910	....	710
700	630	755	815	815	940	1005	....	780
750	655	785	845	845	980	1045	....	810
800	680	815	880	880	1020	1085	....	845
900	730	870	940	940	....	....	....	905
1000	780	935	1000	1000	1165	1240	....	965
1250	890	1065	1130	1130				
1500	980	1175	1260	1260	1450	....	....	1215
1750	1070	1280	1370	1370				
2000	1155	1385	1470	1470	1715	....	....	1405

° Special use only. See Table 310-2(a).
† For dry locations only. See Table 310-2(a).
These ampacities relate only to conductors described in Table 310-2(a).
‡ The ampacities for Types FEP, FEPB, PHH, THHN, and XHHW conductors for sizes AWG 14, 12, and 10 shall be the same as designated for 75°C conductors in this table.
For ambient temperatures over 30°C, see Correction Factors, Note 15.

Be sure to read Notes on pages 625 and 626.

TABLE 310-14 Allowable Ampacities of Insulated Aluminum Conductors
Not More than Three Conductors in Raceway or Cable or Direct Burial (Based on Ambient Temperature of 30° C 86° F)

Size	Temperature rating of conductor. See Table 310-2(a)						
AWG MCM	60°C (140°F)	75°C (167°F)	85°C (185°F)	90°C (194°F)	110°C (230°F)	125°C (257°F)	200°C (392°F)
	Types RUW (12-2), T, TW	Types RH, RHW, RUH (12-2), THW, THWN, XHHW	Types V, MI	Types TA, TBS, SA, AVB, SIS, RHH THHN XHHW †	Types AVA, AVL	Types AI (12-8), AIA	Types A (12-8), AA
12	15	15	25	25 ‡	25	30	30
10	25	25	30	30 ‡	35	40	45
8	30	40	40	40	45	50	55
6	40	50	55	55	60	65	75
4	55	65	70	70	80	90	95
3	65	75	80	80	95	100	115
2 °	75	90	95	95	105	115	130
1 °	85	100	110	110	125	135	150
0 °	100	120	125	125	150	160	180
00 °	115	135	145	145	170	180	200
000 °	130	155	165	165	195	210	225
0000 °	155	180	185	185	215	245	270
250	170	205	215	215	250	270	
300	190	230	240	240	275	305	
350	210	250	260	260	310	335	
400	225	270	290	290	335	360	
500	260	310	330	330	380	405	
600	285	340	370	370	425	440	
700	310	375	395	395	455	485	
750	320	385	405	405	470	500	
800	330	395	415	415	485	520	
900	355	425	455	455			
1000	375	445	480	480	560	600	
1250	405	485	530	530			
1500	435	520	580	580	650		
1750	455	545	615	615			
2000	470	560	650	650	705		

These ampacities relate only to conductors described in Table 310-2(a).

° For three-wire single-phase service, the allowable ampacity of RH, RHH, RHW, and THW aluminum conductors shall be for sizes No. 2-100 amp, No. 1-110 amp, No. 1/0-125 amp, No. 2/0-150 amp, No. 3/0-170 amp, and No. 4/0-200 amp.

† For dry locations only. See Table 310-2(a).

‡ The ampacities for Types RHH, THHN, and XHHW conductors for sizes AWG 12 and 10 shall be the same as designated for 75°C conductors in this table.

For ambient temperatures over 30°C, see Correction Factors, Note 15.

Be sure to read Notes on pages 625 and 626.

TABLE 310-15 **Allowable Ampacities of Insulated Aluminum Conductors**
Single Conductor in Free Air
(Based on Ambient Temperature of 30° C 86° F)

Size	Temperature rating of conductor. See Table 310-2(a)							
AWG MCM	60°C (140°F)	75°C (167°F)	85°C (185°F)	90°C (194°F)	110°C (230°F)	125°C (257°F)	200°C (392°F)	
	Types RUW (12-2), T, TW	Types RH, RHW, RUH (12-2), THW, THWN, XHHW	Types V, MI	Types TA, TBS, SA, AVB, SIS, RHH, THHN, XHHW°	Types AVA, AVL	Types AI (12-8), AIA	Types A (12-8), AA	Bare and covered conductors
12	20	20	30	30 †	40	40	45	30
10	30	30	45	45 †	50	55	60	45
8	45	55	55	55	65	70	80	55
6	60	75	80	80	95	100	105	80
4	80	100	105	105	125	135	140	100
3	95	115	120	120	140	150	165	115
2	110	135	140	140	165	175	185	135
1	130	155	165	165	190	205	220	160
0	150	180	190	190	220	240	255	185
00	175	210	220	220	255	275	290	215
000	200	240	255	255	300	320	335	250
0000	230	280	300	300	345	370	400	290
250	265	315	330	330	385	415	....	320
300	290	350	375	375	435	460	....	360
350	330	395	415	415	475	510	....	400
400	355	425	450	450	520	555	....	435
500	405	485	515	515	595	635	....	490
600	455	545	585	585	675	720	....	560
700	500	595	645	645	745	795	....	615
750	515	620	670	670	775	825	....	640
800	535	645	695	695	805	855	....	670
900	580	700	750	750	....	....	....	725
1000	625	750	800	800	930	990	....	770
1250	710	855	905	905				
1500	795	950	1020	1020	1175	....	....	985
1750	875	1050	1125	1125				
2000	960	1150	1220	1220	1425	....	....	1165

These ampacities relate only to conductors described in Table 310-2(a).
° For dry locations only. See Table 310-2(a).
† The ampacities for Types RHH, THHN, and XHHW conductors for sizes AWG 12 and 10 shall be the same as designated for 75°C conductors in this table.
For ambient temperatures over 30°C, see Correction Factors, Note 15.

Be sure to read Notes on pages 625 and 626.

Notes to Tables 310-12 through 310-15.

Ampacity. The maximum, continuous, ampacities of copper conductors are given in Tables 310-12 and 310-13. The ampacities of aluminum conductors are given in Tables 310-14 and 310-15.

1. Explanation of Tables. For explanation of Type Letters, and for recognized size of conductors for the various conductor insulations, see Sections 310-2 and 310-3. For installation requirements, see Section 310-1 through 310-7, and the various Articles of this Code. For flexible cords see Tables 400-9 and 400-11.

2. Application of Tables. For open wiring on insulators and for concealed knob-and-tube work, the allowable ampacities of Tables 310-13 and 310-15 shall be used. For all other recognized wiring methods, the allowable ampacities of Tables 310-12 and 310-14 shall be used, unless otherwise provided in this Code.

3. Aluminum Conductors. For aluminum conductors, the allowable ampacities shall be in accordance with Tables 310-14 and 310-15.

4. Bare Conductors. Where bare conductors are used with insulated conductors, their allowable ampacities shall be limited to that permitted for the insulated conductors of the same size.

5. Type MI Cable. The temperature limitation on which the ampacities of Type MI cable are based, is determined by the insulating materials used in the end seal. Termination fittings incorporating unimpregnated organic insulating materials are limited to 85°C operation.

6. Ultimate Insulation Temperature. In no case shall conductors be associated together in such a way with respect to the kind of circuit, the wiring method employed, or the number of conductors, that the limiting temperature of the conductors will be exceeded.

7. Use of Conductors with Higher Operating Temperatures. Where the room temperature is within 10 degrees C of the maximum allowable operating temperature of the insulation, it is desirable to use an insulation with a higher maximum allowable operating temperature; although insulation can be used in a room temperature approaching its maximum allowable operating temperature limit if the current is reduced in accordance with the Correction Factors for different room temperatures as shown in the Correction Factor Table, Note 15.

8. More Than Three Conductors in a Raceway or Cable. Tables 310-12 and 310-14 give the allowable ampacities for not more than three conductors in a raceway or cable. Where the number of conductors in a raceway or cable exceeds three, the allowable ampacity of each conductor shall be reduced as shown in the following Table:

Number of Conductors	Tables 310-12 and 310-14, %
4 to 6	to 80
7 to 24	to 70
25 to 42	to 60
43 and above	to 50

Exception No. 1—When conductors of different systems, as provided in Section 300-3, are installed in a common raceway the derating factors shown above apply to the number of

Power and Lighting (Articles 210, 215, 220 and 230) conductors only.

Where the number of conductors in a raceway or cable exceeds three, or where single conductors or multiconductor cables are stacked or bundled without maintaining spacing as required in Article 318 and are not installed in raceways, the individual ampacity of each conductor shall be reduced as shown in the above table.

Exception No. 2—The derating factors of Sections 210-23(b) and 220-2 (second paragraph) do not apply when the above derating factors are also required.

9. Where Type XHHW crosslinked thermosetting polyethylene insulated wire is used in wet locations, the allowable ampacities shall be that of Column 3 in Tables 310-12 through 310-15. Where used in dry locations, the allowable ampacities shall be that of Column 4 in Tables 310-12 through 310-15.

10. Overcurrent   Protection. Where the standard ratings and settings of overcurrent devices do not correspond with the ratings and settings allowed for conductors, the next higher standard rating and setting may be used.

Exception—Except as limited in Section 240-5.

11. Neutral   Conductor. (a) A neutral conductor which carries only the unbalanced current from other conductors, as in the case of normally balanced circuits of three or more conductors, shall not be counted in determining ampacities as provided for in Note 8.

(b) In a 3-wire circuit consisting of two-phase wires and the neutral of a 4-wire, 3-phase WYE connected system, a common conductor carries approximately the same current as the other conductors and shall be counted in determining ampacities as provided in Note 8.

Where the major portion of the load consists of electric discharge lighting there may be harmonic currents present in the neutral conductor which may be equal to the phase currents, thus the neutral could be considered to be a current-carrying conductor.

12. Voltage Drop. The allowable ampacities in Tables 310-12 through 310-15 are based on temperature alone and do not take voltage drop into consideration.

13. Deterioration of Insulation. It should be noted that even the best grades of rubber insulation will deteriorate in time, so eventually will need to be replaced.

14. Aluminum Sheathed Cable. The ampacities of Type ALS cable are determined by the temperature limitation of the insulated conductors incorporated within the cable. Hence the ampacities of aluminum sheathed cable may be determined from the columns in Tables 310-12 and 310-14 applicable to the type of insulated conductors employed within the cable. See Note 9.

15. Correction Factors Ambient temps over 30°C 86°F

°C	°F	60°C (140°F)	75°C (167°F)	85°C (185°F)	90°C (194°F)	110°C (230°F)	125°C (257°F)	200°C (392°F)
40	104	.82	.88	.90	.90	.94	.95	
45	113	.71	.82	.85	.85	.90	.92	
50	122	.58	.75	.80	.80	.87	.89	
55	131	.41	.67	.74	.74	.83	.86	
60	140	. . . .	.58	.67	.67	.79	.83	.91
70	158	. . . .	.35	.52	.52	.71	.76	.87
75	167	. . . .	. . . .	.43	.43	.66	.72	.86
80	176	. . . .	. . . .	.30	.30	.61	.69	.84
90	194	. . . .	. . . .	. . . .	. . . .	.50	.61	.80
100	212	. . . .	. . . .	. . . .	. . . .	. . . .	.51	.77
120	248	. . . .	. . . .	. . . .	. . . .	. . . .	. . . .	.69
140	284	. . . .	. . . .	. . . .	. . . .	. . . .	. . . .	.59

TABLE 310-20(c)   Typical Ambient Temperatures

Location	Temperature	Minimum rating of required conductor insulation
Well-ventilated, normally heated buildings	30°C (86°F)	(See note below)
Buildings with such major heat sources as power stations or industrial processes	40°C(104°F)	75°C(167°F)
Poorly ventilated spaces such as attics	45°C(113°F)	
Furnaces and boiler rooms (min.)	40°C(104°F)	75°C(167°F)
(max.)	60°C(140°F)	90°C(194°F)
Outdoors in shade in air	40°C(104°F)	75°C(167°F)
In thermal insulation	45°C(113°F)	75°C(167°F)
Direct solar exposure	45°C(113°F)	75°C(167°F)
Places above 60°C(140°F)		110°C(230°F)

Note: 60°C for up to and including No. 8 AWG copper and up to and including No. 6 AWG aluminum.   75°C for over No. 8 AWG copper and No. 6 AWG aluminum.

**TABLE 310-21   Simplified Wiring Table
(See Section 310-20 for use)
Conductor Size *—6 or Fewer Conductors
in Raceway or Cable**

Am-peres	Copper				Aluminum			
	Noncontinuous		Continuous		Noncontinuous		Continuous	
	AWG	MCM	AWG	MCM	AWG	MCM	AWG	MCM
15	14		14		12		12	
20	12		12		10		10	
25	10		10		8		8	
30	10		10		8		8	
35	8		8		6		6	
40	8		8		6		6	
45	6		6		4		4	
50	6		6		4		4	
60	4		4		4		4	
70	4		4		3		3	
80	3		3		3		2	
90	3		2		2		1	
100	2		1		1		0	
110	1		0		0		2/0	
125	1		0		2/0		3/0	
150	0		2/0		3/0		4/0	
175	2/0		3/0		4/0			250
200	3/0		4/0			250		300
225	4/0			250		300		350
250		250		300		350		400
300		350		400		400		750
350		400		500		500		1000
400		500		750		750		
450		750		1000		1000		
500		750				1000		
600		1000						

° Neutral conductors shall be treated in accordance with Note 11—Neutral Conductors of Notes to Tables 310-12 through 310-15.

TABLE 430-148   Full-load Currents in Amperes
Single-phase Alternating-current Motors

The following values of full-load currents are for motors running at usual speeds and motors with normal torque characteristics.   Motors built for especially low speeds or high torques may have higher full-load currents, and multispeed motors will have full-load current varying with speed, in which case the name-plate current ratings shall be used.

To obtain full-load currents of 208- and 200-volt motors, increase corresponding 230-volt motor full-load currents by 10 and 15%, respectively.

The voltages listed are rated motor voltages.   Corresponding nominal system voltages are 110 to 120 and 220 to 240.

Horsepower	115 volts	230 volts
$\frac{1}{6}$	4.4	2.2
$\frac{1}{4}$	5.8	2.9
$\frac{1}{3}$	7.2	3.6
$\frac{1}{2}$	9.8	4.9
$\frac{3}{4}$	13.8	6.9
1	16	8
$1\frac{1}{2}$	20	10
2	24	12
3	34	17
5	56	28
$7\frac{1}{2}$	80	40
10	100	50

TABLE 430-150   Full-load Current *
Three-phase AC Motors

Horse-power	Induction type, squirrel-cage and wound rotor, amperes					Synchronous type, unity power factor, amperes †			
	115 volts	230 volts	460 volts	575 volts	2,300 volts	220 volts	440 volts	550 volts	2,300 volts
½	4	2	1	.8					
¾	5.6	2.8	1.4	1.1					
1	7.2	3.6	1.8	1.4					
1½	10.4	5.2	2.6	2.1					
2	13.6	6.8	3.4	2.7					
3	. . .	9.6	4.8	3.9					
5	. . .	15.2	7.6	6.1					
7½	. . .	22	11	9					
10	. . .	28	14	11					
15	. . .	42	21	17					
20	. . .	54	27	22					
25	. . .	68	34	27	. . .	54	27	22	
30	. . .	80	40	32	. . .	65	33	26	
40	. . .	104	52	41	. . .	86	43	35	
50	. . .	130	65	52	. . .	108	54	44	
60	. . .	154	77	62	16	128	64	51	12
75	. . .	192	96	77	20	161	81	65	15
100	. . .	248	124	99	26	211	106	85	20
125	. . .	312	156	125	31	264	132	106	25
150	. . .	360	180	144	37	. . .	158	127	30
200	. . .	480	240	192	49	. . .	210	168	40

For full-load currents of 208- and 200-volt motors, increase the corresponding 230-volt motor full-load current by 10 and 15%, respectively.

° These values of full-load current are for motors running at speeds usual for belted motors and motors with normal torque characteristics. Motors built for especially low speeds or high torques may require more running current, and multispeed motors will have full-load current varying with speed, in which case the name-plate current rating shall be used.

† For 90 and 80% power factor the above figures shall be multiplied by 1.1 and 1.25 respectively.

The voltages listed are rated motor voltages. Corresponding nominal system voltages are 110 to 120, 220 to 240, 440 to 480, and 550 to 600 volts.

TABLE 430-151   Locked-rotor Current Conversion Table

As Determined from Horsepower and Voltage Rating
For Use Only with Section 430-83, Exception No. 3,
and 430-110(b)

Max horsepower rating	Motor locked-rotor current, amperes					
	Single phase		Two- or three-phase			
	115 volts	230 volts	115 volts	230 volts	460 volts	575 volts
½	58.8	29.4	24	12	6	4.8
¾	82.8	41.4	33.6	16.8	8.4	6.6
1	96	48	42	21	10.8	8.4
1½	120	60	60	30	15	12
2	144	72	78	39	19.8	15.6
3	204	102	. . .	54	27	24
5	336	168	. . .	90	45	36
7½	480	240	. . .	132	66	54
10	600	300	. . .	162	84	66
15	. . .	. . .	. . .	240	120	96
20	. . .	. . .	. . .	312	156	126
25	. . .	. . .	. . .	384	192	156
30	. . .	. . .	. . .	468	234	186
40	. . .	. . .	. . .	624	312	246
50	. . .	. . .	. . .	750	378	300
60	. . .	. . .	. . .	900	450	360
75	. . .	. . .	. . .	1,110	558	444
100	. . .	. . .	. . .	1,476	738	588
125	. . .	. . .	. . .	1,860	930	744
150	. . .	. . .	. . .	2,160	1,080	864
200	. . .	. . .	. . .	2,880	1,440	1,152

TABLE 430-152   Maximum Rating or Setting of Motor
Branch-circuit Protective Devices for Motors Marked
with a Code Letter Indicating Locked-rotor KVA

Type of motor	Full-load current, %		
		Circuit-breaker setting	
	Fuse rating	Instantaneous type	Time-limit type
All AC single-phase and polyphase squirrel-cage and synchronous motors with full-voltage, resistor or reactor starting:			
Code Letter A	150	700	150
Code Letter B to E	250	700	200
Code Letter F to V	300	700	250
All AC squirrel-cage and synchronous motors with auto-transformer starting:			
Code Letter A	150	700	150
Code Letter B to E	200	700	200
Code Letter F to V	250	700	200

For certain exceptions to the values specified see Sections 430-52 and 430-54. The values given in the last column also cover the ratings of nonadjustable, time-limit types of circuit breakers which may also be modified as in Section 430-52.

Synchronous motors of the low-torque, low-speed type (usually 450 rpm or lower), such as are used to drive reciprocating compressors, pumps, etc., which start up unloaded, do not require a fuse rating or circuit-breaker setting in excess of 200% of full-load current.

For motors not marked with a Code Letter, see Table 430-153.

TABLE 430-153   Maximum Rating or Setting of Motor Branch-circuit Protective Devices for Motors Not Marked with a Code Letter Indicating Locked-rotor KVA

Type of motor	Full-load current, %		
		Circuit-breaker setting	
	Fuse rating	Instan-taneous type	Time-limit type
Single-phase, all types	300	700	250
Squirrel-cage and synchronous (full-voltage, resistor and reactor starting)	300	700	250
Squirrel-cage and synchronous (auto-trans-former starting)			
Not more than 30 amperes	250	700	200
More than 30 amperes	200	700	200
High-reactance squirrel-cage			
Not more than 30 amperes	250	700	250
More than 30 amperes	200	700	200
Wound-rotor	150	700	150
Direct-current			
Not more than 50 hp	150	250	150
More than 50 hp	150	175	150
Sealed (hermetic type)			
Refrigeration compressor °			
400 KVA locked-rotor or less	175 †	. . .	175 †

For certain exceptions to the values specified see Sections 430-52, and 430-59.   The values given in the last column also cover the ratings of nonadjustable, time-limit types of circuit breakers which may also be modified as in Section 430-52.

Synchronous motors of the low-torque low-speed type (usually 450 rpm or lower) such as are used to drive reciprocating compressors, pumps, etc., which start up unloaded, do not require a fuse rating or circuit-breaker setting in excess of 200% of full-load current.

For motors marked with a Code Letter, see Table 430-152.

° The locked rotor KVA is the product of the motor voltage and the motor locked-rotor current (LRA) given on the motor name-plate divided by 1,000 for single-phase motors, or divided by 580 for 3-phase motors.

† This value may be increased to 225% if necessary to permit starting.

Code Chapter 9

Code Chapter 9 consists of two parts:

A. Tables not shown elsewhere in the Code. Only Tables 1, 1A, and 8 are of sufficient general interest to warrant repeating here. The others are of minor interest or concern projects beyond the scope of this book. However, *parts* of some of these tables have already been shown in the text of this book.

B. Examples. These show how to apply Code rules and will be repeated in the pages following.

TABLE 1 Maximum Number of Conductors in Trade Sizes of Conduit or Tubing—NEW WORK

Col. A = Types RF-2, RFH-2, RH, RHH, RHW, RUH, RUW, T, TF, THW, TW
  XHHW (AWG 14 through 6)
  FEPB (AWG 6 through 2)

Col. B = FEP, THHN, THWN, TFN, PF, PGF
  XHHW (AWG 4 through 2,000 MCM)
  FEPB (AWG 14 through 8)

Derating factors for more than three conductors in raceways, see Note 8, Tables 310-12 through 310-15 (See Sections 300-17, 300-18, 346-6, and 348-6)

Maximum number of conductors in conduit or tubing (based upon per cent conductor fill, Table 3, Chapter 9 for New Work)

Size AWG or MCM	½ in. A	½ in. B	¾ in. A	¾ in. B	1 in. A	1 in. B	1¼ in. A	1¼ in. B	1½ in. A	1½ in. B	2 in. A	2 in. B	2½ in. A	2½ in. B	3 in. A	3 in. B	3½ in. A	3½ in. B	4 in. A	4 in. B	4½ in. A	4½ in. B	5 in. A	5 in. B	6 in. A	6 in. B
18	7	11	12	20	20	33	35	58	49	80	80	131	115	187	176											
16	6	9	10	16	17	27	30	47	41	64	68	106	98	151	150											
14	4	8	6	15	10	24	18	43	25	58	41	96	58	137	90	158	121		155		197					
12	3	6	5	11	8	18	15	32	21	43	34	71	50	102	76		103		132		168					
10	1	4	4	7	7	11	13	20	17	27	29	45	41	65	64	100	86	134	110	172	140		173			
8	1	2	3	4	4	6	7	11	10	16	17	26	25	37	38	58	52	78	67	100	85	127	105	157	152	
6	1	1	1	2	3	4	4	7	6	9	10	16	15	23	23	35	32	47	41	61	52	78	64	96	93	139
4	1	1	1	1	1	2	3	4	5	6	8	9	12	14	18	21	24	29	31	37	40	48	49	59	72	85

3	2	1	0	2/0	3/0	4/0	250	300	350	400	500	600	700	750	800	900	1,000	1,250	1,500	1,750	2,000
72	61	45	38	32	27	23	19	16	15	13	11	9	8	8	7	7	6	5	4	4	3
63	55	42	37	32	27	23	19	16	15	13	11	9	8	8	7	7	6	5	4	4	3
50	42	31	26	22	19	16	13	11	10	9	8	6	6	5	5	4	4	3	3	2	1
44	38	29	25	22	19	16	13	11	10	9	8	6	6	5	5	4	4	3	3	2	1
40	34	25	21	18	15	13	11	9	8	7	6	5	4	4	4	4	3	3	2	2	1
35	31	23	20	18	15	13	11	9	8	7	6	5	4	4	4	4	3	3	2	2	1
31	26	20	16	14	12	10	8	7	6	6	5	4	3	3	3	3	3	2	1	1	1
28	24	18	16	14	12	10	8	7	6	6	5	4	3	3	3	3	3	1	1	1	1
24	20	15	13	11	9	8	6	5	5	4	4	3	3	3	2	2	1	1	1	1	1
21	19	14	12	11	9	8	6	5	5	4	4	3	3	3	2	1	1	1	1	1	1
18	15	11	9	8	7	6	5	4	3	3	3	2	2	1	1		1	1	1	1	
16	14	10	9	8	7	6	5	4	3	3	3	1	1	1	1		1	1	1	1	
12	10	7	6	5	4	3	3	3	2	2	1	1	1	1	1		1	1			
10	9	7	6	5	4	3	3	3	1	1		1	1	1	1		1	1			
8	7	5	4	3	3	2	2	1	1	1		1	1	1	1		1				
7	6	4	4	3	3	2	1	1	1	1		1	1	1	1		1				
5	4	3	2	2	1	1	1	1	1	1											
4	3	3	2	1	1	1	1	1	1	1											
3	3	2	2	1	1	1	1	1	1												
3	3	1	1	1	1	1	1	1	1												
2	1	1	1	1	1																
1	1	1	1	1	1																
1	1	1																			
1	1	1																			

TABLE 1A Maximum Number of Conductors in Trade Sizes of Conduit or Tubing – REWIRING *

Col. A = THW, TW, T, TF, RUH, RUW
(RHH and RHW without outer covering)
XHHW (AWG 14 through 6)
FEPB (AWG 6 through 2)

Col. B = THWN, THHN, FEP, TFN, PF, PGF
XHHW (AWG 4 through 2,000 MCM)
FEPB (AWG 14 through 8)

Derating factors for more than three conductors in raceways, see Note 8, Tables 310-12 through 310-15 (See Sections 300-17, 300-18, 346-6 and 348-6)

Maximum number of conductors in conduit or tubing (based upon per cent conductor fill, Table 3, Chapter 9 for rewiring)

Size AWG or MCM	½ in.		¾ in.		1 in.		1¼ in.		1½ in.		2 in.		2½ in.		3 in.		3½ in.		4 in.		4½ in.		5 in.		6 in.	
	A	B	A	B	A	B	A	B	A	B	A	B	A	B	A	B	A	B	A	B	A	B	A	B	A	B
18	13	19	24	33	38	53	68	93	93	127	152															
16	11	15	19	27	31	43	55	76	75	103	123	170	176													
14	5	13	10	24	16	39	29	69	40	94	65	154	93	164	143		192									
12	4	10	8	18	13	29	24	51	32	70	53	114	76		117		157									
10	4	6	6	11	11	18	19	32	26	44	43	72	61	104	95	160	127		163							
8	1	3	4	6	6	10	11	19	15	26	25	42	36	60	56	93	75	125	96	160	123		152			
6	1	1	2	4	4	6	7	11	10	15	16	25	23	37	36	56	48	76	62	98	79	125	97	154	141	
4	1	1	1	2	3	4	5	7	7	9	12	16	17	22	27	35	36	47	46	60	60	76	73	94	106	136

Size	1	2	3	4	5	6	7	8	9	10	11	12	13	14	15	16	17	18	19	20	21	22	23	24	25	26
3	116	91	80	63	65	51	51	40	39	31	29	23	19	15	13	10	8	6	6	4	3	2	1	1	1	1
2	97	78	67	54	54	44	43	34	33	27	25	20	16	13	11	9	7	5	5	4	3	1	1	1	1	1
1	72	57	50	39	40	32	32	25	25	19	18	14	12	9	8	6	5	4	3	2	1	1	1	1	1	
0	61	48	42	33	34	27	27	21	21	16	15	12	10	8	7	5	4	3	3	2	1	1	1	1		
2/0	51	41	35	28	28	23	22	18	17	14	13	10	8	7	6	4	3	3	2	1	1	1	1	1		
3/0	42	35	29	24	24	19	18	15	14	12	11	9	7	5	5	4	3	2	1	1	1	1				
4/0	35	29	24	20	20	16	15	13	12	10	9	7	6	5	4	3	2	1	1	1	1	1				
250	28	23	20	16	16	13	12	10	9	8	7	6	4	4	3	2	1	1	1	1	1					
300	24	20	17	14	14	11	11	9	8	7	6	5	4	3	3	2	1	1	1	1	1					
350	21	18	15	12	12	10	9	8	7	6	5	4	3	3	2	1	1	1	1	1						
400	19	16	13	11	11	9	8	7	6	5	5	4	3	2	1	1	1	1	1							
500	16	14	11	9	9	7	7	6	5	4	4	3	2	1	1	1	1	1	1							
600	12	11	8	7	7	6	5	5	4	4	3	3	1	1	1	1	1	1								
700	11	10	7	7	6	5	4	4	3	3	2	2	1	1	1	1	1									
750	10	9	7	6	6	5	4	4	3	3	2	2	1	1	1	1										
800	10	9	6	6	5	5	4	4	3	3	1	1	1	1	1	1										
900	9	8	6	5	5	4	4	3	3	2	1	1	1	1	1											
1,000	8	7	5	5	4	4	3	3	2	2	1	1	1	1	1	1										
1,250	6	6	4	4	3	3	2	2	1	1	1	1	1	1	1											
1,500	5	5	3	3	3	3	1	1	1	1	1	1	1	1												
1,750	4	4	3	3	2	2	1	1	1	1	1	1	1	1												
2,000	4	4	3	2	2	2	1	1	1	1	1	1														

° For Types RF-2, RFH-2, RH, RHH, see Column A, Table 1.

TABLE 8   Properties of Conductors

Size AWG	Area, circular mils	Concentric lay stranded conductors		Bare conductors		DC resistance, ohms per 1,000 feet at 25° C, 77° F		
		No. wires	Diam each wire, inches	Diam, inches	Area, square inches °	Copper		Aluminum
						Bare cond.	Tin'd. cond.	
18	1,624	Solid	0.0403	0.0403	0.0013	6.510	6.77	10.9
16	2,583	Solid	0.0508	0.0508	0.0020	4.094	4.25	6.85
14	4,107	Solid	0.0641	0.0641	0.0032	2.575	2.68	4.31
12	6,530	Solid	0.0808	0.0808	0.0051	1.619	1.69	2.71
10	10,380	Solid	0.1019	0.1019	0.0081	1.018	1.06	1.70
8	16,510	Solid	0.1285	0.1285	0.0130	0.641	0.660	1.07
6	26,250	7	0.0612	0.184	0.027	0.410	0.426	0.674
4	41,740	7	0.0772	0.232	0.042	0.259	0.269	0.423
3	52,640	7	0.0867	0.260	0.053	0.205	0.213	0.336
2	66,370	7	0.0974	0.292	0.067	0.162	0.169	0.266
1	83,690	19	0.0664	0.332	0.087	0.129	0.134	0.211
0	105,500	19	0.0745	0.373	0.109	0.102	0.106	0.168
00	133,100	19	0.0837	0.418	0.137	0.0811	0.0844	0.134
000	167,800	19	0.0940	0.470	0.173	0.0642	0.0668	0.105
0000	211,600	19	0.1055	0.528	0.219	0.0509	0.0524	0.0837
	250,000	37	0.0822	0.575	0.260	0.0431	0.0444	0.0708
	300,000	37	0.0900	0.630	0.312	0.0360	0.0371	0.0590
	350,000	37	0.0973	0.681	0.364	0.0308	0.0318	0.0506
	400,000	37	0.1040	0.728	0.416	0.0270	0.0278	0.0443
	500,000	37	0.1162	0.814	0.520	0.0216	0.0225	0.0354
	600,000	61	0.0992	0.893	0.626	0.0180	0.0185	0.0295
	700,000	61	0.1071	0.964	0.730	0.0154	0.0159	0.0253
	750,000	61	0.1109	0.998	0.782	0.0144	0.0148	0.0236
	800,000	61	0.1145	1.031	0.835	0.0135	0.0139	0.0221
	900,000	61	0.1215	1.093	0.938	0.0120	0.0124	0.0197
	1,000,000	61	0.1280	1.152	1.042	0.0108	0.0111	0.0176
	1,250,000	91	0.1172	1.289	1.305	0.00864	0.00890	0.0142
	1,500,000	91	0.1284	1.412	1.566	0.00719	0.00740	0.0118
	1,750,000	127	0.1174	1.526	1.829	0.00617	0.00636	0.0101
	2,000,000	127	0.1255	1.631	2.089	0.00539	0.00555	0.00884

° Area given is that of a circle having a diameter equal to the over-all diameter of a stranded conductor.

The values given in the table are those given in Circular 31 of the National Bureau of Standards except that those shown in the eighth column are those given in Specification B33 of the American Society for Testing Materials.

The resistance values given in the last three columns are applicable only to direct current.   When conductors larger than No. 4/0 are used with alternating current, the multiplying factors in Table 9, Chap. 9, of the Code should be used to compensate for skin effect.

Examples

Selection of Conductors.   In the following examples, the size of conductor has been selected on the basis of the allowable ampacities tabulated in the second column of Table 310-12.   If other types of insulated conductors are used, or if the conductors are run open, or with more than three conductors in a raceway, the size of conductor may vary from those shown.   Tables 310-12 through 310-15 and Notes thereto should be consulted in selecting the size of conductor for a particular installation.

Voltage.   For uniform application of the provisions of Articles 210, 215, and 220 a nominal voltage of 115 and 230 volts shall be used in computing the ampere load on the conductor.

Fractions of an Ampere.   Except where the computations result in a major fraction of an ampere, such fractions may be dropped.

Ranges.   For the computation of the range loads in these examples Column A of Table 220-5 has been used.   For optional methods, see Columns B and C of Table 220-5.

EXAMPLE NO. 1.   Single-family Dwelling

Dwelling has a floor area of 1,500 sq ft exclusive of unoccupied cellar, unfinished attic, and open porches.   It has a 12-kw range.

Computed Load (see Sec. 220-4):
General lighting load:
1,500 sq ft at 3 watts per sq ft = 4,500 watts

Minimum Number of Branch Circuits Required (see Sec. 220-3):
General lighting load:
4,500 ÷ 115 = 39.1 amp or three 15-amp 2-wire circuits; or two 20-amp
2-wire circuits
Small appliance load: two 2-wire 20-amp circuits [Sec. 220-3(b)]
Laundry load: one 2-wire 20-amp circuit [Sec. 220-3(b)]

Minimum Size Feeders Required (see Sec. 220-4)

Computed load	
General lighting	4,500 watts
Small appliance load	3,000 watts
Laundry	1,500 watts
Total (without range)	9,000 watts
3,000 watts at 100%	3,000 watts
9,000 − 3,000 = 6,000 watts at 35% =	2,100 watts
Net computed (without range) .	5,100 watts
Range Load (see Table 220-5) . . .	8,000 watts
Net computed (with range) . .	13,100 watts

For 115/230-volt 3-wire system feeders, 13,100 ÷ 230 = 57 amp

Net computed load exceeds 10 kw, so service conductors shall be 100 amp (see Sec. 230-41, Exception No. 1).

EXAMPLE NO. 1(a).   Single-family Dwelling

Same conditions as Example No. 1, plus addition of one 6-amp 230-volt room air-conditioning unit and three 12-amp 115-volt room air-conditioning units.   See Art. 422, Part F.

From Example No. 1, feeder current is 57 amp (3-wire, 230-volt)

Line A	Neutral	Line B	
57		57	amp from Example No. 1
6		6	one 230-volt air-conditioning motor
12		12	two 115-volt air-conditioning motors
—		12	one 115-volt air-conditioning motor
3		3	25% of largest motor (Sec. 430-24)
78		90	amp per line

EXAMPLE NO. 1(b).   Single-family Dwelling
Optional Calculation for One-Family Dwelling (Sec. 220-7)

Dwelling has a floor area of 1,500 sq ft exclusive of unoccupied cellar, unfinished attic, and open porches.   It has a 12-kw range, a 2.5-kw water heater, a 1.2-kw dishwasher, 9 kw of electric space heating installed in five rooms, a 4.5-kw clothes dryer, and a 6-amp 230-volt room air-conditioning unit.

Air conditioner kw is $6 \times 230 \div 1,000 = 1.38$ kw

1.38 kw is less than the connected load of 9 kw of space heating; therefore, the air-conditioner load need not be included in the service calculation [see Sec. 220-4(k)].

1,500 sq ft at 3 watts.	4.5 kw
Two 20-amp appliance outlet circuits at 1,500 watts each	3.0 kw
Laundry circuit.	1.5 kw
Range (at name-plate rating)	12.0 kw
Water heater	2.5 kw
Dishwasher	1.2 kw
Space heating.	9.0 kw
Clothes dryer	4.5 kw
	38.2 kw

First 10 kw at 100% = 10.00 kw
Remainder at 40% (28.2 kw × .4) = 11.28 kw
Calculated load for service size 21.28 kw = 21,280 watts
$21,280 \div 230 = 92.5$

Therefore, this dwelling may be served by a 100-amp service.

EXAMPLE NO. 1(c).   Single-family Dwelling
Optional Calculation for One-family Dwelling (See Sec. 220-7)

Dwelling has a floor area of 1,500 sq ft exclusive of unoccupied cellar, unfinished attic, and open porches.   It has two 20-amp small appliance circuits, one 20-amp laundry circuit, two 4-kw wall-mounted ovens, one 5.1-kw counter-mounted cooking unit, a 4.5-kw water heater, a 1.2-kw dishwasher, a 4.2-kw combination clothes washer and dryer, six 7-amp 230-volt room air-conditioning units, and a 1.5-kw permanently installed bathroom space heater.

Air conditioning kw calculation:

$$\begin{aligned}
\text{Total amperes } 6 \times 7 &= 42.00 \text{ amp} \\
25\% \text{ of largest motor } .25 \times 7 &= \underline{\ 1.75 \text{ amp}} \\
&\ \ 43.75 \text{ amp}
\end{aligned}$$

$$43.75 \times 230 \div 1,000 = 10.1 \text{ kw of air-conditioner load}$$

Load included at 100%

Air conditioning. . . . . 10.1 kw
Space heater [omit, see Sec. 220-4(k)]

Other load

1,500 sq ft at 3 watts	4.5
Two 20-amp small appliance circuits at 1,500 watts	3.0
Laundry circuit	1.5
2 ovens	8.
1 cooking unit	5.1
Water heater	4.5
Dishwasher	1.2
Washer/dryer	4.2
Total other load	32.0

First 10 kw at 100%	10.0 kw
Remainder at 40% (22.0 kw × .4)	8.8 kw
Total calculated load	28.9 kw = 28,900 watts

$$28,900 \div 230 = 125.6 \text{ amp (service rating)}$$

EXAMPLE NO. 2. Small Roadside Fruitstand with No Show Windows

A small roadside fruitstand with no show windows has a floor area of 150 sq ft. The electrical load consists of general lighting and a 1,000-watt flood-light. There are no other outlets.

Computed load (Sec. 220-4):

General lighting: °
150 sq ft at 3 watts per sq ft × 1.25 = 562 watts
(3 watts per sq ft for stores)
562 watts ÷ 115 = 4.88 amp
One 15-amp 2-wire branch circuit required (Sec. 220-3)

Minimum size service conductor required (Sec. 230-41, Exception No. 2):

Computed load . . . .   562 watts
Floodlight load . . . . 1,000 watts
  Total load . . . . 1,562 watts
  1,562 ÷ 115 × 13.6 amp
Use No. 8 service conductor (Sec. 230-41, Exception No. 2)
Use a 30-amp service switch or breaker (Sec. 230-71)

° See footnote on next page.

EXAMPLE NO. 3.   Store Building

A store 50 by 60 ft, or 3,000 sq ft, has 30 ft of show window.

Computed load (Sec. 220-4):
General lighting load: °
3,000 sq ft at 3 watts per sq ft × 1.25 . . . . 11,250 watts
Show-window lighting load: †
30 ft at 200 watts per ft. . . . . 6,000 watts

Minimum Number of Branch Circuits Required (Sec. 220-3):
General lighting load: ‡ 11,250 ÷ 230 = 49 amp for 3-wire, 115/230 volts;
or 98 amp for 2-wire, 115 volts:
Three 30-amp, 2-wire; and one 15-amp, 2-wire circuits; or
Five 20-amp, 2-wire circuits; or
Three 20-amp, 2-wire, and three 15-amp, 2-wire circuits; or
Seven 15-amp, 2-wire circuits; or
Three 15-amp, 3-wire, and one 15-amp, 2-wire circuits.

Special lighting load (show window): [Secs. 220-2 Exception No. 2 and
220-4(b)]: 6,000 ÷ 230 = 26 amp for 3-wire, 115/230 volts; or 52 amp for 2-
wire, 115 volts:
Four 15-amp, 2-wire circuits; or
Three 20-amp, 2-wire circuits; or
Two 15-amp, 3-wire circuits.

Minimum Size Feeders (or Service Conductors) Required (Sec. 215-2):
For 115/230-volts, 3-wire system:
Ampere load: 49 plus 26 = 75 amp (Sec. 220-2):
Size of each feeder, No. 3
For 115-volt system:
Ampere load: 98 plus 52 = 150 amp (Sec. 220-2):
Size of each feeder, No. 3/0

° The above examples assume that the entire general lighting load is likely
to be used for long periods of time and the load is therefore increased by 25%
in accordance with Sec. 220-2.   The 25% increase is not applicable to any
portion of the load not used for long periods.
† If show-window load computed as per Sec. 220-2, the unit load per outlet
to be increased 25%.
‡ The load on individual branch circuits not to exceed 80% of the branch-
circuit rating [Sec. 210-23(b)].

EXAMPLE NO. 4. Multifamily Dwelling

Multifamily dwelling having a total floor area of 32,000 sq ft with 40 apartments.

Meters in two banks of 20 each and individual subfeeders to each apartment.

One-half of the apartments are equipped with electric ranges of not exceeding 12 kw each.

Area of each apartment is 800 sq ft.

Laundry facilities on premises available to all tenants.   Add no circuit to individual apartment.   Add 1,500 watts for each laundry circuit to house load and add to the example as a "house load."

Computed Load for Each Apartment (Art. 220):
General lighting load:
800 sq ft at 3 watts per sq ft. . . . .   2,400 watts
Special appliance load:
Electric range . . . .   8,000 watts

Minimum Number of Branch Circuits Required for Each Apartment (Sec. 220-3):

General lighting load: $2,400 \div 115 = 21$ amp or two 15-amp, 2-wire circuits or two 20-amp, 2-wire circuits.

Small appliance load: Two 2-wire circuits of No. 12 wire [see Sec. 220-3(b)].

Range circuit: $8,000 \div 230 = 34$ amp or a circuit of two No. 8's and one No. 10 as permitted by Sec. 210-9(c).

Minimum Size Subfeeder Required for Each Apartment (Sec. 215-2):
Computed load (Art. 220):
General lighting load.   2,400 watts
Small appliance load, two 20-amp circuits. . . . .   3,000 watts
Total computed load (without ranges). . . . .   5,400 watts

Application of demand factor:
3,000 watts at 100%.   3,000 watts
2,400 watts at 35%.   840 watts
Net computed load (without ranges). . . . .   3,840 watts

Range load. . . . .   8,000 watts
Net computed load (with ranges) . . . .   11,840 watts

For 115/230-volt, 3-wire system (without ranges):
Net computed load, $3,840 \div 230 = 16.7$ amp
Size of each subfeeder (see Sec. 215-2).

For 115/230-volt, 3-wire system (with ranges):
Net computed load, 11,840 ÷ 230 = 51.5 amp
Size of each ungrounded subfeeder, No. 6.

Neutral subfeeder:

Lighting and small appliance load . . . .   3,840 watts
Range load, 8,000 watts at 70% [see Sec.
220-4(f)] . . . .   5,600 watts
Net computed load (neutral) . . . .   9,440 watts

9,440 ÷ 230 = 41 amp
Size of neutral subfeeder, No. 6.

Minimum Size Feeders Required from Service Equipment to Meter Bank
(for 20 apartments — 10 with ranges):

Total computed load:
Lighting and small appliance load, 20 × 5,400 . . . .   108,000 watts

Application of demand factor:
3,000 watts at 100% . . . .   3,000 watts
105,000 watts at 35% . . . .   36,750 watts
Net computed lighting and small appliance
load . . . .   39,750 watts

Range load, 10 ranges (less than 12 kw; column A,
Table 220-5) . . . .   25,000 watts
Net computed load (with ranges) . . . .   64,750 watts

For 115/230-volt, 3-wire system:
Net computed load, 64,750 ÷ 230 = 282 amp

Size of each ungrounded feeder to each meter bank: 500,000 c.m.

Neutral feeder:
Lighting and small appliance load . . . .   39,750 watts
Range load: 25,000 watts at 70% [see Sec.
220-4(f)] . . . .   17,500 watts
Computed load (neutral) . . . .   57,250 watts

57,250 ÷ 230 = 249 amp

Further demand factor [Sec. 220-4(f)]:
200 amp at 100%   = 200 amp
49 amp at 70%   = 34 amp
Net computed load (neutral) . . . . . .   234 amp

Size of neutral feeder to each meter bank: 300,000 c.m.

Minimum Size Main Feeder (or Service Conductors) Required (for 40 apartments—20 with ranges):

Total computed load:
Lighting and small appliance load, 40 × 5,400 . . . . . .   216,000 watts

Application of demand factor:

3,000 watts at 100%.	3,000 watts
117,000 watts at 35%.	40,950 watts
96,000 watts at 25%.	24,000 watts
Net computed lighting and small appliance	
load	67,950 watts
Range load, 20 ranges (less than 12 kw, column A,	
Table 220-5).	35,000 watts
Net computed load.	102,950 watts

For 115/230-volt, 3-wire system:
Net computed load, $102,950 \div 230 = 448$ amp

Size of each ungrounded main feeder: 1,000,000 c.m.

Neutral feeder:

Lighting and small appliance load	67,950 watts
Range load, 35,000 watts at 70% [see Sec.	
220-4(f)].	24,500 watts
Computed load (neutral)	92,450 watts

$92,450 \div 230 = 402$ amp

Further demand factor [see Section 220-4 (f)]:

200 amp at 100%	= 200 amp
202 amp at 70%	= 141 amp
Net computed load (neutral)	341 amp

Size of neutral main feeder: 600,000 c.m.
See Tables 310-12 through 310-15, Notes 8 and 12.

EXAMPLE NO. 4(a).   Optional Calculation for Multifamily Dwelling

Multifamily dwelling equipped with electric cooking and space heating or air conditioning and having a total floor area of 32,000 sq ft with 40 apartments.

Meters in two banks of 20 each plus house metering and individual subfeeders to each apartment.

Each apartment is equipped with an electric range of 8-kw name-plate rating.

A common laundry facility available to all tenants [Sec. 210-22(b), Exception 1].

Area of each apartment is 800 sq ft.

Computed Load for Each Apartment (Art. 220):

General lighting load:

800 sq ft at 3 watts per sq ft.	2,400 watts
Electric range	8,000 watts
Electric heat 6 kw	6,000 watts
(or air conditioning if larger)	

Minimum Number of Branch Circuits Required for Each Apartment:

General lighting load 2,400 watts $\div$ 115 = 21 amp or two 15-amp 2-wire circuits or two 20-amp 2-wire circuits.

Small appliance load: two 2-wire circuits of No. 12 [see Sec. 220-3(b)]

Range circuit 8,000 watts $\times$ 80% $\div$ 230 = 28 amp on a circuit of three No. 10 as permitted in Column C of Table 220-5

Space heating 6,000 watts $\div$ 230 = 26 amp
No. of circuits (see Sec. 210-24)

Minimum Size Subfeeder Required for Each Apartment (Sec. 215-2):

Computed load (Art. 220):

General lighting load	2,400 watts
Small appliance load, two 20-amp circuits . .	3,000 watts
Total computed load (without range and space heating)	5,400 watts

Application of demand factor:

3,000 watts at 100%	3,000 watts
2,400 watts at 35%	840 watts
Net computed load (without range and space heating)	3,840 watts
Range load	6,400 watts
Space heating Sec 220-4(c)	6,000 watts
Net computed load for individual apartment .	16,240 watts

For 115/230-volt 3-wire system
Net computed load 16,240 $\div$ 230 = 71 amp
Size of each ungrounded subfeeder No. 3

Neutral subfeeder [Sec. 220-4(d)]

Lighting and small appliance load	3,840 watts
Range load 6,400 watts at 70% [see Sec. 220-4(d)].	4,480 watts
Space heating (no neutral) 230 volt	0 watts
Net computed load (neutral)	8,320 watts

8,320 $\div$ 230 = 36 amp
Size of neutral subfeeder No. 8

Minimum Size Feeder Required from Service Equipment to Meter Bank for 20 Apartments:

Total computed load:

Lighting and small appliance load 20 $\times$ 5,400	108,000 watts
Range and space-heating load 20 $\times$ 14,000 .	280,000 watts
Net computed load (20 apartments)	388,000 watts

Net computed using optional calculation (Table 220-9)

  388,000 × .38 . . . . 147,440 watts

147,440 ÷ 230 = 641 amp

Size of each ungrounded feeder to each meter bank two **400 MCM** per phase.

Neutral feeder

  Lighting and small appliance load

    108,000 × 38% (Table 220-9) = ... 41,040 watts

  Range load 8,000 × 20 × 70% = 112,000 watts

    112,000 × 38% [Sec. 220-4(d)] . . . . . . 42,560 watts

                        (Table 220-9)

    Computed load (neutral) . . . . 83,600 watts

83,600 ÷ 230 = 364 amp

Size of neutral feeder to each meter bank one **500 MCM**

Minimum Size Mains Feeder Required (Less House Load) for 40 Apartments:

Total computed load:

  Lighting and small appliance load 40 × 5,400 216,000 watts

  Range and space heating 40 × 14,000 . . . . 560,000 watts

Net computed load (40 apartments) . . . . . . 776,000 watts

Net computed using optional calculation (Table 220-9)

  776,000 × 28% . . . . 216,280 watts

216,280 ÷ 230 = 940 amp

Size of each ungrounded feeder required: two **750 MCM** per phase.

Neutral feeder

  Lighting and small appliance load 40 × 5,400 × 28%

    (Table 220-9) . . . . 60,480 watts

  Range load 224,000 × 28% (Table 220-9) . . 62,720 watts

  Computed load (neutral) Sec. 220-4(d) . . . 123,200 watts

123,200 ÷ 230 = 536 amp

Size of neutral feeder required: two **250 MCM**

Add to obtain size of service conductors, the entire house load including laundry circuit(s) in accordance with applicable section of Art. 220.

**EXAMPLE NO. 5.   Calculation of Neutral Feeder
(See Sec. 220-4)**

The following example illustrates the method of calculating size of neutral feeder for the computed load of a 5-wire, 2-phase system where it is desired to modify the load in accordance with provisions of Sec. 220-4.
An installation consisting of a computed load of 250 amp connected between neutral feeder and each ungrounded feeder.

Neutral Feeder (maximum unbalance of load 250 amp × 140% = 350 amp):

200 amp (first) at 100% = 200 amp
150 amp (excess) at   70% = 105 amp

Computed load   305 amp

Size of neutral feeder: 500,000 c.m.

EXAMPLE NO. 6.   Maximum Demand for Range Loads

Table 220-5, column A, applies to ranges not over 12 kw.   The application of Note 1 to ranges over 12 kw (and not over 27 kw) is illustrated in the following examples:

A. Ranges all of same rating:
Assume 24 ranges each rated 16 kw.

From column A the maximum demand for 24 ranges of 12-kw rating is 39 kw.

16 kw exceeds 12 kw by 4.
5% × 4 = 20% (5% increase for each kw in excess of 12).
39 kw × 20% = 7.8 kw increase.

39 + 7.8 = 46.8 kw: value to be used in selection of feeders.

B. Ranges of unequal rating:
Assume 5 ranges each rated 11 kw.
2 ranges each rated 12 kw.
20 ranges each rated 13.5 kw.
3 ranges each rated 18 kw.

5 × 12   =   60 Use 12 kw for range rated less than 12.
2 × 12   =   24
20 × 13.5 = 270
3 × 18   =   54

408 kw

408 ÷ 30 = 13.6 kw (average to be used for computation)

From column A the demand for 30 ranges of 12-kw rating is 15 + 30 = 45 kw.

13.6 exceeds 12 by 1.6 (use 2).
5% × 2 = 10% (5% increase for each kw in excess of 12).
45 kw × 10% = 4.5 kw increase.
45 + 4.5 = 49.5 kw = value to be used in selection of feeders.

EXAMPLE NO. 7. **Ranges on a 3-phase System**
[Sec. 220-4(d)]

Thirty ranges rated at 12 kw each are supplied by a 3-phase, 4-wire, 120/ 208-volt feeder, 10 ranges on each phase.

As there are 20 ranges connected to each ungrounded conductor, the load should be calculated on the basis of 20 ranges (or in case of unbalance, twice the maximum number between any two phase wires) since diversity applies only to the number of ranges connected to adjacent phases and not the total.

The current in any one conductor will be one-half the total watt load of two adjacent phases divided by the line-to-neutral voltage. In this case, 20 ranges, from Table 220-5, will have a total watt load of 35,000 watts for two phases; therefore, the current in the feeder conductor would be:

$17,500 \div 120 = 146$ amp

On a three-phases basis the load would be:

$3 \times 17,500 = 52,500$ watts

and the current in each feeder conductor—

$$\frac{52,500}{208 \times 1.73} = 146 \text{ amp}$$

EXAMPLE NO. 8.   Motors, Conductors, and Overcurrent Protection

(See Sections 430-22, 430-24, 430-32, 430-52, 430-62, and Tables 310-12, 430-150, 430-152, and 430-153)

Determine the size of copper conductors, the motor-running overcurrent protection, the branch-circuit protection, and the feeder protection, for one 25-hp squirrel-cage induction motor (full-voltage starting, service factor 1.15, Code letter F), and two 30-hp wound-rotor induction motors (40°C rise), on a 460-volt, 3-phase, 60-cycle-per-second supply.

Conductor Sizes

The full-load current of the 25-hp motor is 34 amp (Table 430-150). A full-load current of 34 amp × 1.25 = 42.5 amp (Sec. 430-22) requires a No. 6 conductor with 60°C insulation (Table 310-12). The full-load current of the 30-hp motor is 40 amp (Table 430-150). A full-load current of 40 amp × 1.25 = 50 amp (Sec. 430-22) requires a No. 6 conductor with 60°C insulation (Table 310-12).

The feeder conductor capacity will be 125% of 40 plus 40 plus 34, or 124 amp (Sec. 430-24). In accordance with Table 310-12, this would require a No. 0, 60°C feeder conductor.

Note: For conductors with 60°C insulation run open in air, or for conductors with temperature ratings other than 60°C, see Tables 310-12 through 310-15.

Overcurrent Protection

Running.   The 25-hp motor, with full-load current of 34 amp must have running overcurrent protection of not over 42.5 amp. The 30-hp motor with full-load current of 40 amp must have running overcurrent protection of not over 50 amp.

Branch Circuit.   The branch circuit of the 25-hp motor must have branch-circuit overcurrent protection of not over 300% for a fuse (Table 430-153) or 3.00 × 34 = 102 amp. The nearest standard fuse which does not exceed this maximum value is 100 amp.

For the 30-hp motor the branch-circuit overcurrent protection is 150% (Table 430-152) or 1.50 × 40 = 60 amp. Where the maximum value of overcurrent protection is not sufficient to start the motor the value for a fuse may be increased to 400% [Sec. 430-52, Exception (a)].

Feeder Circuit.   The maximum rating of the feeder overcurrent protection device is based on the sum of the largest branch-circuit protective device (100-amp fuse) plus the sum of the full-load currents of the other motors, or 100 plus 40 plus 40 = 180 amp. The nearest standard fuse which does not exceed this value is 175 amp.

Index